Lecture Notes in Computer Science 7844

Commenced Publication in 1973
Founding and Former Series Editors:
Gerhard Goos, Juris Hartmanis, and Jan van Leeuwen

T0202596

Editorial Board

David Hutchison
Lancaster University, UK
Takeo Kanade
Carnegie Mellon University, Pittsburgh, PA, USA
Josef Kittler
University of Surrey, Guildford, UK
Jon M. Kleinberg
Cornell University, Ithaca, NY, USA
Alfred Kobsa
University of California, Irvine, CA, USA
Friedemann Mattern
ETH Zurich, Switzerland
John C. Mitchell
Stanford University, CA, USA
Moni Naor
Weizmann Institute of Science, Rehovot, Israel
Oscar Nierstrasz
University of Bern, Switzerland
C. Pandu Rangan
Indian Institute of Technology, Madras, India
Bernhard Steffen
TU Dortmund University, Germany
Madhu Sudan
Microsoft Research, Cambridge, MA, USA
Demetri Terzopoulos
University of California, Los Angeles, CA, USA
Doug Tygar
University of California, Berkeley, CA, USA
Gerhard Weikum
Max Planck Institute for Informatics, Saarbruecken, Germany

Elvira Albert (Ed.)

Logic-Based Program Synthesis and Transformation

22nd International Symposium, LOPSTR 2012 Leuven, Belgium, September 18-20, 2012 Revised Selected Papers

Volume Editor

Elvira Albert
Complutense University of Madrid
Department of Computer Science (DSIC)
Séneca Avenue 2
28040 Madrid, Spain
E-mail: elvira@fdi.ucm.es

ISSN 0302-9743 e-ISSN 1611-3349
ISBN 978-3-642-38196-6 e-ISBN 978-3-642-38197-3
DOI 10.1007/978-3-642-38197-3
Springer Heidelberg Dordrecht London New York

Library of Congress Control Number: 2013936976

CR Subject Classification (1998): D.1.6, D.2.4-5, F.4.1, G.2.2, F.3, I.2.2-4, D.3.1, F.1.1

LNCS Sublibrary: SL 1 – Theoretical Computer Science and General Issues

© Springer-Verlag Berlin Heidelberg 2013

This work is subject to copyright. All rights are reserved, whether the whole or part of the material is concerned, specifically the rights of translation, reprinting, re-use of illustrations, recitation, broadcasting, reproduction on microfilms or in any other way, and storage in data banks. Duplication of this publication or parts thereof is permitted only under the provisions of the German Copyright Law of September 9, 1965, in its current version, and permission for use must always be obtained from Springer. Violations are liable to prosecution under the German Copyright Law.
The use of general descriptive names, registered names, trademarks, etc. in this publication does not imply, even in the absence of a specific statement, that such names are exempt from the relevant protective laws and regulations and therefore free for general use.

Typesetting: Camera-ready by author, data conversion by Scientific Publishing Services, Chennai, India

Printed on acid-free paper

Springer is part of Springer Science+Business Media (www.springer.com)

Preface

This volume contains a selection of the papers presented at the 22nd International Symposium on Logic-Based Program Synthesis and Transformation (LOPSTR 2012) held during September 18-20, 2012, in Leuven. LOPSTR 2012 was co-located with PPDP 2012, the 14th International ACM SIGPLAN Symposium on Principles and Practice of Declarative Programming.

Previous LOPSTR symposia were held in Odense (2011), Hagenberg (2010), Coimbra (2009), Valencia (2008), Lyngby (2007), Venice (2006 and 1999), London (2005 and 2000), Verona (2004), Uppsala (2003), Madrid (2002), Paphos (2001), Manchester (1998, 1992, and 1991), Leuven (1997), Stockholm (1996), Arnhem (1995), Pisa (1994), and Louvain-la-Neuve (1993). Information about the conference can be found at: http://costa.ls.fi.upm.es/lopstr12/.

The aim of the LOPSTR series is to stimulate and promote international research and collaboration in logic-based program development. LOPSTR traditionally solicits contributions, in any language paradigm, in the areas of specification, synthesis, verification, analysis, optimization, specialization, security, certification, applications and tools, program/model manipulation, and transformational techniques. LOPSTR has a reputation for being a lively, friendly forum for presenting and discussing work in progress. Formal proceedings are produced only after the symposium so that authors can incorporate this feedback in the published papers.

In response to the call for papers, 27 contributions were submitted from 19 different countries. The Program Committee accepted four full papers for immediate inclusion in the formal proceedings, and nine more papers were accepted after a revision and another round of reviewing. In addition to the 13 contributed papers, this volume includes the abstracts of the invited talks by two outstanding researchers: one LOPSTR invited talk by Tom Schrijvers (University of Ghent, Belgium) and a joint PPDP-LOPSTR invited talk by Jürgen Giesl (RWTH Aachen, Germany).

I want to thank the Program Committee members, who worked diligently to produce high-quality reviews for the submitted papers, as well as all the external reviewers involved in the paper selection. I am very grateful the LOPSTR 2012 Conference Co-chairs, Daniel De Schreye and Gerda Janssens, and the local organizers for the great job they did in preparing the conference. I also thank Andrei Voronkov for his excellent EasyChair system that automates many of the tasks involved in chairing a conference. Finally, I gratefully acknowledge the sponsors of LOPSTR: The Association for Logic Programming and Fonds voor Wetenschappelijk Onderzoek Vlaanderen.

March 2013 Elvira Albert

Organization

Program Committee

Elvira Albert	Complutense University of Madrid, Spain
Sergio Antoy	Portland State University, USA
Demis Ballis	University of Udine, Italy
Henning Christiansen	Roskilde University, Denmark
Michael Codish	Ben-Gurion University of the Negev, Israel
Danny De Schreye	K.U. Leuven, Belgium
Esra Erdem	Sabanci University, Istanbul
Maribel Fernandez	King's College London, UK
Carsten Fuhs	University College London, UK
John Gallagher	Roskilde University, Denmark
Robert Glück	University of Copenhagen, Denmark
Miguel Gomez-Zamalloa	Complutense University of Madrid, Spain
Rémy Haemmerlé	Technical University of Madrid, Spain
Geoff Hamilton	Dublin City University, Ireland
Reiner Hähnle	Technical University of Darmstadt, Germany
Gerda Janssens	K.U. Leuven, Belgium
Isabella Mastroeni	University of Verona, Italy
Kazutaka Matsuda	University of Tokyo, Japan
Paulo Moura	Universidade da Beira Interior, Portugal
Johan Nordlander	Luleå University of Technology, Sweden
Andrey Rybalchenko	Technische Universität München, Germany
Kostis Sagonas	Uppsala University, Sweden
Francesca Scozzari	Università "G. D'Annunzio" di Chieti, Italy
Valerio Senni	Universtà di Roma "Tor Vergata", Italy
German Vidal	Technical University of Valencia, Spain

Additional Reviewers

Amtoft, Torben
De Angelis, Emanuele
Fioravanti, Fabio
Grossniklaus, Michael
Hentschel, Martin
Nigam, Vivek
Nogueira, Vitor
Riesco, Adrian

Table of Contents

Symbolic Evaluation Graphs and Term Rewriting — A General Methodology for Analyzing Logic Programs*
(Abstract)

Jürgen Giesl[1], Thomas Ströder[1], Peter Schneider-Kamp[2], Fabian Emmes[1], and Carsten Fuhs[3]

[1] LuFG Informatik 2, RWTH Aachen University, Germany
[2] Dept. of Mathematics and Computer Science, University of Southern Denmark
[3] Dept. of Computer Science, University College London, UK

There exist many powerful techniques to analyze *termination* and *complexity* of *term rewrite systems* (TRSs). Our goal is to use these techniques for the analysis of other programming languages as well. For instance, approaches to prove termination of definite logic programs by a transformation to TRSs have been studied for decades. However, a challenge is to handle languages with more complex evaluation strategies (such as Prolog, where predicates like the *cut* influence the control flow).

We present a general methodology for the analysis of such programs. Here, the logic program is first transformed into a *symbolic evaluation graph* which represents all possible evaluations in a finite way. Afterwards, different analyses can be performed on these graphs. In particular, one can generate TRSs from such graphs and apply existing tools for termination or complexity analysis of TRSs to infer information on the termination or complexity of the original logic program.

More information can be found in the full paper [1].

Reference

1. Giesl, J., Ströder, T., Schneider-Kamp, P., Emmes, F., Fuhs, C.: Symbolic evaluation graphs and term rewriting — a general methodology for analyzing logic programs. In: Proc. PPDP 2012, pp. 1–12. ACM Press (2012)

* Supported by the DFG under grant GI 274/5-3, the DFG Research Training Group 1298 (*AlgoSyn*), and the Danish Council for Independent Research, Natural Sciences.

E. Albert (Ed.): LOPSTR 2012, LNCS 7844, p. 1, 2013.
© Springer-Verlag Berlin Heidelberg 2013

An Introduction to Search Combinators

Tom Schrijvers[1], Guido Tack[2], Pieter Wuille[1,3],
Horst Samulowitz[4], and Peter J. Stuckey[5]

[1] Universiteit Gent, Belgium
{tom.schrijvers,pieter.wuille}@ugent.be
[2] National ICT Australia (NICTA) and Monash University, Victoria, Australia
guido.tack@monash.edu
[3] Katholieke Universiteit Leuven, Belgium
pieter.wuille@cs.kuleuven.be
[4] IBM Research, New York, USA
samulowitz@us.ibm.com
[5] National ICT Australia (NICTA) and University of Melbourne, Victoria, Australia
pjs@cs.mu.oz.au

Abstract. The ability to model search in a constraint solver can be an essential asset for solving combinatorial problems. However, existing infrastructure for defining search heuristics is often inadequate. Either modeling capabilities are extremely limited or users are faced with a general-purpose programming language whose features are not tailored towards writing search heuristics. As a result, major improvements in performance may remain unexplored.

This article introduces *search combinators*, a lightweight and solver-independent method that bridges the gap between a conceptually simple modeling language for search (high-level, functional and naturally compositional) and an efficient implementation (low-level, imperative and highly non-modular). By allowing the user to define application-tailored search strategies from a small set of primitives, search combinators effectively provide a rich *domain-specific language* (DSL) for modeling search to the user. Remarkably, this DSL comes at a low implementation cost to the developer of a constraint solver.

1 Introduction

Search heuristics often make all the difference between effectively solving a combinatorial problem and utter failure. Heuristics make a search algorithm efficient for a variety of reasons, e.g., incorporation of domain knowledge, or randomization to avoid heavy-tailed runtimes. Hence, the ability to swiftly design search heuristics that are tailored towards a problem domain is essential for performance. This article introduces search combinators, a versatile, modular, and efficiently implementable language for expressing search heuristics.

1.1 Status Quo

In CP, much attention has been devoted to facilitating the modeling of combinatorial problems. A range of high-level modeling languages, such as OPL [1], Comet [2],

E. Albert (Ed.): LOPSTR 2012, LNCS 7844, pp. 2–16, 2013.
© Springer-Verlag Berlin Heidelberg 2013

or Zinc [3], enable quick development and exploration of problem models. But there is substantially less support for high-level specification of accompanying search heuristics. Most languages and systems, e.g. ECLiPSe [4], Gecode [5], Comet [2], or MiniZinc [6], provide a set of predefined heuristics "off the shelf". Many systems also support user-defined search based on a general-purpose programming language (e.g., all of the above systems except MiniZinc). The former is clearly too confining, while the latter leaves much to be desired in terms of productivity, since implementing a search heuristic quickly becomes a non-negligible effort. This also explains why the set of predefined heuristics is typically small: it takes a lot of time for CP system developers to implement heuristics, too – time they would much rather spend otherwise improving their system.

1.2 Contributions

In this article we show how to resolve this stand-off between solver developers and users, by introducing a domain-specific modular search language based on combinators, as well as a modular, extensible implementation architecture.

For the User, we provide a modeling language for expressing complex search heuristics based on an (extensible) set of primitive combinators. Even if the users are only provided with a small set of combinators, they can already express a vast range of combinations. Moreover, using combinators to program application-tailored search is vastly more productive than resorting to a general-purpose language.

For the System Developer, we show how to design and implement modular combinators. The modularity of the language thus carries over directly to modularity of the implementation. Developers do not have to cater explicitly for all possible combinator combinations. Small implementation efforts result in providing the user with a lot of expressive power. Moreover, the cost of adding one more combinator is small, yet the return in terms of additional expressiveness can be quite large.

The technical challenge is to bridge the gap between a conceptually simple search language and an efficient implementation, which is typically low-level, imperative and highly non-modular. This is where existing approaches are weak; either the expressiveness is limited, or the approach to search is tightly tied to the underlying solver infrastructure.

The contribution is therefore the novel design of an expressive, high-level, compositional search language with an equally modular, extensible, and efficient implementation architecture.

1.3 Approach

We overcome the modularity challenge by implementing the primitives of our search language as *mixin* components [7]. As in Aspect-Oriented Programming [8], mixin components neatly encapsulate the *cross-cutting behavior* of primitive search concepts, which are highly entangled in conventional approaches. Cross-cutting means that a mixin component can interfere with the behavior of its sub-components (in this case, sub-searches). The combination of encapsulation *and* cross-cutting behavior is essential

s ::= prune	| ifthenelse($cond, s_1, s_2$)
prunes the node	perform s_1 until $cond$ is false, then perform s_2
| base_search($vars, var\text{-}select, domain\text{-}split$)	| and($[s_1, s_2, \ldots, s_n]$)
label	perform $s1$, on success $s2$ otherwise fail, …
| let(v, e, s)	| or($[s_1, s_2, \ldots, s_n]$)
introduce new variable v with initial value e, then perform s	perform $s1$, on termination start $s2$, …
	| portfolio($[s_1, s_2, \ldots, s_n]$)
| assign(v, e)	perform $s1$, if not exhaustive start $s2$, …
assign e to variable v and succeed	| restart($cond, s$)
| post(c, s)	restart s as long as $cond$ holds
post constraint c at every node during s	

Fig. 1. Catalog of primitive search heuristics and combinators

for systematic reuse of search combinators. Without this degree of modularity, minor modifications require rewriting from scratch.

An added advantage of mixin components is extensibility. We can add new features to the language by adding more mixin components. The cost of adding such a new component is small, because it does not require changes to the existing ones. Moreover, experimental evaluation bears out that this modular approach has no significant overhead compared to the traditional monolithic approach. Finally, our approach is solver-independent and therefore makes search combinators a potential standard for designing search.

This article provides on overview of the work on search combinators, that appeared in the proceedings of the 17th International Conference on Principles and Practice of Constraint Programming (CP) 2011 [9] and the Special Issue on Modeling and Reformulation of the Constraints journal [10].

2 High-Level Search Language

This section introduces the syntax of our high-level search language and illustrates its expressive power and modularity by means of examples. The rest of the article then presents an architecture that maps the modularity of the language down to the implementation level.

The search language is used to define a *search heuristic*, which a *search engine* applies to each node of the search tree. For each node, the heuristic determines whether to continue search by creating child nodes, or to prune the tree at that node. The queuing strategy, i.e., the strategy by which new nodes are selected for further search (such as depth-first traversal), is determined separately by the search engine, it is thus orthogonal to the search language. The search language features a number of primitives, listed in the catalog of Fig. 1. These are the building blocks in terms of which more complex heuristics can be defined, and they can be grouped into *basic heuristics* (base_search and prune), *combinators* (ifthenelse, and, or, portfolio, and restart), and *state management* (let, assign, post). This section introduces the three groups of primitives in turn.

For many users, the given primitives will represent a simple and at the same time sufficiently expressive language that allows them to implement complex, problem-specific search heuristics. The examples in this section show how versatile this base language is. However, we emphasize that the catalog of primitives is open-ended. Advanced users may need to add new, problem-specific primitives, and Sect. 3 explains how the language implementation explicitly supports this.

The concrete syntax we chose for presentation uses simple nested terms, which makes it compatible with the *annotation* language of MiniZinc [6]. However, other concrete syntax forms are easily supported (e.g., we support C++ and Haskell).

2.1 Basic Heuristics

Let us first discuss the two basic primitives, base_search and prune.

base_search. The most widely used method for specifying a basic heuristic for a constraint problem is to define it in terms of a *variable selection* strategy which picks the next variable to constrain, and a *domain splitting* strategy which splits the set of possible values of the selected variable into two (or more) disjoint sets.

The CP community has spent a considerable amount of work on defining and exploring many such variable selection and domain splitting heuristics. The provision of a flexible language for defining new basic searches is an interesting problem in its own right, but in this article we concentrate on search combinators that combine and modify basic searches.

To this end, our search language provides the primitive base_search(*vars*, *var-select*, *domain-split*), which specifies a systematic search. If any of the variables *vars* are still not fixed at the current node, it creates child nodes according to *var-select* and *domain-split* as variable selection and domain splitting strategies respectively.

Note that base_search is a CP-specific primitive; other kinds of solvers provide their own search primitives. The rest of the search language is essentially solver-independent. While the solver provides few basic heuristics, the search language adds great expressive power by allowing these to be combined arbitrarily using combinators.

prune. The second basic primitive, prune, simply cuts the search tree below the current node. Obviously, this primitive is useless on its own, but we will see shortly how prune can be used together with combinators.

2.2 Combinators

The expressive power of the search language relies on combinators, which combine search heuristics (which can be basic or themselves constructed using combinators) into more complex heuristics.

and/or. Probably the most widely used combination of heuristics is *sequential composition*. For instance, it is often useful to first label one set of problem variables before starting to label a second set. The following heuristic uses the and combinator to first label all the xs variables using a first-fail strategy, followed by the ys variables with a different strategy:

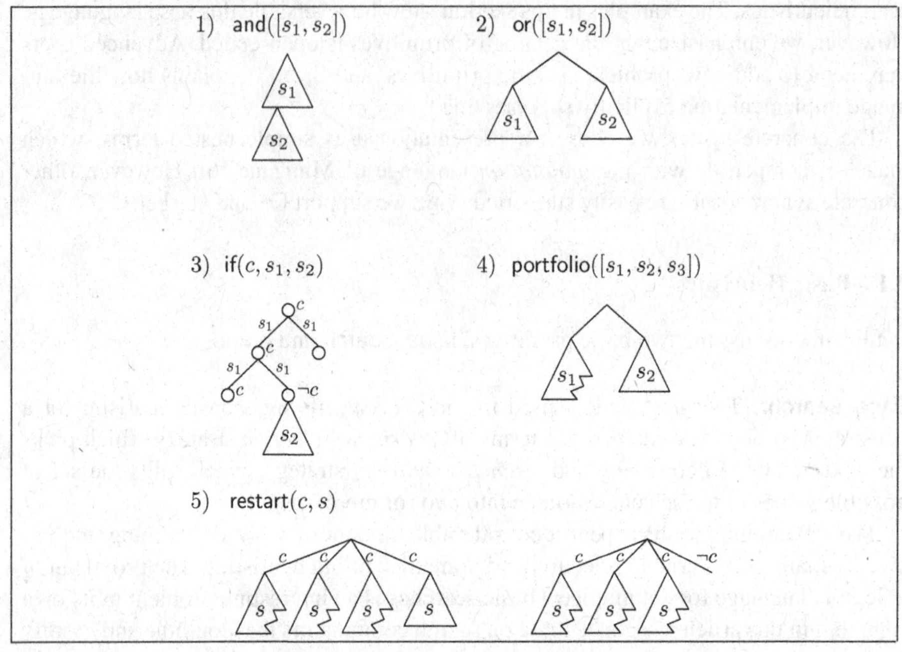

Fig. 2. Primitive combinators

```
and([base_search(xs, firstfail, min),
     base_search(ys, smallest, max)])
```

As you can see in Fig. 1, the and combinator accepts a list of searches $s_1, \ldots, s_n$, and performs their and-sequential composition. And-sequential means, intuitively, that solutions are found by performing *all* the sub-searches sequentially down one branch of the search tree, as illustrated in Fig. 2.1.

The dual combinator, or($[s_1, \ldots, s_n]$), performs a disjunctive combination of its sub-searches – a solution is found using *any* of the sub-searches (Fig. 2.2), trying them in the given order.

Statistics and ifthenelse. The ifthenelse combinator is centered around a conditional expression *cond*. As long as *cond* is true for the current node, the sub-search s_1 is used. Once *cond* is false, s_2 is used for the complete subtree below the current node (see Fig. 2.3).

We do not specify the *expression language* for conditions in detail, we simply assume that it comprises the typical arithmetic and comparison operators and literals that require no further explanation. It is notable though that the language can refer to the constraint variables and parameters of the underlying model. Additionally, a condition may refer to one or more *statistics* variables. Such statistics are collected for the duration of a subsearch until the condition is met. For instance ifthenelse($\text{depth} < 10, s_1, s_2$)

maintains the search depth statistic during subsearch s_1. At depth 10, the ifthenelse combinator switches to subsearch s_2.

We distinguish two forms of statistics: *Local statistics* such as depth and discrepancies express properties of individual nodes. *Global statistics* such as number of explored nodes, encountered failures, solution, and time are computed for entire search trees.

It is worthwhile to mention that developers (and advanced users) can also define their own statistics, just like combinators, to complement any predefined ones. In fact, Sect. 3 will show that statistics can be implemented as a *subtype* of combinators that can be queried for the statistic's value.

Abstraction. Our search language draws its expressive power from the combination of primitive heuristics using combinators. An important aspect of the search language is *abstraction*: the ability to create new combinators by effectively defining macros in terms of existing combinators.

For example, we can define the limiting combinator $\mathsf{limit}(cond,s)$ to perform s while condition *cond* is satisfied, and otherwise cut the search tree using prune:

$$\mathsf{limit}(cond,s) \equiv \mathsf{ifthenelse}(cond,s,\mathsf{prune})$$

The $\mathsf{once}(s)$ combinator, well-known in Prolog as `once/1`, is a special case of the limiting combinator where the number of solutions is less than one. This is simply achieved by maintaining and accessing the solutions statistic:

$$\mathsf{once}(s) \equiv \mathsf{limit}(\mathsf{solutions} < 1, s)$$

Exhaustiveness and portfolio/restart. The behavior of the final two combinators, portfolio and restart, depends on whether their sub-search was *exhaustive*. Exhaustiveness simply means that the search has explored the entire subtree without ever invoking the prune primitive.

The $\mathsf{portfolio}([s_1,\ldots,s_n])$ combinator performs s_1 until it has explored the whole subtree. If s_1 was exhaustive, i.e., if it did not call prune during the exploration of the subtree, the search is finished. Otherwise, it continues with $\mathsf{portfolio}([s_2,\ldots,s_n])$. This is illustrated in Fig. 2.4, where the subtree of s_1 represents a non-exhaustive search, s_2 is exhaustive and therefore s_3 is never invoked.

An example for the use of portfolio is the $\mathsf{hotstart}(cond,s_1,s_2)$ combinator. It performs search heuristic s_1 while condition *cond* holds to initialize global parameters for a second search s_2. This heuristic can for example be used to initialize the widely applied *Impact* heuristic [11]. Note that we assume here that the parameters to be initialized are maintained by the underlying solver, so we omit an explicit reference to them.

$$\mathsf{hotstart}(cond,s_1,s_2) \equiv \mathsf{portfolio}([\mathsf{limit}(cond,s_1),s_2])$$

The $\mathsf{restart}(cond,s)$ combinator repeatedly runs s in full. If s was not exhaustive, it is restarted, until condition *cond* no longer holds. Fig. 2.5 shows the two cases, on the left terminating with an exhaustive search s, on the right terminating because *cond* is no longer true.

The following implements random restarts, where search is stopped after 1000 failures and restarted with a random strategy:

```
restart(true, limit(failures < 1000, base_search(xs, randomvar, randomval)))
```

Clearly, this strategy has a flaw: If it takes more than 1000 failures to find the solution, the search will never finish. We will shortly see how to fix this by introducing user-defined search variables.

The prune primitive is the only source of non-exhaustiveness. Combinators propagate exhaustiveness in the obvious way:

- and($[s_1,\ldots,s_n]$) is exhaustive if all s_i are
- or($[s_1,\ldots,s_n]$) is exhaustive if all s_i are
- portfolio($[s_1,\ldots,s_n]$) is exhaustive if one s_i is
- restart($cond, s$) is exhaustive if the last iteration is
- ifthenelse($cond, s_1, s_2$) is exhaustive if, whenever $cond$ is true, then s_1 is, and, whenever $cond$ is false, then s_2 is.

2.3 State Access and Manipulation

The remaining three primitives, let, assign, and post, are used to access and manipulate the state of the search:

- let(v, e, s) introduces a new search variable v with initial value of the expression e and visible in the search s, then continues with s. Note that search variables are distinct from the decision variables of the model.
- assign(v, e): assigns the value of the expression e to search variable v and succeeds.
- post(c, s): provides access to the underlying constraint solver, posting a constraint c at every node during s. If s is omitted, it posts the constraint and immediately succeeds.

These primitives add a great deal of expressiveness to the language, as the following examples demonstrate.

Random Restarts: Let us reconsider the example using random restarts from the previous section, which suffered from incompleteness because it only ever explored 1000 failures. A standard way to make this strategy complete is to increase the limit geometrically with each iteration:

```
geom_restart(s) ≡ let(maxfails, 100,
                      restart(true, portfolio([limit(failures < maxfails, s),
                                               and([assign(maxfails, maxfails * 1.5),
                                                    prune])])))
```

The search initializes the search variable *maxfails* to 100, and then calls search s with *maxfails* as the limit. If the search is exhaustive, both the portfolio and the restart combinators are finished. If the search is not exhaustive, the limit is multiplied by 1.5, and the search starts over. Note that assign succeeds, so we need to call prune afterwards in order to propagate the non-exhaustiveness of s to the restart combinator.

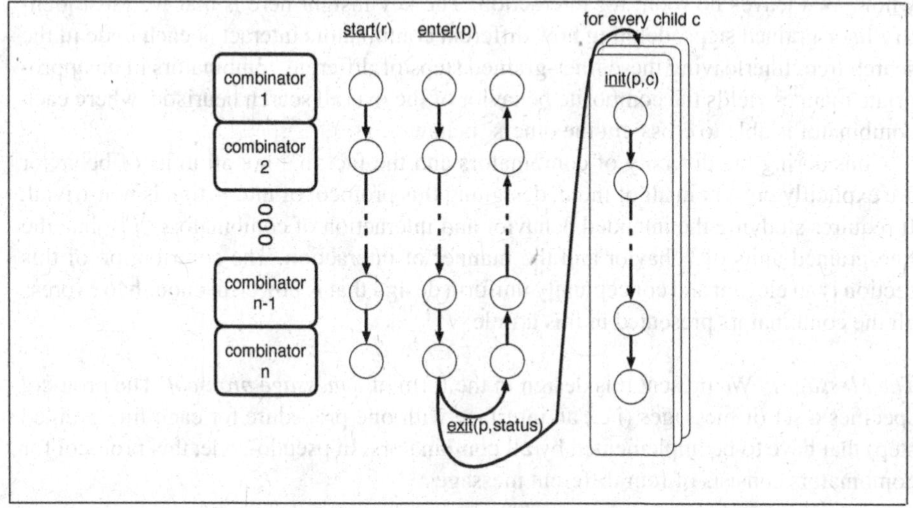

Fig. 3. The modular message protocol

Other Heuristics. Many more heuristics can be implemented with the primitive combinators: branch-and-bound, restarting branch-and-bound, limited discrepancy search, iterative deepening, dichotomic search, ... See [10] for the details of these heuristics.

3 Modular Combinator Design

The previous section caters for the user's needs, presenting a high-level modular syntax for our combinator-based search language. To cater for advanced users' and system developers' needs, this section goes beyond modularity of syntax, introducing modularity of *design*.

Modularity of design is the one property that makes our approach practical. Each combinator corresponds to a separate module that has a meaning and an implementation independent of the other combinators. This enables us to actually realize the search specifications defined by modular syntax.

Solver independence is another notable property of our approach. While a few combinators access solver-specific functionality (e.g., base_search and post), the approach as such and most combinators listed in Fig. 1 are in fact generic (solver- and even CP-independent); their design and implementation is reusable.

In the following we explain our design in detail by means of code implementations of most of the primitive combinators we have covered in the previous section.

3.1 The Message Protocol

To obtain a modular design of search combinators we step away from the idea that the behavior of a search combinator, like the and combinator, forms an indivisible whole; this leaves no room for interaction. The key insight here is that we must identify finer-grained steps, defining how different combinators interact at each node in the search tree. Interleaving these finer-grained steps of different combinators in an appropriate manner yields the composite behavior of the overall search heuristic, where each combinator is able to cross-cut the others' behavior.

Considering the diversity of combinators and the fact that not all units of behavior are explicitly present in all of them, designing this protocol of interaction is non-trivial. It requires studying the intended behavior and interaction of combinators to isolate the fine-grained units of behavior and the manner of interaction. The contribution of this section is an elegant and conceptually uniform design that is powerful enough to express all the combinators presented in this article.

The Messages. We present this design in the form of a *message protocol*. The protocol specifies a set of messages (i.e., an interface with one procedure for each fine-grained step) that have to be implemented by all combinators. In pseudo-code, this protocol for combinators consists of four different messages:

```
protocol combinator
  start(rootNode);
  enter(currentNode);
  exit(currentNode,status);
  init(parentNode,childNode);
```

The protocol concerns the *dynamic* behavior of a search combinator. A single static occurrence of a search combinator in a search heuristic may have zero or more dynamic *life cycles*. During a life cycle, the combinator observes and influences the search of a particular subtree of the overall search tree.

- The message `start(rootNode)` starts up a new life cycle of a combinator for the subtree rooted at `rootNode`. The typical implementation of this message allocates and initializes data for the life cycle.
- The message `enter(currentNode)` notifies the combinator of the fact that the node `currentNode` of its subtree is currently active. At this point the combinator may for instance decide to prune it.
- The message `exit(currentNode,status)` informs the combinator that the currently active node `currentNode` is a leaf node of its subtree. The node's status is one of **`failure`**, **`success`** or **`abort`** which denote respectively an inconsistent node, a solution and a pruned node.
- The message `init(parentNode,childNode)` registers with the combinator the node `childNode` as a child node of the currently active node `parentNode`.

Typically, during a life cycle, a combinator sees every node three times. The first time the node is included in the life cycle, either as a root with `start` or as the child of another node with `init`. The second time the node is processed with `enter`. The last time the node processing has determined that the node is either a leaf with `exit` or the parent of one or more other nodes with `init`.

The Nodes. All of the message signatures specify one or two search tree *nodes* as parameters. Each such node keeps track of a solver `State` and the information associated by combinators to that `State`.

We observe three different access patterns of nodes:

1. In keeping with the solver independence stipulated above, we will see that most combinators only query and update their associated information and do not access the underlying solver `State` at all.
2. Restarting-based combinators, like `restart` and `portfolio`, copy nodes. This means copying the solver's `State` representation and all associated information for later restoration.
3. Finally, selected solver-specific combinators like `base_search` do perform solver-specific operations on the underlying `State`, like querying variable domains and posting constraints.

The Calling Hierarchy. In addition to the message signatures, the protocol also stipulates in what order the messages are sent among the combinators (see Fig. 3). While in general a combinator composition is tree-shaped, the processing of any single search tree node p only involves a stack of combinators. For example, given $\mathsf{or}([\mathsf{and}_1([s_1,s_2]), \mathsf{and}_2([s_3,s_4])])$,[1] p is included in life cycles of $[\mathsf{or},\mathsf{and}_1,s_1]$, $[\mathsf{or},\mathsf{and}_1,s_2]$, $[\mathsf{or},\mathsf{and}_2,s_3]$ or $[\mathsf{or},\mathsf{and}_2,s_4]$. We also say that the particular stack is *active* at node p. The picture shows this stack of active combinators on the left.

Every combinator in the stack has both a *super*-combinator above and a *sub*-combinator below, except for the *top* and the *bottom* combinators. The bottom is always a basic heuristic (base_search, prune, assign, or post). The important aspect to take away from the picture is the direction of the four different messages, either top-down or bottom-up.

The protocol initializes search by sending the <u>`start`</u>`(root)` message, where `root` is the root of the overall search tree, to the topmost combinator. This topmost combinator decides what child combinator to forward the message to, that child combinator propagates it to one of its children and so on, until a full stack of combinators is initialized.

Next, starting from the root node, nodes are processed in a loop. The <u>`enter`</u>`(`**`node`**`)` message is passed down through the stack of combinator stack to the primitive heuristic at the bottom, which determines whether the node is a leaf or has children. In the former case, the primitive heuristic passes the <u>`exit`</u>`(`**`node`**`, status)` message up. In the latter case, it passes the <u>`init`</u>`(`**`node`**`, child)` message down from the top for each child. These child nodes are added to the queue that fuels the loop. At any point, intermediate combinators can decide not to forward messages literally, but to alter them instead (e.g., to change the status of a leaf from **`success`** to **`abort`**), or to initiate a different message flow (e.g. to start a new subtree).

3.2 Basic Setup

Before we delve into the interesting search combinators, we first present an example implementation of the basic setup consisting of a base search (base_search) and a

[1] The left and right and are subscripted to distinguish them.

search engine (dfs). This allows us to express overall search specifications of the form: dfs(base_search(*vars*,*var-select*,*domain-split*)).

Base Search. We do not provide full details on a base_search combinator, as it is not the focus of this article. However, we will point out the aspects relevant to our protocol.

The first line of base_search's implementation expresses two facts. Firstly, it states that the base_search implements the **combinator** protocol. Secondly, its constructor has three parameters (vars, var-select, domain-select) that can be referred to in its message implementations.

In the enter message, the node's solver state is propagated. Subsequently, the condition isLeaf(c,vars) checks whether the solver state is unsatisfiable or there are no more variables to assign. If either is the case, the exit status (respectively **failure** or **success**) is sent to the parent combinator. For now, the parent combinator is just the search engine, but later we will see how how other combinators can be inserted between the search engine and the base search.

If neither is the case, the search branches depending on the variable selection and domain splitting strategies. This involves creating a child node for each branch, determining the variable and value for that child and posting the assignment to the child's state. Then, the **top** combinator (i.e., the engine) is asked to initialize the child node. Finally the child node is pushed onto the search queue.

```
combinator base_search(vars,var-select,domain-select)
  enter(c):
    c.propagate
    if isLeaf(c,vars)
      parent.exit(c,leafstatus(c))
    pos = ...          // from vars based on var-select
    for each child: // based on domain-select
      val = ...        // based on domain-select
      child.post(vars[pos]=val)
      top.init(c,child)
      queue.push(child)
```

Note that, as the base_search combinator is a base combinator, its exit message is immaterial (there is no child heuristic of base_search that could ever call it). The start and init messages are empty. Many variants on and generalizations of the above implementation are possible.

Depth-First Search Engine. The engine dfs serves as a pseudo-combinator at the **top** of a combinator expression heuristic and serves as the heuristic's immediate parent as well. It maintains the **queue** of nodes, a stack in this case. The search starts from a given root node by starting the heuristic with that node and then entering it. Each time a node has been processed, new nodes may have been pushed onto the queue. These are popped and entered successively.

```
combinator dfs(heuristic)
  start(root):
    top=this
    heuristic.parent=this
    queue=new stack()
    heuristic.start(root)
    heuristic.enter(root)
    while not queue.empty
      heuristic.enter(queue.pop())

  init(n,c):
    heuristic.init(n,c)
```

The engine's exit message is empty, the enter message is never called and the init message delegates initialization to the heuristic.

Other engines may be formulated with different queuing strategies.

3.3 Combinator Composition

The idea of search combinators is to augment a base_search. We illustrate this with a very simple print combinator that prints out every solution as it is found. For simplicity we assume a solution is just a set of constraint variables *vars* that is supplied as a parameter. Hence, we obtain the basic search setup with solution printing with:

dfs(print(*vars*,base_search(*vars*, *strategy*)))

Print. The print combinator is parametrized by a set of variables vars and a search combinator child. Implicitly, in a composition, that child's parent is set to the print instance. The same holds for all following search combinators with one or more children.

The only message of interest for print is exit. When the exit status is **success**, the combinator prints the variables and propagates the message to its parent.

```
combinator print(vars,child)
  exit(c,status):
    if status==success
      print c.vars
    parent.exit(c,status)
```

The other messages are omitted. Their behavior is default: they all propagate to the child. The same holds for the omitted messages of following unary combinators.

We refer to [10] for the definitions of all the primitive search combinators.

4 Modular Combinator Implementation

The message-based combinator approach lends itself well to different implementation strategies. In the following we briefly discuss two diametrically opposed approaches we have explored:

Dynamic composition implements combinators as objects that can be combined arbitrarily at runtime. It therefore acts like an *interpreter*. This is a lightweight implementation, it can be ported quickly to different platforms, and it does not involve a compilation step between the formulation and execution of a search heuristic.

Static composition uses a code generator to translate an entire combinator expression into executable code. It is therefore a *compiler* for search combinators. This approach lends itself better to various kinds of analysis and optimization.

As both approaches are possible, combinators can be adapted to the implementation choices of existing solvers. Experimental evaluation [10] has shown that both approaches have competitive performance.

4.1 Dynamic Composition

To support dynamic composition, we have implemented our combinators as C++ classes whose objects can be allocated and composed into a search specification at runtime. The protocol events correspond to virtual method calls between these objects. For the delegation mechanism from one object to another, we explicitly encode a form of dynamic inheritance called *open recursion* or *mixin inheritance* [7]. In contrast to the OOP inheritance built into C++ and Java, this mixin inheritance provides two essential abilities: 1) to determine the inheritance graph *at runtime* and 2) to use multiple copies of the same combinator class at different points in the inheritance graph. In contrast, C++'s built-in static inheritance provides neither.

The C++ library currently builds on top of the Gecode constraint solver [5]. However, the solver is accessed through a layer of abstraction that is easily adapted to other solvers (e.g., we have a prototype interface to the Gurobi MIP solver). The complete library weighs in at around 2500 lines of code, which is even less than Gecode's native search and branching components.

4.2 Static Composition

In a second approach, also on top of Gecode, we statically compile a search specification to a tight C++ loop. Again, every combinator is a separate module independent of other combinator modules. A combinator module now does not directly implement the combinator's behavior. Instead it implements a code generator (in Haskell), which in turn produces the C++ code with the expected behavior.

Hence, our search language compiler parses a search specification, and composes (in mixin-style) the corresponding code generators. Then it runs the composite code generator according to the message protocol. The code generators produce appropriate C++ code fragments for the different messages, which are combined according to the protocol into the monolithic C++ loop. This C++ code is further post-processed by the C++ compiler to yield a highly optimized executable.

As for dynamic composition, the mixin approach is crucial, allowing us to add more combinators without touching the existing ones. At the same time we obtain with the press of a button several 1000 lines of custom low-level code for the composition of just a few combinators. In contrast, the development cost of hand crafted code is prohibitive.

As the experiments in the next section will show, compiling the entire search specification into an optimised executable achieves better performance than dynamic composition. However, the dynamic approach has the big advantage of not requiring a compilation step, which means that search specifications can be constructed at runtime, as exemplified by the following application.

4.3 Further Implementations

We are in the process of implementing the search combinators approach on three more platforms:

MiniZinc. As a proof of concept and platform for experiments, we have integrated search combinators into a complete MiniZinc toolchain:[2] The toolchain comprises a pre-compiler, which is necessary to support arbitrary expressions in annotations, such as the condition expressions for an ifthenelse. The expressions are translated into standard MiniZinc annotations that are understood by the FlatZinc interpreter. We extended the Gecode FlatZinc interpreter to parse the search combinator annotation and construct the corresponding heuristic using the Dynamic Composition approach described above.

Prolog. Our Tor library [13] implements a subset of the search message protocol in Prolog. The library is currently available for SWI-Prolog [14] and B-Prolog [15], and extends the capabilities of their respective finite domain solver libraries. Among others, it provides all the search heuristics of ECLiPSe Prolog's [4] `search/6` predicate, but in a fully compositional way. The library implements the dynamic approach supplemented with load-time program specialization.

Scala. Desouter [16] has implemented a preliminary library of search combinators for Scala [17] on the Java Virtual Machine. His implementation exploits Scala's built-in mixin mechanism (called *traits*) to further factorize the combinator implementations. The library's current backend is the JaCoP solver [18].

5 Conclusion

Search combinators provide a powerful high-level language for modeling complex search heuristics. To make this approach useful in practice, the architecture matches the modularity of the language with the modularity of the implementation. This relieves system developers from a high implementation cost and yet, as experiments show, imposes no runtime penalty.

Acknowledgements. NICTA is funded by the Australian Government as represented by the Department of Broadband, Communications and the Digital Economy and the Australian Research Council. This work was partially supported by Asian Office of Aerospace Research and Development grant 10-4123.

[2] The source code including examples can be downloaded from `http://www.gecode.org/flatzinc.html`

References

1. Van Hentenryck, P., Perron, L., Puget, J.F.: Search and strategies in OPL. ACM TOCL 1(2), 285–315 (2000)
2. Van Hentenryck, P., Michel, L.: Constraint-Based Local Search. MIT Press (2005)
3. Marriott, K., Nethercote, N., Rafeh, R., Stuckey, P., Garcia de la Banda, M., Wallace, M.: The design of the Zinc modelling language. Constraints 13(3), 229–267 (2008)
4. Schimpf, J., Shen, K.: ECLiPSe – From LP to CLP. Theory and Practice of Logic Programming 12(1-2), 127–156 (2012)
5. Schulte, C., et al.: Gecode, the generic constraint development environment (2009) `http://www.gecode.org/` (accessed November 2012)
6. Nethercote, N., Stuckey, P.J., Becket, R., Brand, S., Duck, G.J., Tack, G.: MiniZinc: Towards a standard CP modelling language. In: Bessière, C. (ed.) CP 2007. LNCS, vol. 4741, pp. 529–543. Springer, Heidelberg (2007)
7. Cook, W.R.: A denotational semantics of inheritance. PhD thesis, Brown University (1989)
8. Kiczales, G., Lamping, J., Menhdhekar, A., Maeda, C., Lopes, C., Loingtier, J.-M., Irwin, J.: Aspect-oriented programming. In: Akşit, M., Matsuoka, S. (eds.) ECOOP 1997. LNCS, vol. 1241, pp. 220–242. Springer, Heidelberg (1997)
9. Schrijvers, T., Tack, G., Wuille, P., Samulowitz, H., Stuckey, P.J.: Search combinators. In: Lee, J. (ed.) CP 2011. LNCS, vol. 6876, pp. 774–788. Springer, Heidelberg (2011)
10. Schrijvers, T., Tack, G., Wuille, P., Samulowitz, H., Stuckey, P.: Search combinators. Constraints, 1–37 (2012)
11. Refalo, P.: Impact-based search strategies for constraint programming. In: Wallace, M. (ed.) CP 2004. LNCS, vol. 3258, pp. 557–571. Springer, Heidelberg (2004)
12. Perron, L.: Search procedures and parallelism in constraint programming. In: Jaffar, J. (ed.) CP 1999. LNCS, vol. 1713, pp. 346–361. Springer, Heidelberg (1999)
13. Schrijvers, T., Triska, M., Demoen, B.: Tor: Extensible search with hookable disjunction. In: Principles and Practice of Declarative Programming, PPDP 2012. ACM (2012)
14. Wielemaker, J., Schrijvers, T., Triska, M., Lager, T.: SWI-Prolog. Theory and Practice of Logic Programming 12(1-2), 67–96 (2012)
15. Zhou, N.F.: The language features and architecture of B-Prolog. Theory and Practice of Logic Programming 12(1-2), 189–218 (2012)
16. Desouter, B.: Modular Search Heuristics in Scala. Master's thesis, Ghent University (2012) (in Dutch)
17. Cremet, V., Garillot, F., Lenglet, S., Odersky, M.: A core calculus for Scala type checking. In: Královič, R., Urzyczyn, P. (eds.) MFCS 2006. LNCS, vol. 4162, pp. 1–23. Springer, Heidelberg (2006)
18. Kuchcinski, K., Szymanek, R.: JaCoP - Java Constraint Programming solver (2012), `http://www.jacop.eu/` (accessed November 2012)

A Declarative Pipeline Language for Complex Data Analysis

Henning Christiansen, Christian Theil Have,
Ole Torp Lassen, and Matthieu Petit

Research group PLIS: Programming, Logic and Intelligent Systems
Department of Communication, Business and Information Technologies
Roskilde University, P.O. Box 260, 4000 Roskilde, Denmark
{henning,cth,otl,petit}@ruc.dk

Abstract. We introduce BANpipe – a logic-based scripting language designed to model complex compositions of time consuming analyses. Its declarative semantics is described together with alternative operational semantics facilitating goal directed execution, parallel execution, change propagation and type checking. A portable implementation is provided, which supports expressing complex pipelines that may integrate different Prolog systems and provide automatic management of files.

1 Introduction

Computations for biological sequence processing are often complex compositions of time consuming analyses, including calls to external resources and databases. The expected output, intermediate results, and original input data are often huge files. To facilitate such computations, we introduce a new declarative scripting language called BANpipe. The name BANpipe is in reference to Bayesian Annotation Networks (BANs) [4,11], which BANpipe is designed to support. BANpipe is not limited to BANs, but it is a general and extensible pipeline programming language which can be adapted for different scenarios. The language supports complex pipelines of Prolog programs, PRISM [19] models, and other types of programs through rules which specify dependencies between files.

BANpipe rules express dependencies between symbolically represented files that are automatically mapped to the underlying filesystem. A dependency is specified in terms of a function from (symbolic) input files to (symbolic) output files. The implementation of this function is delegated to a predicate in a Prolog or PRISM program. BANpipe handles the calling of these predicates (in separate processes) and converts between symbolical identifiers and real files in the filesystem. The symbolic filenames may include logic variables, and dependency rules may include guards which enable advanced control mechanisms.

Execution of a pipeline script is goal directed, where only the desired result is specified, and the system then executes the programs necessary to achieve the result as entailed by the dependencies in the script. Computations entailed by multiple goals are only performed once and subsequently shared by the goals.

E. Albert (Ed.): LOPSTR 2012, LNCS 7844, pp. 17–34, 2013.
© Springer-Verlag Berlin Heidelberg 2013

Incremental change propagation is supported, so that only the necessary files are recomputed in case one or more program or input data files are modified.

BANpipe is implemented in Logtalk [12], a portable language based on Prolog. BANpipe supports seamless integration of programs written for different Prolog systems and runs on multiple operating systems. It is designed to be familiar to the logic programming community, and scripts in BANpipe can utilize the full power of Logtalk/Prolog and associated libraries to automate complex workflows. The BANpipe is designed to be extensible. It is easy to define a new specialized semantics, as well as extending many of the core components of the system such as the mechanisms to handle tracking and organization of generated files. The system is extended with semantics that enables parallel execution and advanced debugging facilities (type checking, interactive tracing and visualization of the pipeline structure). These facilities are particularly useful when working complex pipelines that may be very time-consuming to run.

Overview of the Paper

Section 2 gives an informal introduction to the language and its semantics, and Section 3 explains the declarative semantics in depth. Because of the way the semantics are presented, proofs in this and the following sections often turn out to be trivial. A number of operational semantics are described in Section 4, including sequential operational semantics, change propagation semantics and parallel operational semantics. Following this, Section 5 introduces an alternative semantics for type inference. Examples of BANpipe scripts from the domain of biological sequence analysis are given in Section 6. Section 7 discusses the implementation of BANpipe. Related work is discussed in Section 8. Conclusions and future research directions are presented in Section 9.

2 Syntax and Informal Semantics of BANpipe

The BANpipe language consists of scripts for composing collections of programs that work on data files. We add another layer on top of the traditional file system, so that an arbitrary ground Prolog term can be used for identifying a file. We assume a *(local) file environment* that maps such file identifiers into files, via their real file name in the local file system or a remote resource. The syntax inherits Prolog's notions of terms and variables (written with capital letters). A Prolog constant whose name indicates a protocol is treated as a URL. The files for other terms are determined through the local file environment. For example:

- `'file:///a/b/c/datafile'`: refers to the file with the real name `datafile` in the `/a/b/c` directory in the local file system,
- `'http://a.b.c/file'`: refers to a file referenced using the http protocol.
- `f(7)`: may refer to a file in the local file system.

To simplify usage, we refer to Prolog terms expected to denote files as *file names*.

The programs referred to in BANpipe scripts are placed in modules, and a program defines a function from zero or more files to one or more files; a program may take options as additional arguments that modify the function being calculated. Programs are referred to in the body of BANpipes *dependency rules*, exemplified as follows.

```
file1, file2 <- m::prog([file3,file4],op1,op2).        (1)
```

Here `prog` is a program in module `m`, taking two files `file3` and `file4`, plus options `op1` and `op2` as input. The rule states how two output files `file1` and `file2` depend on `file3`, `file4`, namely being the result of applying the function (or *task*) given by `m::prog([–,–],op1,op2)`.

File names in rules can be parameterized as shown in this rule:

```
f(N) <- m::prog(g(N)).        (2)
```

For any ground instance of `N`, this rule explains the dependency between two files, e.g., between `f(7)` and `g(7)` or `f(h(a))` and `g(h(a))`. Rules can be recursive as shown in the following example.

```
f(0) <- file::get('file:///data').        (3)
f(N) <- N > 0, N1 is N-1 | m::prog(f(N1)).        (4)
```

Here, rule (3) applies a built-in module for simple file handling including the `get` facility that provides a copy of a file as shown; rule (4) includes a *guard*, which may precede the program call and is used for rule selection and for instantiating variables not given by the matching in the head. The recursion works as expected: the evaluation of a query `f(2)` involves the calculation of files named `f(2)`, `f(1)` and `f(0)` from the local file with the real name `data`.

A BANpipe *script* is a sequence of rules defined by the following syntax, extended with any definitions of Prolog predicates that are used in rule guards.

$$\langle\text{rule}\rangle ::= \langle\text{head}\rangle \texttt{ <- } \{\langle\text{guard}\rangle \texttt{|}\} \langle\text{body}\rangle \tag{5}$$

$$\langle\text{head}\rangle ::= \langle\text{file}\rangle_1 \texttt{,} \ldots \texttt{,} \langle\text{file}\rangle_m \quad m \geq 1 \tag{6}$$

$$\langle\text{file}\rangle ::= \textit{any Prolog term, as described above} \tag{7}$$

$$\langle\text{body}\rangle ::= \langle\text{program call}\rangle \tag{8}$$

$$\langle\text{guard}\rangle ::= \textit{sequence of one or more Prolog calls} \tag{9}$$

$$\langle\text{program call}\rangle ::= \langle\text{module}\rangle \texttt{::} \langle\text{program name}\rangle \texttt{(} \langle\text{file}\rangle_1 \texttt{,} \ldots \texttt{,} \langle\text{file}\rangle_n \texttt{,} \textit{options}\texttt{)} \quad n \geq 0 \tag{10}$$

The following restrictions must hold for any BANpipe script:

- File names given as URLs cannot occur in rule heads.
- When a rule head contains file names with variables, any file name in that head must contain the same variables.
- The evaluation of a guard must always terminate.

A *query* is a finite set of *goals*, each of which is a ground term, supposed to represent a file name. The selection of a rule for the evaluation of a goal must be unique. We capture the essential properties in the following definition and explain afterwards how such a selection function is implemented in practice.

Definition 1. *A selection function for a BANpipe script S is a partial function σ_S from non-URL ground Prolog terms to ground instances of rules of S such that if*

$$\sigma_S(f) = (f_1^{out}, \ldots, f_n^{out} \texttt{ <- } guard \mid m :: prog(f_1^{in}, \ldots, f_m^{in}, opts)) \tag{11}$$

then guard evaluates to true, and it holds that

$$f = f_i^{out} \;\; \textit{for some } i = 1, \ldots, n, \tag{12}$$

$$\sigma_S(f_i^{out}) = \sigma_S(f) \;\; \textit{for all } i = 1, \ldots, n. \tag{13}$$

Any such instance is called a selection instance *for S.*

To simplify notation later, we may leave out the guard when referring to a selected instance, as it has made its duty for testing and variable instantiation, once the selection is effectuated. Condition (12) states that the chosen rule is actually relevant for f, and condition (13) indicates that whenever a rule is applied, it calculates the unique results for all files mentioned in its head, independently of which request for a file that triggered the rule.

In the implemented system, the rules are checked in the order they appear in the script and the file names in their heads from left to right. If such a head file name unifies with the given f, and the guard succeeds, the rule is a candidate for selection. However, if the execution of a guard leads to a runtime error or does not instantiate all remaining variables in the rule, the search stops and no rule is selected. Condition (13) of Def. 1 is undecidable, but it is straightforward to define sufficient conditions that can be checked syntactically; we do not consider this topic further here. In the implemented system, it is not checked.

The evaluation of a query Q can be done in a standard recursive way, which will be described in more detail in Section 4.

2.1 Defining Programs and Modules

As mentioned, the tasks activated from a BANpipe script are defined by programs that are grouped into named modules. How these modules are structured is not required for the understanding of the BANpipe script language, so we give here just a brief overview. A module m must contain a designated interface file that implements each task through a Prolog predicate of the following form,

$$task\texttt{([}in\text{-}file_1, \ldots, in\text{-}file_n\texttt{]}, opts, \texttt{[}out\text{-}file_1, \ldots, out\text{-}file_m\texttt{])} \tag{14}$$

that will be matched by a program call $m\texttt{::}task(\cdots)$ in a script as described above. In accordance with the precise semantics specified below, the file names (whether URLs or arbitrary Prolog ground terms) encountered by the script are mapped into references to actual files, which then are given to task predicates that access the files through standard input/output built-ins.

The interface file may contain all the code that implements the tasks but, typically, a module contains a number of source files shared by the different tasks. The execution of a task is done by a Prolog system or the PRISM system.

The system includes a sort of dynamic types that are specified in the interface files and not visible in the BANpipe scripts. Besides the predicate that implements a particular task, each task is declared using a fact on the form,

$$\texttt{task}(\mathit{task}\texttt{(}\texttt{[}\mathit{Type}_1^{in},\ldots,\mathit{Type}_n^{in}\texttt{]}\texttt{,}\mathit{options}\texttt{,}\texttt{[}\mathit{Type}_1^{out},\ldots,\mathit{Type}_m^{out}\texttt{]}\texttt{)}\texttt{)} \tag{15}$$

Each $Type_i^{in/out}$ are Prolog terms – this is described in more detail in Section 5. The *options* argument is a list of valid options (functors) and their default values.

3 Declarative Semantics of BANpipe

Raw data and results of analyses are represented as data files. No assumptions are made about the structure of those files, and we assume an unspecified domain

$$\mathit{DataFile} \tag{16}$$

including a $\perp$ element, indicating an unsuccessful result.

Program calls in a script denote *tasks* that are mappings from a (perhaps empty) sequence of data files into another sequence of data files. Thus

$$\mathit{Task} = \sum_{i=0,1,\ldots;j=1,2,\ldots} \mathit{Task}_{i,j} \tag{17}$$

$$\mathit{Task}_{i,j} = \mathit{DataFile}^i \to \mathit{DataFile}^j \tag{18}$$

Tasks must be strict: if any input argument is $\perp$, the result is $\perp$. A task may also result in $\perp$ reflecting a runtime error or a Prolog failure.

Definition 2. *A* program semantics *is a function* $[\![-]\!]$ *from triples of module name, program name, and ground values for possible option parameters into tasks. For module mod, program prog (with n input and m output files), and option values opts, this function is indicated as*

$$[\![\mathit{mod} :: \mathit{prog}(\mathit{opts})]\!] \in \mathit{Task}_{n,m}. \tag{19}$$

Ground file names are used as synonyms for variables ranging over the domain *DataFile*; for a ground file name f, the corresponding unique variable, called a *file variable*, is denoted $\widehat{f}$, and this notation is extended to sets, $\widehat{F} = \{\widehat{f} \mid f \in F\}$; whenever f is a URL, $\widehat{f}$ is called a *URL variable.* (Partial) answers to queries are represented below as substitution for file variables into *DataFile* and are typically indicated by the letter Φ with possible subscripts. We recognize a special form of *URL substitutions* for URL variables only. For ease of notation, a URL substitution is assumed to provide a value for any URL variable which might be $\perp$. The notation Φ_0 typically refers to a URL substitution. A substitution Φ is considered equivalent to the set of equations $\{\widehat{f} \doteq d \mid \Phi(\widehat{f}) = d\}$.[1]

The declarative meaning of a BANscript is given by a recursive systems of equations defined as follows.

Definition 3. *Given a BANscript S, a* defining equation for *a non-URL file name f is of the form*

$$\langle \widehat{f_1^{out}}, \ldots, \widehat{f_m^{out}} \rangle \doteq [\![mod :: prog(opts)]\!] \langle \widehat{f_1^{in}}, \ldots, \widehat{f_n^{in}} \rangle \tag{20}$$

where $f = \widehat{f_i^{out}}$ for some $i = 1, \ldots, m$, and S has a selection instance for f,

$$f_1^{out}, \ldots, f_m^{out} \texttt{<-} \, m :: prog(f_1^{in}, \ldots, f_m^{in}, opts). \tag{21}$$

Given such S and $[\![-]\!]$, the defining set of equations for a query *Q, denoted Eq(Q, S) is defined as the smallest set E of defining equations such that*

- *E contains a defining equation for any $q \in Q$,*
- *for any equation in E whose righthand side contains a non-URL variable $\widehat{f}$, E contains a defining equation for f.*

We say that a BANscript S is well-behaved *for a query Q if Eq(Q, S) exists, is finite, and contains no circularities.*[2]

Notice that $Eq(Q, S)$ is defined independently of program semantics, so this definition is equally relevant for a standard semantics as for different abstract semantics reflecting different program properties.

The solution to a set of equations is given as usual, as a substitution that maps variables to values, such that the left and right hand sides of each equation become identical when all functions are evaluated. Whenever a script S is well-behaved for a query Q, and Φ_0 is a URL substitution for the URL variables of $Eq(Q, S)$, there exists a unique solution for $Eq(Q, S) \cup \Phi_0$. To prove this, we first convert each equation on tuples into equations of the form $\widehat{f_i} \doteq \ldots$ by projections, and then notice that all variables in the righthand sides can be eliminated in a finite number of steps. Condition (13) of Def. 1 ensures that the resulting set of equations with variable-free righthand sides is unique. We can thus define:

Definition 4. *Let $[\![-]\!]$ be a program semantics, S a BANscript which is well-behaved for a query Q and Φ_0 a URL substitution, and let Φ be the solution to $Eq(Q, S) \cup \Phi_0$. The* answer *to Q (with respect to S, $[\![-]\!]$ and Φ_0) is the restriction of Φ to $\widehat{Q}$; the substitution Φ is referred to as the* full answer *to Q.*

The query is failed *whenever the solution assigns $\perp$ to any variable in $\widehat{Q}$.*

[1] We use the symbol "$\doteq$" to distinguish equations that are explicit syntactic objects from the normal use of "=" as meta-notation.

[2] "No circularities" can be formalized by separating variables into disjoint, indexed strata, such that for an equation $\cdots V \cdots = \cdots V' \cdots$, that the stratum number for V' is always lower than the stratum number for V.

Alternatively, this semantics could have been formulated in terms of a fixed point or a least model, which is straightforward due to the well-behavedness property.

Well-behavedness is obviously an undecidable property as an arbitrary Turing machine can be encoded through recurrence of variables in the terms that represent file names. In practice, this can be checked by symbolic execution of the script using depth-first search with a depth limit.

4 Operational Semantics

We present a number of alternative operational semantics for BANpipe as abstract algorithms.

4.1 Bottom-Up Operational Semantics with Memoization

The following algorithm defines an operational semantics that works in a bottom-up fashion, calculating all involved files from scratch. Each file needed to obtain the final results is evaluated exactly once, even if used in different program calls.

Algorithm 1. Bottom-up operational semantics for BANscript
Input: *A query Q, a BANscript S, program semantics $[\![-]\!]$ and initial substitution Φ_0;*
Output: *A substitution;*

$\Phi := \Phi_0$;
while $Eq(Q, S)$ *contains an equation*
$\langle \widehat{f_1^{out}}, \ldots, \widehat{f_m^{out}} \rangle \doteq [\![P]\!] \langle \widehat{f_1^{in}}, \ldots, \widehat{f_n^{in}} \rangle$
for which $\Phi(\widehat{f_i^{out}})$ *is undefined for all* $i = 1, \ldots, m$,
and $\Phi(\widehat{f_j^{in}})$ *is defined for all* $j = 1, \ldots, n$
do $\Phi := \Phi[\widehat{f_1^{out}}/df_1, \ldots, \widehat{f_m^{out}}/df_m]$
where $\langle df_1, \cdots, df_m \rangle = [\![P]\!] \langle \Phi(\widehat{f_1^{in}}), \ldots, \Phi(\widehat{f_n^{in}}) \rangle$;
return Φ;

Theorem 1. *Given a program semantics $[\![-]\!]$, a BANscript S which is well-behaved for a query Q and a URL substitution Φ_0, Algorithm 1 returns the full answer Φ to Q. The solution to Q is a found as a the restriction of Φ to $\widehat{Q}$.*

Having the algorithm to return the full substitution produced, makes it possible to use it also for incremental maintenance of solutions, cf. Section 4.2 below.

Sketch of Proof. Each step performed in the while loop in Algorithm 1 corresponds to a variable elimination step in $Eq(Q, S) \cup \Phi_0$. Furthermore, each such step that processes an equation of the form $\langle \widehat{f_1^{out}}, \ldots, \widehat{f_m^{out}} \rangle \doteq [\![P]\!] \langle \widehat{f_1^{in}}, \ldots, \widehat{f_n^{in}} \rangle$, will bind variables $\widehat{f_1^{out}}, \ldots, \widehat{f_m^{out}}$ to their final values in the resulting solution. □

This abstract operational semantics can be transformed into a running implementation by adding suitable data structures for representing the defining equations and the file environment (appearing in the algorithm as substitutions).

Algorithm 1 can also be applied for non-standard program semantics to calculate program properties, which is used below for predicting change propagation, Section 4.2, and type inference, Section 5.

4.2 Operational Semantics for Incremental Change Propagation

When one or more data files, or programs producing data files, are modified, the current file substitution may become inconsistent with the script. As BANpipe is intended for programs with substantial execution time, the number of program runs needed to reestablish consistency should be reduced to a minimum.

An alternative program semantics is used for measuring change propagation, based on the domain $\mathit{DataFile}^{prop} = \{\mathit{changed}, \mathit{unchanged}, \bot\}$. The matching program semantics $[\![-]\!]^{prop}$ is defined as follows: whenever the program *prog* (with n input and m output files) in module *mod* has been modified, or one of its input arguments (x_i below) has the value *changed*, we set

$$[\![\mathit{mod} :: \mathit{prog}(\mathit{opts})]\!]^{prop}\langle x_1, \ldots, x_n\rangle = \underbrace{\langle \mathit{changed}, \ldots, \mathit{changed}\rangle}_{m \text{ times}}; \tag{22}$$

otherwise (i.e., program not modified, input $= \langle \mathit{unchanged}, \ldots, \mathit{unchanged}\rangle$), the program call returns $\langle \mathit{unchanged}, \ldots, \mathit{unchanged}\rangle$.

We consider the difference between two substitutions Φ^{before} and Φ^{after}, intended to represent correct values for all file variables before and after the modification. This is characterized by a *propagation substitution* defined as follows.

$$\mathit{Diff}(\Phi^{before}, \Phi^{after})(\widehat{f}) = \begin{cases} \mathit{unchanged} & \text{whenever } \Phi^{before}(\widehat{f}) = \Phi^{after}(\widehat{f}) \\ \mathit{changed} & \text{otherwise} \end{cases} \tag{23}$$

Changes in the set of URL files (i.e., where the ultimate input comes from) is characterized by a propagation substitution for URL variables.

The following algorithm that predicts which files that need to be re-evaluated, is used as a helper for Algorithm 3, below.

Algorithm 2. Change prediction for BANscript
Input: *A query Q, a BANscript S and URL change substitution* Φ_0^{prop};
Output: *A substitution of variables into* $\{\mathit{changed}, \mathit{unchanged}\}$;

Φ^{prop} := **run** *Algorithm 1 for Q, S,* $[\![-]\!]^{prop}$ *and* Φ_0^{prop};
return Φ^{prop};

We state the following weak correctness statement for Algorithm 2.

Theorem 2 (Soundness of the change prediction algorithm). *Let Φ_0^{before} and Φ_0^{after} be URL substitutions for the same set of variables into DataFile, and assume two program semantics $[\![-]\!]^{before}$ and $[\![-]\!]^{after}$. Let, furthermore,*

Φ_1^{before} be the full answer for Q wrt S, $[\![-]\!]^{before}$ and Φ_0^{before},
Φ_1^{after} the full answer for Q wrt S, $[\![-]\!]^{after}$ and Φ_0^{after}, and
Φ_1^{prop} the result of running Algorithm 2 for Q, S and $Diff(\Phi_0^{before}, \Phi_0^{after})$.

Then it holds for any ground file name f, that if $\Phi_1^{prop}(\widehat{f}) = unchanged$, then $\Phi_1^{before}(\widehat{f}) = \Phi_2^{before}(\widehat{f})$.

Sketch of Proof. According to theorem 1, Φ_1^{before} (resp. Φ_1^{after}) can be characterized as the result of running Algorithm 1 for Q, S, $[\![-]\!]^{before}$ and Φ_0^{before} (resp. $[\![-]\!]^{after}$ and Φ_0^{after}). We can thus construct three synchronized runs of algorithm 1, calculating Φ_1^{before}, Φ_1^{after} and Φ_1^{prop} selecting the same equations in the same order. The theorem is easily shown by induction over these runs. □

The algorithm for incremental maintenance of the solution, when input data or programs are modified, is specified as follows. It uses Algorithm 2 to identify which files to recompute; their values are set to undefined in the current file substitution, and a run of Algorithm 1 starting from this substitution leads to correctly updated file substitutions with as few program calls as possible.

Algorithm 3. Incremental maintenance for BANscript
Input: *A query Q, a BANscript S, a (revised) program semantics $[\![-]\!]^{after}$, a substitution Φ_1 produced by Algorithm 1 from some Q, S, $[\![-]\!]^{before}$ and Φ_0^{before}, and a (revised) URL substitution Φ_0^{after};*
Output: *A substitution;*

Φ^{prop} := **run** *Algorithm 2 for Q, S and $Diff(\Phi_0^{before}, \Phi_0^{after})$;*
$\Phi_2 := \Phi_1 \setminus \{ (\widehat{f}/\Phi_1(\widehat{f})) \mid \Phi^{prop}(\widehat{f}) = changed\}$;
ϕ_3 := **run** *Algorithm 1 for Q, S, $[\![-]\!]^{after}$ and Φ_2;*
return Φ_3;

We leave out the correctness statement, which is straightforward to formulate and follows easily from the previous theorems.

4.3 A Parallel Operational Semantics

BANpipe scripts are obvious candidates for parallel execution, as we can illustrate with this fragment of a script.

```
f0 <- m0::p0(f1,f2,f3).            f2 <- m2::p2('file:///data').
f1 <- m1::p1('file:///data').      f3 <- file::get('http://serv/remoteF').
```

Here `f1` and `f2` can be computed independently in parallel, and at the same time `f3` can be downloaded from the internet. When they all have finished, `m0::p0` can start running, taking as input the files thus produced, but not before.

A parallel operational semantics ia described as modification of Algorithm 1. We assume a *task manager* that maintains a queue of defining equations ready to be executed. Whenever sufficient resources are available, e.g., a free processor plus a suitable chunk of memory, it takes an equation from the queue and start its evaluation in a new process. The task manager receives messages of the form

enqueue(e), e being a defining equation,

and sends messages back of the form

finished($e, df_1, df_2, \ldots$), e being a defining equation, $df_1, df_2, \ldots \in \mathit{DataFile}$.

Such a message should guarantee that the task referred to in e has been applied correctly in order to produce the resulting file values $df_1, df_2, \ldots$ according to the standard semantics $[\![-]\!]$. A parallel operational semantics can now be given by the following abstract algorithm.

Algorithm 4. Parallel operational semantics for BANscript
Input: *A query Q, a BANscript S, program semantics $[\![-]\!]$ and initial substitution Φ_0;*
Output: *A substitution;*

$\Phi := \Phi_0$;
$E := Eq(Q, S)$; // *Equations not yet enqueued*
$F := \emptyset$; // *Equations that have been processed*
while $F \neq Eq(Q, S)$ **do**
  while *there is an* $e \in E$ *of the form*
  $$\langle \widehat{f_1^{out}}, \ldots, \widehat{f_m^{out}} \rangle \doteq [\![P]\!] \langle \widehat{f_1^{in}}, \ldots, \widehat{f_n^{in}} \rangle$$
  for which $\Phi(f_i^{out})$ *is undefined for all* $i = 1, \ldots, m$, *and* $\Phi(f_j^{in})$ *is defined for all* $j = 1, \ldots, n$
  do
    enqueue(e);
    $E := E \setminus \{e\}$;
  await message finished($e', df_1, df_2, \ldots$);
  $\Phi := \Phi[\widehat{f_1'^{out}}/df_1, \ldots, \widehat{f_{m'}'^{out}}/df_{m'}]$
    where $e' = (\langle \widehat{f_1'^{out}}, \ldots, \widehat{f_{m'}'^{out}} \rangle \doteq \cdots)$;
  $F := F \cup \{e'\}$;
return Φ;

Correctness is straightforward as this algorithm performs exactly the same file assignments as algorithm 1. We refrain from a formal exposition.

This algorithm is implemented in our system for a multicore computer, but it should also work for other architectures such as grids and clusters. Algorithms 3 and 4 can be combined into a system that monitors the user's actions and, after each editing, automatically initiates the necessary program runs in order to restore consistency. This will also involve stopping active processes, whose input files have become outdated.

5 Types and Type Inference for BANpipe Scripts

The system includes a dynamic type system such that, for a given program call, the output files are assigned types based on the types of the input files. These types are programmer-defined and may not indicate anything about the internal structure of the file, but provide a general mechanism for checking and inferring aspects of scripts. It is up to the programmer to associate a meaning with the types, and they are used by the system to check and infer specified aspects which may, e.g., be file formats.

Types are not visible in a script, but are managed through optional declarations in the interface files (cf. Section 2.1) and are checked separately by a symbolic execution of the program as explained in the following.

A type can be any Prolog term. A URL file has a default type `file`, which may be coerced into a more specific type by a variant of the `file::get` task, illustrated as follows.

```
f <- file::get(['http://server/file.html'],type(text(html))).
```

A task specified with a task declaration as in (15) that includes an polymorphic type declaration specifying requirements for its input and output files. Each $\mathit{Type}_i^{in/out}$ and *options* are Prolog terms, possibly with variables, and any variable in a Type_j^{out} must occur in some Type_k^{in} or *options*. Thus, if the types for n actual input files plus actual option values simultaneously unifies with $\mathit{Type}_1^{in}, \ldots, \mathit{Type}_n^{in}, \mathit{opts}$, unique ground instances are created for $\mathit{Type}_1^{out}, \ldots, \mathit{Type}_m^{out}$, which are then assigned as types for m output files.

Correct typing of a well-behaved BANscript S with respect to a given query Q are formalized through a type semantics $[\![-]\!]^{type}$. For each task with n input and m output files, assuming a type declaration as in (15) above, we define,

$$[\![\mathit{mod}::\mathit{prog}(\mathit{opts})]\!]^{type}\langle x_1,\ldots,x_n\rangle = \langle \mathit{Type}_1^{out},\ldots,\mathit{Type}_m^{out}\rangle\rho$$
$$\text{where } \rho \text{ is the unifier of } \langle x_1,\ldots,x_n,\mathit{opts}\rangle \quad (24)$$
$$\text{and } \langle \mathit{Type}_1^{in},\ldots,\mathit{Type}_n^{in},\mathit{options}\rangle$$

If the mentioned unifier does not exists, the result is instead $\langle\bot,\ldots,\bot\rangle$.

We can now use Algorithm 1 with this semantics for type checking a BANpipe script with respect to a given query as a symbolic execution of the program. The initial substitution Φ_0^{type} maps any URL variable to the type `file`.

Algorithm 5. Type inference for BANscript
Input: *A query Q, a BANscript S;*
Output: *A substitution of variables into types;*

Φ^{type} := **run** *Algorithm 1 for Q, S,* $[\![-]\!]^{type}$ *and* Φ_0^{type};
return Φ^{type};

If, in the resulting substitution, any file is mapped to $\bot$, we say that type checking of S failed for Q; otherwise type checking succeeded. Taking $[\![-]\!]^{type}$ as definition of correct typing, correctness of this algorithm is a consequence of theorem 1.

6 Examples

We exemplify BANpipe using examples drawn from biological sequence analysis and machine learning. In the first example we present a simple gene prediction pipeline and in the second example we show how such a pipeline can be extended with recursive rules used to implement self-training.

6.1 A Basic Gene Prediction Pipeline

The following is an example of a simple gene prediction pipeline, corresponding to experiments previously reported [4]. The premise is to train a gene finder expressed as a PRISM model using some of the known genes of the *Escherischia Coli* genome (the training set) and verify its prediction accuracy on a different set of known genes (the test set). First we get some initial data files; namely a genome sequence (`fasta_seq`) and a list of reference genes (`genes_ptt`),

```
fasta_seq <- file::get(['ftp://ftp.ncbi.nih.gov/.../NC_000913.fna']).
genes_ptt <- file::get(['ftp://ftp.ncbi.nih.gov/.../NC_000913.ptt']).
```

The fetched files are parsed into suitable format (Prolog facts), and we then extract all open reading frames (`orfs`) from the genome, and divide them into a training set and a test set,

```
genome <- fasta::parse([fasta_seq]).
genes(reference) <- ptt::parse([genes_ptt]).
orfs <- sequence::extract_orfs([genome]).
orfs(training_set), orfs(test_set) <- file::random_split([orfs],seed(42)).
```

Slightly simplified, open reading frames are subsequences of the genome which may contain genes. The `random_split` program divides the `orfs` randomly into the two files `orfs(training_set)` and `orfs(test_set)`. The process is deterministic due to the `seed(42)` option, i.e. it will split `orfs` in the same way if it is rerun. Next, we extract known genes corresponding to each set,

```
genes(Set) <- ranges::intersect([genes(reference),orfs(Set)]).
```

The `ranges` module contains tasks which deal with files containing particular facts which, besides representing sub-sequences, also includes their positions in a genome. The `intersect` task finds all facts from `genes(reference)` where the represented sub-sequences are completely overlapped[3] by a member of `orfs(Set)`. It is used here to find the reference genes that belong to some `Set`, i.e., either the `training_set` or the `test_set`. This concludes the preparation of data files and we turn to the rules for the gene finder:

```
params <- genefinder::learn([genes(training_set)]).
predictions <- genefinder::predict([orfs(test_set), params]).
report <- accuracy::measures([genes(test_set), predictions]).
```

The `genefinder` module contains a PRISM based gene finder and the `learn` task bootstraps and invokes PRISMs machine learning procedure from the facts in the file `genes(traning_set)`. The resulting `params` file is a parameterization for the PRISM model. Using this parameterization, the task `predict` probabilistically predicts which orfs in `orfs(test_set)` represent genes, resulting in the file `predictions`. Finally, the accuracy of the `predictions` are evaluated with regards to the reference genes, `genes(test_set)`, by the task `accuracy::measures` which calculates, e.g., sensitivity and specificity.

6.2 Self-training

Self-training has been demonstrated to yield improved gene prediction accuracy [2]. A self-training gene finder can be expressed by mutually recursive rules,

```
known_genes <- ...
self_learn(1) <- genefinder::learn([known_genes]).
self_learn(N) <- N > 1, N1 is N-1 | genefinder::learn([predict(N1)]).
predict(N) <- genefinder::predict([self_learn(N)]).
```

To elaborate: `known_genes` (obtained somehow) is the starting point for training and `self_learn(1)` is the parameter file resulting from training on `known_genes`. The second `self_learn(N)` rule is the recursive case, which learns parameters from the predictions of the previous iteration. The goal `predict(N)` produces a set of gene predictions based on the parameters of obtained from `self_learn(N)`. For instance, the goal `predict(100)` corresponds to predictions after 100 iterations of self-training.

The example is simple and elegant, but it is not fully satisfactory. Typically, we are interested in termination when learning converges – not after a predefined number of steps. This demonstrates a limitation of BANpipe – we cannot check for convergence in the guard, since the guard does not have direct access of generated files. It is possible to work around this issue by, e.g., letting the task itself check for convergence[4] , but this would not be elegant.

[3] For the biologically inclined; we require this overlap to be in the same reading frame.
[4] This could be done with extra input files with the accuracy and results of previous iteration. If accuracy of the previous iteration is good enough, the task would just output the same files as done in the previous iteration.

7 Implementation

The system was originally implemented directly in PRISM, but the latest implementation is in Logtalk to ensure portability across a wider range of Prolog systems and operating systems. The implementation in Logtalk is considerably simpler and more closely resembles the algorithms given in this paper. In particular, the use of parameterized objects [13] makes it easy to plug in alternative semantics. The implementation also contains several debugging facilities implemented as alternative semantics. It is possible to combine components from several Prolog systems in the same script and the full power of Logtalk and its libraries is available within scripts. The implementation is designed to be extendable by users. It is available from `http://banpipe.org`.

7.1 Task Invocation

When the system needs to invoke a task, it first looks up the interface file of the module where the task is defined. The interface is a plain Prolog file from which BANpipe extracts information about tasks in the module and how to invoke them. A optional declaration, `invoke_with/1`, in the interface file is used to specify that tasks in the module should be run within a particular Prolog system. The argument of `invoke_with/1` identifies a particular Logtalk object (an invoker) which is responsible for invocation of the Prolog/PRISM process. The purpose of an invoker object is to launch the system of choice in the directory of the module, load the interface file (along with possible dependencies) and finally to call a goal corresponding to the task to be invoked.

This goal is constructed from the relevant rule in the script – it has a functor identical to the name of the task and three arguments: The first argument is a list of (real) filenames, which is derived from the symbolic input filenames. The second argument is a list of options: The list of options along with their default values are read from the task declaration in the interface file, but options given in the corresponding rule of the script override (replace) default options. The third argument is a list of file names to be written when the goal completes.

7.2 File Maintenance

BANpipe keeps track of generated files using a transactional *file index*, the operations of which are defined through a Logtalk protocol. The file index defines a many-to-many relation; the combination of module name, task name, input files and options is used as a unique key to identify generated files.

Before the invocation of a task, a lookup in the file index determines if a compatible previously generated file is available and can be returned immediately. If not, a transaction is initiated which allocates unique output filenames to be associated with the task call. If the task completes and writes all its associated output files, the transaction is committed. Otherwise, if the task fails, the transaction rolls back and the filename allocations are relinquished.

Note that we do not have destructive updates as with, e.g., `make`. Files are never overwritten when dependencies change or task definitions change. Changes are not detected automatically, but must be indicated in the task declaration of a module, e.g., with a `version` option with a new default value. Such changes result in different unique keys for relevant task calls and hence distinct output files, which can propagate upwards as changed dependencies. As consequence, we have full revision history of previously computed result files (this can be important for traceability in data analysis experiments). This approach is slightly different from Algorithm 3, which works with destructive updates.

7.3 Semantics

The bottom-up semantics (Algorithm 1) is implemented as a parametric Logtalk object, that can be parameterized with a *semantics object*. This is an object adhering to a Logtalk protocol that requires it to supply the operation `apply/3`. The arguments to `apply/3` is a matched rule, a corresponding task and a result argument. Different semantics objects with alternative implementations of `apply/3` can then be used to specialize the implementation of Algorithm 1 for different purposes. For instance, with the execution semantics, `apply/3` will run a task and unify the result argument to a list of names of generated output files, whereas for type checking semantics the result will be a list of types.

Debugging Facilities are realized through specialized semantics objects. *Type checking* is implemented using a particular semantics object that can serve as parameter to our generic implementation of Algorithm 1. This exactly mimics how typechecking semantics is formally defined in section 5. The typechecking semantics object does not run the tasks, but performs the type unification as detailed in section 5.

A call graph is a graph data structure which, for a particular file request, contains a node for file that may transitively be a dependency file and edges between such files represent task calls. We generate call graphs using a semantics object that with Algorithm 1 builds a tree of all task calls involved when requesting a filename. A post-processing of this tree to identify identical nodes yields a call graph. The system includes functionality to pretty-print the graph.

The *tracer semantics* is a facility which adds interactive tracing. It is implemented as a semantics object which can be parameterized with any other semantics object. With such a parameterization, interactive tracing is available with any other mode of execution (semantics object).

Parallel Execution is implemented through a scheduler – corresponding to Algorithm 5 – which picks tasks that are ready to run by utilizing a call graph. Ready tasks are orphans in the call graph. For each enqueued task, a new thread[5]

[5] We use portable high-level multi-threading constructs provided by Logtalk [14].

is started which runs Algorithm 1. When the thread completes, the job is removed from the queue and also the call graph. If files for the scheduled task are available, Algorithm 1 immediately returns.

8 Related Work

Computational pipelines are ubiquitous. The classic example is Unix pipes, which feed the output of one program into another. Declarative pipeline languages with non-procedural semantics goes back at least to the make utility [7]. The safeness of make-file based incremental recompilation has later been proved through a specification of its semantics [9]. There are several contemporary and very powerful derivatives of the make utility such as Ant (Java) [1], Rake (Ruby) [24], SCons (Python) [10] and plmake (Prolog) [15]. These frameworks are very expressive and integrate with their "host" languages which gives developers almost unlimited capabilities. None of them, however, have available formal semantics.

The importance of pipelines for biological sequence analysis has been acknowledged [17] and there are a variety of biological pipeline languages to choose from, e.g., EGene [6], BioPipe [8], DIYA [21], and SKAM [16]. The first three are configured using a static XML format with limited expressivity. DIYA targets annotation of bacterial genomes which has also been a motivating case for us. SKAM is Prolog based like BANpipe and allows interaction with Prolog, but does not have well-defined semantics.

Many systems have features comparable to those of BANpipe, e.g., automatic parallelization, type checking and advanced control mechanisms, although one rarely finds all of these features in the same system. BANpipe is, to our knowledge, the only system which provides all these features in a declarative language with well-defined semantics.

Another distinct difference and advantage of BANpipe over other systems which represent dependencies between files, is that the symbolic filenames represented by Prolog terms are independent from the actual filenames. Different scripts using different symbolic file names may share computed files.

The family of concurrent logic languages [20] has syntactical and semantic similarities to BANpipe. Rules have guards and the successful execution of a guard implies a committed choice to evaluate the body of a rule. In *flat* variants of the these languages, guards are restricted to a predefined set of predicates as opposed to arbitrary used defined predicates. BANpipe allows user defined predicates, but for the semantics to be well-defined, these are subject to obvious restrictions, e.g. they should terminate. BANpipe rules have a single goal in the rule body, whereas concurrent logic languages typically have conjunctions of goals. These languages execute in parallel and synchronize computations by suspension of unification, which may be subject to certain restrictions. For instance, in Guarded Horn Clauses [23], the guard is not allowed to bind variables in the head and the body may not bind variables in the guard. BANpipe have more restricted assumptions; the guard never binds variables in the head and the body never binds variables in the head or the guard.

9 Conclusions

BANpipe is a declarative logic language for modeling of computational pipelines with time consuming analyses. It has well-defined abstract and operational semantics, and is extended for change propagation, parallelism and type checking.

There are currently two implementations of BANpipe. The original PRISM-based implementation has been used to run fairly large pipelines for biological sequence analysis. The current implementation in Logtalk is simpler, better documented and more suitable for use in other domains. No evaluation of the performance either implementation has been done, since the computations done by tasks are expected to dwarf the overhead of the pipeline. Obviously, parallelization reduce computation time under certain conditions. In our parallelization strategy, only dependencies dictate which tasks can run simultaneously. It would be interesting to explore how further constraints can be integrated, for instance with respect to available memory.

Many contemporary approaches to data analysis, particularly for big data, exploit distributed processing and it would be obvious also to include support for distributed computing in BANpipe. This would involve distributing files among nodes and deciding what nodes should run what tasks. This could be either explicitly defined by the script (with suitable syntactic extensions) or automatically managed using a bag-of-tasks approach.

Perhaps inspiration can be drawn from distributed query processing systems such as Scope [3], Pig [18] and Hive [22], which are based on distributed map-reduce [5]. The purpose of these systems are different than BANpipe, but they also share some characteristics, i.e., a declarative specification of desired result and caching of intermediary computations. A main difference, though, is that computations are much more course-grained in BANpipe – it delegates the actual processing of data to tasks.

Acknowledgement. This work is part of the project "Logic-statistic modeling and analysis of biological sequence data" funded by the NABIIT program under the Danish Strategic Research Council.

References

1. Apache ant, `http://ant.apache.org/` (accessed November 30, 2012)
2. Lomsadze, A., Besemer, J., Borodovsky, M.: Genemarks: a self-training method for predicition of gene starts in microbial genomes. Implications for finding sequence motifs in regulatory regions. Nucleic Acids Research 29, 2607–2618 (2001)
3. Chaiken, R., Jenkins, B., Larson, P.Å., Ramsey, B., Shakib, D., Weaver, S., Zhou, J.: Scope: easy and efficient parallel processing of massive data sets. Proceedings of the VLDB Endowment 1(2), 1265–1276 (2008)
4. Christiansen, H., Have, C.T., Lassen, O.T., Petit, M.: Bayesian Annotation Networks for Complex Sequence Analysis. In: Technical Communications of the 27th International Conference on Logic Programming, ICLP 2011. Leibniz International Proceedings in Informatics (LIPIcs), vol. 11, pp. 220–230. Schloss Dagstuhl–Leibniz-Zentrum fuer Informatik (2011)

5. Dean, J., Ghemawat, S.: Mapreduce: simplified data processing on large clusters. Communications of the ACM 51(1), 107–113 (2008)
6. Durham, A.M., Kashiwabara, A.Y., Matsunaga, F.T.G., Ahagon, P.H., Rainone, F., Varuzza, L., Gruber, A.: Egene: a configurable pipeline generation system for automated sequence analysis. Bioinformatics 21(12), 2812–2813 (2005)
7. Feldman, S.I.: Make – A program for maintaining computer programs. Software – Practice and Experience 9(3), 255–265 (1979)
8. Hoon, S., Ratnapu, K.K., Chia, J.M., Kumarasamy, B., Juguang, X., Clamp, M., Stabenau, A., Potter, S., Clarke, L., Stupka, E.: Biopipe: A flexible framework for protocol-based bioinformatics analysis. Genome Research, 1904–1915 (2003)
9. Jørgensen, N.: Safeness of make-based incremental recompilation. In: Eriksson, L.-H., Lindsay, P.A. (eds.) FME 2002. LNCS, vol. 2391, pp. 126–145. Springer, Heidelberg (2002)
10. Knight, S.: Building software with scons. Computing in Science & Engineering 7(1), 79–88 (2005)
11. Lassen, O.T.: Compositionality in probabilistic logic modelling for biological sequence analysis. PhD thesis, Roskilde University (2011)
12. Moura, P.: Logtalk - Design of an Object-Oriented Logic Programming Language. PhD thesis, Department of Computer Science, University of Beira Interior, Portugal (September 2003)
13. Moura, P.: Programming patterns for logtalk parametric objects. In: Abreu, S., Seipel, D. (eds.) INAP 2009. LNCS (LNAI), vol. 6547, pp. 52–69. Springer, Heidelberg (2011)
14. Moura, P., Crocker, P., Nunes, P.: Multi-threading programming in Logtalk. In: Abreu, S., Costa, V.S. (eds.) Proceedings of the 7th Colloquium on Implementation of Constraint LOgic Programming Systems, pp. 87–101. University of Oporto, Oporto (2007)
15. Mungall, C.: Make-like build system based on prolog, `https://github.com/cmungall/plmake` (accessed November 30, 2012)
16. Mungall, C.: Skam - skolem assisted makefiles, `http://skam.sourceforge.net/` (accessed November 30, 2012)
17. Noble, W.S.: A quick guide to organizing computational biology projects. PLoS Comput. Biol. 5(7), e1000424 (2009)
18. Olston, C., Reed, B., Srivastava, U., Kumar, R., Tomkins, A.: Pig latin: a not-so-foreign language for data processing. In: Proceedings of the 2008 ACM SIGMOD International Conference on Management of Data, pp. 1099–1110. ACM (2008)
19. Sato, T., Kameya, Y.: Prism: a language for symbolic-statistical modeling. In: International Joint Conference on Artificial Intelligence, vol. 15, pp. 1330–1339 (1997)
20. Shapiro, E.: The family of concurrent logic programming languages. ACM Computing Surveys 21(3), 412 (1989)
21. Stewart, A.C., Osborne, B., Read, T.D.: Diya: a bacterial annotation pipeline for any genomics lab. Bioinformatics 25, 962–963 (2009)
22. Thusoo, A., Sarma, J.S., Jain, N., Shao, Z., Chakka, P., Anthony, S., Liu, H., Wyckoff, P., Murthy, R.: Hive: a warehousing solution over a map-reduce framework. Proceedings of the VLDB Endowment 2(2), 1626–1629 (2009)
23. Ueda, K.: Guarded horn clauses. Technical Report TR-103, ICOT, Tokyo (1985)
24. Weirich, J.: Rake – ruby make, `http://rake.rubyforge.org/` (accessed November 30, 2012)

Semantic Code Clones in Logic Programs

Céline Dandois* and Wim Vanhoof

University of Namur - Faculty of Computer Science
21 rue Grandgagnage, 5000 Namur, Belgium
{cda,wva}@info.fundp.ac.be

Abstract. In this paper, we study what is a semantic code clone pair in a logic program. Unlike our earlier work, that focused on simple syntactic equivalence for defining clones, we propose a more general approximation based on operational semantics and transformation rules. This new definition captures a wider set of clones, and allows to formally define the conditions under which a number of refactorings can be applied.

1 Introduction

Code duplication, also called *code cloning*, occurs intuitively when two or more source code fragments have an identical or sufficiently similar computational behavior, independent of them being textually equal or not. Those fragments are described as *duplicated* or *cloned*. Clone detection has received a substantial amount of attention in recent years [15], but there is no standard definition for what constitutes a clone and the latter's definition is often bound to a particular detection technique [14]. Not unsurprisingly, most definitions and the associated detection techniques are based on somehow comparing the syntactical structure of two code fragments as a rough approximation of their semantics. Some examples are the recent abstract syntax-tree based approaches for Erlang [7] and Haskell [2], as well as our own work [4] in the context of logic programming. These syntax-based approaches suffer from a number of limitations. On the one hand, syntactical equivalence is too basic a characterization for defining, *in a simple and uniform way*, the conditions under which a number of refactorings that aim at removing duplication from a logic program [17,22] can be applied. On the other hand, looking at syntactical equivalence only, they are unable to classify as cloned slightly different computations that yield nevertheless the same result such as, in a logic programming setting, the goals $X\ is\ 10$ and $X\ is\ 5*2$. Or, in other words, they cannot capture *semantic equivalence* of code fragments, even in cases where such equivalence could rather simply be established.

As a somewhat contrived but prototype example, representative of various real code clones, consider the following predicate definitions:

$$
\begin{array}{ll}
take\_mult\_n(0, L, [], 1) & \leftarrow \\
take\_mult\_n(N, [], [], 1) & \leftarrow N > 0 \\
take\_mult\_n(N, [X|Xs], [X|Ys], P) & \leftarrow N > 0,\ prev(N_1, N), \\
& \qquad take\_mult\_n(N_1, Xs, Ys, P_1),\ P\ is\ X{*}P_1
\end{array}
$$

* F.R.S.-FNRS Research Fellow.

E. Albert (Ed.): LOPSTR 2012, LNCS 7844, pp. 35–50, 2013.
© Springer-Verlag Berlin Heidelberg 2013

$add\_ij(L, I, 0, 0) \leftarrow$
$add\_ij([], I, J, 0) \leftarrow J > 0$
$add\_ij([A|As], 1, J, S) \leftarrow J > 0,\ J_1\ is\ J\text{-}1,\ add\_ij(As, 1, J_1, S_1),\ S\ is\ A{+}S_1$
$add\_ij([A|As], I, J, S) \leftarrow I > 1,\ J \geq I,\ I_1\ is\ I\text{-}1,\ J_1\ is\ J\text{-}1,\ add\_ij(As, I_1, J_1, S)$

The predicate $prev$ is simply defined as $prev(N_1, N) \leftarrow N_1\ is\ N\text{-}1$. An atom of the form $take\_mult\_n(N, L_1, L_2, P)$ succeeds if the list L_2 contains the N first naturals of the list L_1 and if P is the product of those naturals. On the other hand, an atom $add\_ij(L, I, J, S)$ succeeds if S equals the sum of those naturals in the list L between the positions I and J. While syntactically very different, it is clear that *take_mult_n* and *add_ij* share some common functionality that may be worthwhile to exploit in a refactoring setting. Indeed, both predicates partition a list and apply some operation on one element of this partition. Obviously, this functionality is hard to detect based on syntactically comparing the atoms in the definitions of *take_mult_n* and *add_ij*, as would be the case by our previous work [4] or similar syntax-based techniques.

In the next section, we define the notion of a *semantic* clone pair in a logic programming setting. While semantic equivalence is in general undecidable, we provide in the third section an approximation of our definition, based on well-known program transformations. This approximation allows to formally capture the relation between cloned code fragments such as the functionality shared by *take_mult_n* and *add_ij* from above. Thereby, it goes far beyond syntactical equivalence and considerably extends our previous work on the subject [4]. Finally, in the last section, we compare our work to the related literature, we expose a simple procedure for refactoring a particular case of duplication captured by our definitions, and we give some ideas for further research.

2 Defining Semantic Clones

In what follows, we consider an extended form of logic language including higher-order constructs [9,3], and we assume the reader to be familiar with the basic logic programming concepts [1]. Inspired from the language HiLog [3], the syntax that we will use is based on a countably infinite set $\mathcal{V}$ of *variables* and a countable set $\mathcal{N}$ of arityless *relation names*. The latter set contains what is called functors and predicate names in first-order logic (and by the way, it eliminates the distinction between those entities), including built-in ones such as the unification operator "=". The set $\mathcal{T}$ of *terms* satisfies the following conditions: $\mathcal{V} \cup \mathcal{N} \subseteq \mathcal{T}$ and $\forall x, t_1, \ldots, t_n \in \mathcal{T} : x(t_1, \ldots, t_n) \in \mathcal{T} (n \geq 1)$. A term is thus built by any logical symbol preceding any finite number of arguments. As such, *add*, $add(X)$, $add(X, Y, Z)$, $add(X)(Y, Z)$ and $W(X, Y, Z)$ are valid terms. A *(positive or negative) atom* is a term. A *clause* of *arity* $n (n \geq 0)$ is a construct of the form $C = H \leftarrow B_1, \ldots, B_s$, where H, the *head* of the clause (denoted $hd(C)$), is an atom of the particular form $h(t_1, \ldots, t_n)$ where $h \in \mathcal{N}$ and $t_1, \ldots, t_n \in \mathcal{T}$, and $B_1, \ldots, B_s (s \geq 0)$, the *body* of the clause (denoted $bd(C)$), is a conjunction of atoms, also called a *goal*. A *predicate* is defined by a sequence of clauses whose heads share the same predicate symbol and arity. Finally, a *program* is formed by a set of predicate definitions. To ease the notation, we will use the following auxiliary notations and functions. A sequence of objects $o_1, \ldots, o_n$ will sometimes simply be represented by $\overline{o}$ (for example, a predicate is typically denoted $\overline{C}$). For a sequence of n objects $\overline{o}$ and a set of naturals $S \subseteq \{1, \ldots, n\}$, we use $\overline{o}_{|S}$ to denote the restriction of $\overline{o}$ to the objects with an index in S. The function rel takes a predicate $\overline{C}$ as argument and returns its relation name and arity in the form p/n. For a goal G which is a subgoal of a clause $H \leftarrow L, G, R$ with L and R possibly empty subgoals, $uvars(G)$ denotes the set of variables from G that are shared with L and/or R. Finally, the function $vars$ returns the set of variables contained in a given object and the function dom returns the domain of a given function.

As usual, a *substitution* is defined as a mapping from variables to terms. In what follows, we will only use *idempotent* substitutions, i.e. substitutions $\theta = \{X_1/t_1, \ldots, X_k/t_k\}$ where the variables $\overline{X}$ do not appear free in the terms $\overline{t}$, or in other words, where we have that $\theta\theta = \theta$ [10]. As an example, $\{X/Y, Y/Z, Z/a\}$ is not considered a valid substitution but $\{X/a, Y/a, Z/a\}$ is. A *variable renaming*, or simply *renaming*, is a particular case of idempotent substitution, from variables to variables. As usual, $\theta\sigma$ denotes the composition of substitutions θ and σ. We will use the operator $\simeq_\rho$ to represent the standard *variance* between terms, according to some renaming ρ. Slightly abusing notation, we will use $\simeq_\rho^{\{p/q\}}$ to represent the fact that a predicate definition is a variant of another, according to some renaming ρ and to the renaming of each occurrence of the predicate symbol p by q.

Our main definitions will be based on the operational semantics of logic programming [1,9,3], and more precisely, on the concept of computed answer substitution [8,1]. The computed answer substitution (CAS) semantics may be expressed as the following function: $CAS : P \times Q \rightarrow (\mathcal{P}(Subst), \subseteq)$, where P is the set of programs, Q is the set of queries, and $(\mathcal{P}(Subst), \subseteq)$ is the powerset of the set of substitutions ordered by set inclusion [12]. This function is thus such that, given a program P and a query Q, $CAS(P, Q) = \{\theta \mid \text{there exists an SLD-derivation of } Q \text{ in } P \text{ with computed answer substitution } \theta\}$. An empty set stands for a failed query. Note that for SLD-resolution, we adopt the standard clause and atom selection rules of Prolog, respectively top-down and left-to-right.

We define a *code fragment* in a predicate $\overline{C}$ as either (1) the entire predicate definition of $\overline{C}$, or (2) a strict subsequence of clauses belonging to the predicate definition of $\overline{C}$, or (3) a subgoal in some clause body of $\overline{C}$. Given a code fragment F and assuming that $newp$ is a unique relation name not used in the program at hand, we define the *predicate definition relative to* F, denoted $pdr(F)$, as follows: if F is a predicate definition, then $pdr(F)$ is F itself ; if F is a strict subsequence of clauses with $rel(F) = p/m$, then $pdr(F)$ is such that $F \simeq_\emptyset^{\{p/newp\}} pdr(F)$; finally, if F is a subgoal of a clause, then $pdr(F)$ is defined by $newp(uvars(F)) \leftarrow F$. Note that both cases 1 and 2 consider a sequence of clauses but the predicate definitions relative to those types of code fragment are obtained differently. Indeed, in case 1, contrary to case 2, information about the relation represented by the fragment is already known and do not need to be constructed by extraction. We define the *arity of a code fragment* F, denoted F/n, as the arity of $pdr(F)$. The following definition allows to relate the semantics of two code fragments.

Definition 1. *Given two code fragments F_1 in $\overline{C}_1$ and F_2 in $\overline{C}_2$ from a program P with $rel(pdr(F_1)) = p/m$ and $rel(pdr(F_2)) = q/n$,* F_1 computes a subrelation *of F_2, denoted $F_1 \sqsubseteq F_2$, if and only if for each call $p(\overline{t})$ to $pdr(F_1)$ with $CAS((P \setminus \{\overline{C}_2\}) \cup \{pdr(F_1)\}, p(\overline{t})) = \{\theta_1, \ldots, \theta_k\}$, there exists a call to $pdr(F_2)$, say $q(\overline{X})\sigma$, such that:*

- *$CAS((P \setminus \{\overline{C}_1\}) \cup \{pdr(F_2)\}, q(\overline{X})\sigma) = \{\theta'_1, \ldots, \theta'_k\}$, and*
- *either $k = 0$ (both sets are empty) or there exists a bijective mapping between the substitutions of $\{\theta_1, \ldots, \theta_k\}$ and those of $\{\theta'_1, \ldots, \theta'_k\}$, and*
- *for each pair of corresponding substitutions θ_i and $\theta'_i (1 \leq i \leq k)$, we have that $\overline{t}\theta_i \simeq_\rho \overline{t}\sigma\theta'_i$ with $vars(\overline{t}) \cap dom(\rho) = \emptyset$.*

Intuitively, $F_1 \sqsubseteq F_2$ means that all goals that can be (dis)proven by the relation represented by F_1 can equally be (dis)proven, at least in some way, by the relation represented by F_2. More precisely, for every call $p(\overline{t})$ to $pdr(F_1)$ there exists a call $q(\overline{X})\sigma$ to $pdr(F_2)$ having the same procedural behavior, i.e. either both calls fail or they succeed producing the same solutions, but not necessarily in the same order nor with the same frequency. In other words, in case of success of the query $Q = p(\overline{t})$, the query $Q' = q(\overline{X})\sigma$ must at least contain the variables of Q that are instantiated by the resolution of Q and the resolution of Q' must instantiate these variables in the same way. Corresponding non-ground answers may be variants of each other, as long as the associated variable renaming does not include the variables from Q remaining free, i.e. concerns only the new variables introduced during the resolution of Q.

Note that part of the bindings produced by the call $p(\overline{t})$ might be present in the instantiation of the call to $q(\overline{X})\sigma$ (in the σ), hence the need to compose its computed answers with σ for the comparison of variance. Furthermore, both code fragments are taken to be independent, i.e. one being not defined in function of the other. This explains the exclusion of $\overline{C}_2$ from the program for resolution of the call to $pdr(F_1)$ and vice versa. Finally, F_1 and F_2 are not necessarily code fragments of the same arity nor granularity (a predicate definition may compute a subrelation of a goal, for example) and that both code fragments either may belong to two different predicate definitions or to the same predicate definition.

Example 1. Consider the well-known predicate definitions of $\overline{rev1}$, implementing naive reverse, and of $\overline{rev2}$, implementing reverse with an accumulator:

$rev1([], []) \leftarrow$
$rev1([X|Xs], Y) \leftarrow rev1(Xs, Z),\ append(Z, [X], Y)$

$rev2([], B, B) \leftarrow$
$rev2([A|As], B, C) \leftarrow rev2(As, [A|B], C)$

We may state that $\overline{rev1}$ computes a subrelation of $\overline{rev2}$ since for each call to $\overline{rev1}$ of the form $rev1(t_1, t_2)$ for terms t_1 and t_2, there exists a call to $\overline{rev2}$, namely $rev2(t_1, [], t_2)$, such that both calls either fail or result in the same set of computed answer substitutions. To illustrate the role of the condition $vars(\bar{t}) \cap dom(\rho) = \emptyset$ in Definition 1, let take the call $rev1([X], Y)$ which produces the unique solution $\{Y/[X]\}$. The call $rev2([X], [], Y)$ satisfies the definition of subrelation, as written just above, but not the call $rev2([A], [], Y)$ since it instantiates Y to $[A]$ and a renaming from X to A is not allowed. Note that $\overline{rev2} \sqsubseteq \overline{rev1}$ is not true since, given the call $rev2(X, Y, [a, b, c])$, for example, it does not exist a call to $rev1$ able to provide the solutions for X and Y.

Note that, without loss of generality, we considered two entire predicate definitions in our example, but the concept of subrelation can be applied just as easily to other types of code fragment since it is based in all cases on the predicate definitions relative to the compared fragments.

Returning to the example from the introduction, neither $\overline{take\_mult\_n} \sqsubseteq \overline{add\_ij}$ nor $\overline{add\_ij} \sqsubseteq \overline{take\_mult\_n}$. However, it is possible to define a third predicate, say $\overline{g}$, such that $\overline{take\_mult\_n} \sqsubseteq \overline{g}$ and $\overline{add\_ij} \sqsubseteq \overline{g}$, disclosing consequently the common functionality that exists between $\overline{take\_mult\_n}$ and $\overline{add\_ij}$. This is exactly the basic idea behind our definition of a code clone pair.

Definition 2. *Given a code fragment F, two code fragments F_1 and F_2 form an F-*code clone pair *if and only if $F_1 \sqsubseteq F$ and $F_2 \sqsubseteq F$.*

Intuitively, this definition means that the code fragment F can be seen as generalizing both F_1 and F_2 in such a way that it can be used to compute both the relations computed by F_1 and F_2. Back to our running example, such a possible generalization of $\overline{take\_mult\_n}$ and $\overline{add\_ij}$ is the predicate $\overline{g}$ defined as:

$g(Base, Pred, L, I, 0, Base, 1) \leftarrow$
$g(Base, Pred, [], I, J, Base, 1) \leftarrow J > 0$
$g(Base, Pred, [A|As], 1, J, R_1, R_2) \leftarrow J > 0,\ J_1\ is\ J\text{-}1,\ g(Base, Pred, As, 1, J_1, V, W),$
$\quad Pred(A, V, R_1),\ R_2\ is\ A{*}W$
$g(Base, Pred, [A|As], I, J, R_1, R_2) \leftarrow I > 1,\ J \geq I,\ I_1\ is\ I\text{-}1,\ J_1\ is\ J\text{-}1,$
$\quad g(Base, Pred, As, I_1, J_1, R_1, R_2)$

An atom of the form $g(Base, Pred, L, I, J, R_1, R_2)$ succeeds if R_1 results from combining those naturals of the list L between the positions I and J by an operator $Pred$ whose right-neutral element is $Base$, and if R_2 is the product of those naturals. Regarding the semantics, we effectively have that $\overline{g}$ allows to compute both $\overline{take\_mult\_n}$ and $\overline{add\_ij}$ since a generic call of the form $take\_mult\_n(t_1, t_2, t_3, t_4)$ is equivalent to the call $g([], list, t_2, 1, t_1, t_3, t_4)$ and a call $add\_ij(t_1, t_2, t_3, t_4)$ is equivalent to the call $g(0, add, t_1, t_2, t_3, t_4, \_)$, for terms t_1, t_2, t_3 and t_4, and predicates $\overline{list}$ and $\overline{add}$ defined as follows: $list(X, Xt, [X|Xt]) \leftarrow$ and $add(X, Y, Z) \leftarrow Z\ is\ X{+}Y$. Some ideas for building the above generic calls to $\overline{g}$ in an algorithmic way are given in the last section of this paper.

Returning to the definition of an F-code clone pair, two particular cases may be deduced. First, if $F_1 \sqsubseteq F_2$, then F_1 and F_2 form an F_2-code clone pair, given that by definition $F_2 \sqsubseteq F_2$. The predicates $\overline{rev1}$ and $\overline{rev2}$ belong to this category. Secondly, if we have both $F_1 \sqsubseteq F_2$ and $F_2 \sqsubseteq F_1$, then F_1 and F_2 are fully semantically equivalent, in the sense that every call to $pdr(F_1)$ could be replaced by a call to $pdr(F_2)$, and vice versa. For example, this could be the case for two predicates sorting a list in increasing order – one implementing quicksort, the other implementing bubble sort – or for two predicates representing the relations "greater than" and "smaller than".

As the examples show, the proposed notion of an F-code clone pair captures a larger subset of cloned code fragments than would any definition based on syntactical comparison, such as our previous work [4]. Obviously, the notion of an F-code clone pair is not, in general, computable, but, to the best of our knowledge, this is the first attempt of formal definition for logic programs. As we will argue in the next section, it can also be approximated by using well-known program transformations.

3 Approximating Semantic Clones

3.1 Basic Definitions

Even if *computing* F-code clone pairs in the sense outlined above is beyond the scope of the present work, a first step consists in providing an approximation of F-code clone pairs that is verifiable by program analysis. To that end, we will formalize how two predicate definitions, one computing a subrelation of the other, can be related by program transformation. As a starting point, we borrow from the work of Pettorossi and Proietti [12] the notion of transformation sequence, adapted to our needs.

Definition 3. *Let $\mathcal{R}$ be a set of program transformation rules. Given a logic program P and a predicate $\overline{C}$ defined over the language of P, an $\mathcal{R}$-transformation* sequence of $P \cup \{\overline{C}\}$ *is a finite sequence of programs $P_0 \cup \{\overline{C}_0\}, \ldots, P_k \cup \{\overline{C}_k\}$ where $P_0 = P$ and $\overline{C}_0 = \overline{C}$, and $\forall i (0 < i \leq k) : P_i \cup \{\overline{C}_i\}$ is obtained by the application of one rule from $\mathcal{R}$ on $P_{i-1} \cup \{\overline{C}_{i-1}\}$.*

Given our interest in approximating the notion of subrelation, we restrict our attention to $\mathcal{R}$-transformation sequences that are $\sqsubseteq$-correct.

Definition 4. *A program transformation rule from $\mathcal{R}$ is* $\sqsubseteq$-correct *if and only if its application on any given program $P \cup \{\overline{C}\}$ produces a program $P \cup \{\overline{C}'\}$ such that $\overline{C}' \sqsubseteq \overline{C}$ (i.e. $\overline{C}'$ computes a subrelation of $\overline{C}$). An $\mathcal{R}$-transformation sequence of the form $P_0 \cup \{\overline{C}_0\}, \ldots, P_k \cup \{\overline{C}_k\}$ is* $\sqsubseteq$-correct *if and only if (1) $\forall i (0 < i \leq k) : P_0 \subseteq P_i$ and (2) $\overline{C}_k \sqsubseteq \overline{C}_0$ with respect to P_k.*

Concerning $\sqsubseteq$-correctness of an $\mathcal{R}$-transformation sequence, the first condition means that the original program P_0 remains unchanged throughout the transformation process and thus, that its semantics is preserved. The second condition means that the last derived version of the original predicate $\overline{C}_0$ computes a subrelation of $\overline{C}_0$. This highlights our objective that the transformation process focuses on transforming a single predicate definition.

Recall that we consider a logic program as a set of predicate definitions, not as a set of individual clauses. If we use only $\sqsubseteq$-correct transformation rules and if, at each transformation step i $(0 \leq i < k)$, the chosen rule is applied to a single predicate from the program $(P_i \setminus P_0) \cup \{\overline{C}_i\}$, then the $\mathcal{R}$-transformation sequence is $\sqsubseteq$-correct by construction. Indeed, the first condition ensures that $\overline{C}_k \sqsubseteq \overline{C}_{k-1} \sqsubseteq \ldots \sqsubseteq \overline{C}_0$ and thus $\overline{C}_k \sqsubseteq \overline{C}_0$ since the concept of subrelation can easily be proved transitive, while the second condition ensures that $\forall i$ $(0 < i \leq k) : P_0 \subseteq P_i$. Building an $\mathcal{R}$-transformation sequence in such a way implies that at each transformation step i $(0 \leq i < k)$, either $P_{i+1} = P_i$ (the predicate $\overline{C}_i$ is transformed), or $\overline{C}_{i+1} = \overline{C}_i$ (either one new predicate is created in P_i or one predicate in $P_i \setminus P_0$, i.e. previously created during the transformation process, is transformed). This fact is explained because the predicate $\overline{C}$ becoming $\overline{C}'$ in the definition of $\sqsubseteq$-correctness of a transformation rule does not necessarily correspond to the $\overline{C}_i$ becoming $\overline{C}_{i+1}$ in the definition of $\sqsubseteq$-correctness of an $\mathcal{R}$-transformation sequence. Instead, it corresponds to the single predicate selected in the program $(P_i \setminus P_0) \cup \{\overline{C}_i\}$ to be transformed. Note that other kinds of rules than $\sqsubseteq$-correct ones could be used to obtain a $\sqsubseteq$-correct $\mathcal{R}$-transformation sequence but we will not detail this possibility.

Finally, we are able to approximate the notion of subrelation.

Definition 5. *Given two code fragments F_1 and F_2 from a program P with $rel(pdr(F_1)) = p/m$ and $rel(pdr(F_2)) = q/n$, and given a set $\mathcal{R}$ of program transformation rules, F_1* computes an $\mathcal{R}$-subrelation *of F_2, denoted $F_1 \sqsubseteq^{\mathcal{R}} F_2$, if and only if there exists a $\sqsubseteq$-correct $\mathcal{R}$-transformation sequence of $P \cup \{pdr(F_2)\}$, say of k steps, such that $pdr(F_2)_k \simeq_{\rho}^{\{q/p\}} pdr(F_1)$.*

At the level of code fragments, demonstrating $F_1 \sqsubseteq^{\mathcal{R}} F_2$ boils thus down to deriving the predicate definition relative to the more specific code fragment F_1 from the predicate definition relative to the more general code fragment F_2, by means of a $\sqsubseteq$-correct $\mathcal{R}$-transformation sequence.

3.2 A First Instantiation of $\mathcal{R}$

One obvious and useful instantiation of $\mathcal{R}$ is to use the well-known program transformation rules *Definition introduction*, *Unfolding*, *In-situ folding* and *Deletion of clauses with finitely failed body* defined by Tamaki and Sato [19,20] and by Pettorossi and Proietti [12,13]. These rules can trivially be adapted to our notation, and their definition can be found in Appendix. All these rules are proved totally correct with respect to CAS semantics (i.e. CAS semantics-preserving) and thus also $\sqsubseteq$-correct. We propose two other rules, which allow to specialize the predicate definition under transformation with respect to a given call, and to remove superfluous arguments.

Definition 6. ***[R5] Specialization.*** *Let $P \cup \{\overline{C}\}$ be a program and A a call to $\overline{C}$. The* Specialization *rule transforms the program into $P \cup \{\overline{C}'\}$, where $\overline{C}'$ is the predicate obtained by (once) unfolding the body atom A in the clause $A \leftarrow A$.*

The specialization rule is trivially $\sqsubseteq$-correct, i.e. $\overline{C}'$ computes a subrelation of $\overline{C}$, since for every call to $\overline{C}'$, there exists a call to $\overline{C}$, notably A, computing the same set of answers.

Definition 7. ***[R6] Argument removal.*** *Let $P \cup \{\overline{C}\}$ be a program. The* Argument removal *rule transforms the program into $P \cup \{\overline{C}'\}$, where $\overline{C}'$ is the predicate obtained from $\overline{C}$ by removing, both from the clause heads and from each recursive call in the clause bodies, an argument having the same constant value everywhere in $\overline{C}$ or an argument represented by an anonymous variable, i.e. having only one occurrence in each clause.*

An argument which is a constant or an anonymous variable does not influence the bindings created by any call to the predicate. Consequently, each answer computed by a call to the transformed predicate is a subset of an answer computed by the original predicate (the latter possibly including the binding of a variable to the constant value being removed by the transformation).

Definition 8. *The set $\mathcal{R}_s$ is constituted of the transformation rules Definition introduction, Unfolding, In-situ folding, Deletion of clauses with finitely failed body, Specialization and Argument removal.*

As illustrated below, the set $\mathcal{R}_s$ allows to characterize an important subset of subrelations, namely those in which the subrelation can be obtained from the more general predicate by partially evaluating [16,5] a call to the latter.

Example 2. Consider the following two predicate definitions:
$add1\&sqr([], []) \leftarrow$
$add1\&sqr([A|As], [B|Bs]) \leftarrow N\ is\ A{+}1,\ B\ is\ N{*}N,\ add1\&sqr(As, Bs)$
$map(P, [], []) \leftarrow$
$map(P, [X|Xs], [Y|Ys]) \leftarrow P(X, Y),\ map(P, Xs, Ys)$

A call $add1\&sqr(L_1, L_2)$ transforms a list L_1 into L_2 by replacing each element x by $(x{+}1)^2$, while a call $map(P, L_1, L_2)$ transforms L_1 into L_2 by applying the binary predicate P to each element. It can be easily verified that $\overline{add1\&sqr} \sqsubseteq^{\mathcal{R}_s} \overline{map}$. Indeed, there exists a $\sqsubseteq$-correct $\mathcal{R}_s$-transformation sequence in which $\overline{add1\&sqr}$ is derived from $\overline{map}$:

$$(P_0 \cup \{\overline{map}_0\}), (P_1 \cup \{\overline{map}_0\}), (P_1 \cup \{\overline{map}_1\}), (P_1 \cup \{\overline{map}_2\}), (P_1 \cup \{\overline{map}_3\})$$

where $P_0 = \{\overline{add1\&sqr}\}$ and $\overline{map}_0 = \overline{map}$; P_1 is obtained from P_0 by introducing the predicate $op(A, B) \leftarrow N\ is\ A{+}1,\ B\ is\ N{*}N$; $\overline{map}_1$ is the following predicate, obtained by specializing $\overline{map}_0$ with respect to the call $map(op, A, B)$:
$map(op, [], []) \leftarrow$
$map(op, [X|Xs], [Y|Ys]) \leftarrow op(X, Y),\ map(op, Xs, Ys)$

and $\overline{map}_2$ is obtained by unfolding, in $\overline{map}_1$, the call to $\overline{op}$:
$map(op, [], []) \leftarrow$
$map(op, [X|Xs], [Y|Ys]) \leftarrow N\ is\ X{+}1,\ Y\ is\ N{*}N,\ map(op, Xs, Ys)$

Finally, $\overline{map_3}$ is obtained by removing the superfluous argument $\overline{op}$ from the definition of $\overline{map_2}$:
$map([], []) \leftarrow$
$map([X|Xs], [Y|Ys]) \leftarrow N \ is \ X{+}1, \ Y \ is \ N{*}N, \ map(Xs, Ys)$

It can be easily seen that the obtained predicate is a variant of $\overline{add1\&sqr}$.

3.3 Towards a More Involved Instantiation of $\mathcal{R}$

In order to capture a larger class of subrelations (hence, clone pairs), we will now define two more transformation rules. As the rules in $\mathcal{R}_s$, they change the definition of the given predicate and possibly its arity, but not its name.

We first introduce the concept of a *slice*, similarly to the definition proposed by Vasconcelos and Aragão [23]. A *slice* of a program $P \cup \{\overline{C}\}$ is a program $P \cup \{\overline{C}'\}$ where the predicate $\overline{C}'$ is obtained from $\overline{C}$ by removing a (possibly empty) subset of its clauses and removing, from each remaining clause, a (possibly empty) subset of the atoms therein. In what follows, we restrict our attention to slices that are *correct* with respect to a given set of argument positions.

Definition 9. *Let the program $P \cup \{\overline{C}'\}$ be a slice of $P \cup \{\overline{C}\}$ with $rel(\overline{C}) = rel(\overline{C}') = p/m$, and $\Pi \subseteq \{1, \ldots, m\}$ be a set of argument positions of $\overline{C}$. The slice $P \cup \{\overline{C}'\}$ is* correct with respect to Π *if and only if for each call $p(\overline{t})$ to $\overline{C}$, with $CAS(P \cup \{\overline{C}\}, p(\overline{t})) = \{\theta_1, \ldots, \theta_k\}$, the following holds:*

- *$CAS(P \cup \{\overline{C}'\}, p(\overline{t})) = \{\theta'_1, \ldots, \theta'_k\}$, and*
- *either $k = 0$ (both sets are empty) or there exists a bijective mapping between the substitutions of $\{\theta_1, \ldots, \theta_k\}$ and those of $\{\theta'_1, \ldots, \theta'_k\}$, and*
- *for each pair of corresponding substitutions θ_i and $\theta'_i (1 \leq i \leq k)$, we have that $\overline{t}_{|\Pi}\theta_i \simeq_\rho \overline{t}_{|\Pi}\theta'_i$ with $vars(\overline{t}_{|\Pi}) \cap dom(\rho) = \emptyset$.*

In other words, a slice is correct with respect to a set of argument positions Π if it computes the same answers as those computed by the original program, at least as far as the argument positions in Π are concerned. Correct slices may be computed by existing techniques [23,6].

Example 3. The following predicate definition, executing list traversal, is a correct slice with respect to $\Pi = \{1\}$ of the predicate $\overline{add1\&sqr}$ from Example 2:
$add1\&sqr([], []) \leftarrow$
$add1\&sqr([A|As], [B|Bs]) \leftarrow add1\&sqr(As, Bs)$

Definition 10. ***[R7] Slicing.*** *Let $P \cup \{\overline{C}\}$ be a program, Π a set of argument positions and $P \cup \{\overline{C}'\}$ a correct slice of the program with respect to Π. The* Slicing *rule transforms the program $P \cup \{\overline{C}\}$ into $P \cup \{\overline{C}'\}$.*

By the definition of a correct slice, we have that $\overline{C}' \sqsubseteq \overline{C}$, i.e. $\overline{C}'$ computes a subrelation of $\overline{C}$. As a transformation rule, slicing is related to the merging refactoring for functional programs [2,7] and to the field of skeletons and techniques for synthesizing logic programs [18].

The following rule allows to change the order of the arguments of a predicate. It is obviously $\sqsubseteq$-correct.

Definition 11. *[R8] **Argument permutation.** Let $P \cup \{\overline{C}\}$ be a program with $rel(\overline{C}) = p/m$, and $\phi : \{1,\ldots,m\} \to \{1,\ldots,m\}$ a bijective mapping. The* Argument permutation *rule transforms the program $P \cup \{\overline{C}\}$ into $P \cup \{\overline{C}'\}$ by replacing every atom of the form $p(t_1,\ldots,t_m)$ (clause head or recursive call) by an atom of the form $p(t_{\phi(1)},\ldots,t_{\phi(m)})$.*

We can now define a more involved set of transformation rules. As the following examples will show, this set is sufficient to characterize the predicates $\overline{take\_mult\_n}$ and $\overline{add\_ij}$ from the introduction as subrelations of the more general predicate $\overline{g}$, thereby characterizing them as a $\overline{g}$-clone pair.

Definition 12. *The set $\mathcal{R}_a$ is constituted of the transformation rules in $\mathcal{R}_s$ together with Slicing and Argument permutation.*

Example 4. Returning to our running example, by rule R1, we may introduce the definition of $\overline{list}$. By R5, with the query $g([], list, L, 1, J, R_1, R_2)$, we obtain the predicate definition of $\overline{g}_1$:

$$
\begin{array}{ll}
g([], list, L, 1, 0, [], 1) & \leftarrow \\
g([], list, [], 1, J, [], 1) & \leftarrow J > 0 \\
g([], list, [A|As], 1, J, R_1, R_2) & \leftarrow J > 0,\ J_1 \textit{ is } J\text{-}1,\ g([], list, As, 1, J_1, V, W), \\
& \quad\ list(A, V, R_1),\ R_2 \textit{ is } A{*}W \\
g([], list, [A|As], 1, J, R_1, R_2) & \leftarrow 1 > 1,\ J \geq I,\ I_1 \textit{ is } I\text{-}1,\ J_1 \textit{ is } J\text{-}1, \\
& \quad\ g([], list, As, I_1, J_1, R_1, R_2)
\end{array}
$$

We may delete, by R4, the last clause since the evaluation of the goal $1 > 1$ fails, and by R6, the 1st, 2nd and 4th arguments, constant, to obtain $\overline{g}_5$:

$$
\begin{array}{ll}
g(L, 0, [], 1) & \leftarrow \\
g([], J, [], 1) & \leftarrow J > 0 \\
g([A|As], J, R_1, R_2) & \leftarrow J > 0,\ J_1 \textit{ is } J\text{-}1,\ g(As, J_1, V, W),\ list(A, V, R_1),\ R_2 \textit{ is } A{*}W
\end{array}
$$

By R2 with respect to the atom $list(A, V, R_1)$, we obtain $\overline{g}_6$:

$$
\begin{array}{ll}
g(L, 0, [], 1) & \leftarrow \\
g([], J, [], 1) & \leftarrow J > 0 \\
g([A|As], J, [A|V], R_2) & \leftarrow J > 0,\ J_1 \textit{ is } J\text{-}1,\ g(As, J_1, V, W),\ R_2 \textit{ is } A{*}W
\end{array}
$$

By R3, using the predicate definition of $\overline{prev}$, we obtain $\overline{g}_7$:

$$
\begin{array}{ll}
g(L, 0, [], 1) & \leftarrow \\
g([], J, [], 1) & \leftarrow J > 0 \\
g([A|As], J, [A|V], R_2) & \leftarrow J > 0,\ prev(J_1, J),\ g(As, J_1, V, W),\ R_2 \textit{ is } A{*}W
\end{array}
$$

Lastly, by R8, with argument mapping $\{(1,2),(2,1),(3,3),(4,4)\}$, we get $\overline{g}_8$:

$g(0, L, [], 1) \leftarrow$
$g(J, [], [], 1) \leftarrow J > 0$
$g(J, [A|As], [A|V], R_2) \leftarrow J > 0,\ prev(J_1, J),\ g(As, J_1, V, W),\ R_2\ is\ A{*}W$

We have that $\overline{take\_mult\_n}$ is a variant of $\overline{g}_8$ taking into account the replacement of the relation name $\overline{g}$ by $\overline{take\_mult\_n}$. It follows that the predicate $\overline{take\_mult\_n}$ computes an $\mathcal{R}_a$-subrelation of $\overline{g}$.

Example 5. It is possible to establish in a similar way that $\overline{add\_ij} \sqsubseteq^{\mathcal{R}_a} \overline{g}$. The trickiest part of the demonstration consists in the application of rule R7. After introducing the definition of $\overline{add}$ by rule R1 and applying rule R5 with the query $g(0, add, L, I, J, R_1, R_2)$, we obtain the following predicate definition for $\overline{g}_1$:

$g(0, add, L, 1, 0, 0, 1) \leftarrow$
$g(0, add, [], 1, J, 0, 1) \leftarrow J > 0$
$g(0, add, [A|As], 1, J, R_1, R_2) \leftarrow J > 0,\ J_1\ is\ J\text{-}1,\ g(0, add, As, 1, J_1, V, W),$
$\qquad add(A, V, R_1),\ R_2\ is\ A{*}W$
$g(0, add, [A|As], 1, J, R_1, R_2) \leftarrow 1 > 1,\ J \geq I,\ I_1\ is\ I\text{-}1,\ J_1\ is\ J\text{-}1,$
$\qquad g(0, add, As, I_1, J_1, R_1, R_2)$

Then, by R7 and the set of argument positions $\{1, \ldots, 6\}$, we obtain the correct slice $\overline{g}_2$:

$g(0, add, L, 1, 0, 0, 1) \leftarrow$
$g(0, add, [], 1, J, 0, 1) \leftarrow J > 0$
$g(0, add, [A|As], 1, J, R_1, R_2) \leftarrow J > 0,\ J_1\ is\ J\text{-}1,\ g(0, add, As, 1, J_1, V, W),\ add(A, V, R_1)$
$g(0, add, [A|As], 1, J, R_1, R_2) \leftarrow 1 > 1,\ J \geq I,\ I_1\ is\ I\text{-}1,\ J_1\ is\ J\text{-}1,$
$\qquad g(0, add, As, I_1, J_1, R_1, R_2)$

The atom $R_2\ is\ A{*}W$ may be deleted because it succeeds or fails in the same way as the other atoms of the clause, in particular as the atom $add(A, V, R_1)$ which always succeeds if A is a natural and fails otherwise.

Finally, by suppressing the 1st, 2nd and 7th superfluous arguments by R6 and unfolding the atom $add(A, V, R_1)$ by R2, we may conclude that $\overline{add\_ij}$ computes an $\mathcal{R}_a$-subrelation of $\overline{g}$.

3.4 Return to Clone Pairs

We may state that an $\mathcal{R}$-subrelation can be seen as approximating a subrelation.

Proposition 1. *Given two code fragments F_1 and F_2 from a program P, if F_1 computes an $\mathcal{R}$-subrelation of F_2, then F_1 also computes a subrelation of F_2.*

This result stems from the fact that, by Definition 5, the predicate $\overline{C}$ derived by the transformation sequence originating in $pdr(F_2)$ computes a subrelation of $pdr(F_2)$ while being, at the same time, a variant of $pdr(F_1)$ (modulo a relation name substitution). Note that the converse of the proposition is not necessarily true, at least referring to our set of rules $\mathcal{R}_a$. Indeed, for example, even if the predicate $\overline{rev1}$ computes a subrelation of $\overline{rev2}$, it computes no $\mathcal{R}_a$-subrelation of $\overline{rev2}$. This could be explained because the way of computing the reverse list is inherently different in both predicate definitions, they implement two different algorithms and the one of $\overline{rev1}$ can possibly not be derived from the one of $\overline{rev2}$ by the transformation rules from $\mathcal{R}_a$[1].

We can now also approximate the notion of an F-code clone pair by building upon an $\mathcal{R}$-transformation sequence.

Definition 13. *Given a code fragment F and given a set $\mathcal{R}$ of program transformation rules, two code fragments F_1 and F_2 form an* $(F,\mathcal{R})$-clone pair *if and only if $F_1 \sqsubseteq^{\mathcal{R}} F$ and $F_2 \sqsubseteq^{\mathcal{R}} F$.*

Example 6. As shown by Examples 4 and 5, the predicates $\overline{take\_mult\_n}$ et $\overline{add\_ij}$ form a $(\overline{g},\mathcal{R}_a)$-clone pair.

Given our definition of an F-clone pair, the following result is immediate:

Proposition 2. *Given a code fragment F, given a set $\mathcal{R}$ of program transformation rules and given two code fragments F_1 and F_2 from a program P, if F_1 and F_2 form an $(F,\mathcal{R})$-clone pair, then they also form an F-clone pair.*

4 Refactoring and Ongoing Work

Let us first examine the relevance of our definitions of an F-clone pair and an $(F,\mathcal{R})$-clone pair in the context of refactorings that aim at removing duplicated (cloned) code from a program. Some basic refactorings exist in the literature of logic programming. Pettorossi and Proietti propose two rules concerning exact duplication [12]: "Deletion of Duplicate Clauses" and "Deletion of Duplicate Goals". Those rules allow to replace a sequence of clauses C, C in a predicate definition by the clause C and to replace a goal G, G in the body of a clause by the goal G. Serebrenik et al. present two refactorings for Prolog programs [17]. The first one aims at identifying identical subsequences of goals in different predicate bodies, in order to extract them into a new predicate. The detection phase is said to be comparable to the problem of determining longest common subsequences. The second refactoring aims at eliminating copy-pasted predicates, with identical definitions. However, to limit the search complexity, only predicates with identical names in different modules are considered duplicated. Vanhoof and Degrave expose the idea of a more complex refactoring [21,22]: generalization of two predicate definitions into a higher-order one. Those refactorings are frequently studied for other programming languages [14], and we may point out in particular another kind of generalization realized thanks to a merging operator [2]. Our definitions of clone pairs allow to embody all above refactorings in a single simple schema, under some conditions.

A particular case of subrelation occurs when there exists a general call to $pdr(F_2)$, let say K, such that each call to $pdr(F_1)$ could be replaced by an

[1] In the case of $\overline{rev1}$ and $\overline{rev2}$, this particular relation between both predicates *could* be proven but this requires using properties other than the rules in $\mathcal{R}_a$ [18,11].

instance of this general call. It means that, according to Definition 1, each call to $pdr(F_1)$ of the form $p(\overline{X})\sigma$ has the same procedural behaviour than the call $K\sigma$. In such a situation, F_1 may be written in terms of F_2, in the following way:

1. if F_2 is not a predicate definition, then add $pdr(F_2)$ to P as a new predicate.
2. supposing $rel(pdr(F_2)) = newq/n$, if F_2 is a subgoal of a clause body, then it simply becomes the atom $newq(uvars(F_2))$; if F_2 is a subsequence of clauses with $rel(F_2) = q/n$, then it is replaced by a new clause $q(Y_1, \ldots, Y_n) \leftarrow newq(Y_1, \ldots, Y_n)$.
3. add other new predicates to P if needed
4. if F_1 is a subgoal of a clause body, then it simply becomes the atom K ; if F_1 is a subsequence of clauses or a predicate definition with $rel(F_1) = p/m$, then it is replaced by a new clause $p(X_1, \ldots, X_m) \leftarrow K$.

Consequently, when two code fragments F_1 and F_2 form an F-code clone pair because they are both linked to F by this particular case of subrelation, it is possible to remove the duplication by simply applying the items 1 and 2 of the above procedure for the fragment F and the items 3 and 4 for F_1 and F_2.

Example 7. All previous examples benefit from this particularity, which allows us to write, assuming the creation of the new predicates $\overline{op}$, $\overline{list}$, $\overline{add}$ and $\overline{g}$:
$rev1(X, Y) \leftarrow rev2(X, [], Y)$
$add1\&sqr(L_1, L_2) \leftarrow map(op, L_1, L_2)$
$take\_mult\_n(N, L_1, L_2, P) \leftarrow g([], list, L_1, 1, N, L_2, P)$
$add\_ij(L, I, J, S) \leftarrow g(0, add, L, I, J, S, \_)$

This kind of rewriting is however not possible for all subrelations. For example, consider the simple predicate $\overline{mult}$ computing the product of two naturals: $mult(X, Y, Z) \leftarrow Z\ is\ X{*}Y$. This implementation using the built-in predicate "*is*" implies that every call to $\overline{mult}$ where the two first arguments are not ground fails. This property allows to show that $\overline{mult}$ computes a subrelation of our predicate $\overline{add\_ij}$. Indeed, to every call of the form $mult(t_1, t_2, t_3)$ with t_1, t_2 ground naturals, we may associate a call $add\_ij([t_4, t_5], 1, 2, t_3)$ with t_4, t_5 ground naturals such that $t_1 * t_2 = t_4 + t_5$. However, it is not possible to write $mult(X, Y, Z) \leftarrow add\_ij([A, B], 1, 2, Z)\sigma$ because there exists no substitution σ making an appropriate link between X, Y and A, B.

This particular case of subrelation (hence, of F-clone pair) is not computable, but it also may be approximated. Indeed, it is possible to state a set $\mathcal{R}$ of cautiously chosen program transformation rules such that if F_1 computes an $\mathcal{R}$-subrelation of F_2, then it also computes this particular kind of subrelation of F_2. Concretely, each rule may be associated with a simple operation on a given call, such that at the end of the $\mathcal{R}$-transformation sequence, the call K is constructed. Both sets $\mathcal{R}_s$ and $\mathcal{R}_a$ verify this property. Thus, given an $\mathcal{R}_a$-transformation sequence allowing to prove an $(F, \mathcal{R}_a)$-code clone pair, it is possible to refactor the corresponding duplication.

As for some concluding remarks, note that our definitions of clone pairs generalize the definition of a purely syntactically structural clone as the one in our previous work [4] by considering variance with respect to an $\mathcal{R}$-transformation sequence. Indeed, a purely syntactical clone pair between code fragments F_1 and F_2 corresponds to an $(F, \mathcal{R})$-clone pair where the code fragment F is either F_1 or F_2 and where no transformation rule has to be applied.

The price to pay for the generality of the definition relies of course in the complexity needed for verifying whether two code fragments are related by an $\mathcal{R}$-subrelation. Also note that even if the verification could be completely automated, doing so might turn out to be far from trivial since deciding which transformation rule to apply and how to parametrize it may reveal to be quite complex. An interesting topic of ongoing work is how to *find* a generalization of two code fragments suspected of forming an $(F, \mathcal{R})$-clone pair. Developing an algorithm for detecting (a relevant subset) of $(F, \mathcal{R})$-clone pairs will be an important step in identifying semantic clones [15].

Acknowledgements. We warmly thank the reviewers for their thought-provoking remarks. We particularly thank A. Pettorossi and M. Proietti for their enriching discussions.

References

1. Apt, K.: Logic programming. In: Handbook of Theoretical Computer Science. Formal Models and Sematics, vol. B, pp. 493–574. Elsevier (1990)
2. Brown, C., Thompson, S.: Clone detection and elimination for Haskell. In: Proceedings of the 2010 SIGPLAN Workshop on Partial Evaluation and Program Manipulation, PEPM 2010, pp. 111–120. ACM (2010)
3. Chen, W., Kifer, M., Warren, D.: HiLog: A foundation for higher-order logic programming. Journal of Logic Programming 15(3), 187–230 (1993)
4. Dandois, C., Vanhoof, W.: Clones in logic programs and how to detect them. In: Vidal, G. (ed.) LOPSTR 2011. LNCS, vol. 7225, pp. 90–105. Springer, Heidelberg (2012)
5. Leuschel, M.: Advanced Techniques for Logic Program Specialisation. Ph.D. thesis, Katholieke Universiteit Leuven (1997)
6. Leuschel, M., Vidal, G.: Forward slicing by conjunctive partial deduction and argument filtering. In: Sagiv, M. (ed.) ESOP 2005. LNCS, vol. 3444, pp. 61–76. Springer, Heidelberg (2005)
7. Li, H., Thompson, S.: Clone detection and removal for Erlang/OTP within a refactoring environment. In: Proceedings of the 2009 SIGPLAN Workshop on Partial Evaluation and Program Manipulation, PEPM 2009, pp. 169–178. ACM (2009)
8. Lloyd, J.W.: Foundations of Logic Programming, 2nd edn. Springer (1987)
9. Nadathur, G., Miller, D.: Higher-order logic programming. In: Handbook of Logic in Artificial Intelligence and Logic Programming, vol. 5, pp. 499–590. Oxford University Press (1998)
10. Palamidessi, C.: Algebraic properties of idempotent substitutions. In: Paterson, M.S. (ed.) Automata, Languages and Programming. LNCS, vol. 443, pp. 386–399. Springer, Heidelberg (1996)
11. Pettorossi, A., Proietti, M., Senni, V.: Constraint-based correctness proofs for logic program transformations. Tech. Rep. 24, IASI-CNR (2011)

12. Pettorossi, A., Proietti, M.: Transformation of logic programs. In: Handbook of Logic in Artificial Intelligence and Logic Programming, vol. 5, pp. 697–787. Oxford University Press (1998)
13. Pettorossi, A., Proietti, M.: Synthesis and transformation of logic programs using unfold/fold proofs. The Journal of Logic Programming 41, 197–230 (1999)
14. Roy, C.K., Cordy, J.R.: A survey on software clone detection research. Tech. rep. (2007)
15. Roy, C.K., Cordy, J.R., Koschke, R.: Comparison and evaluation of code clone detection techniques and tools: A qualitative approach. Science of Computer Programming 74(7), 470–495 (2009)
16. Sahlin, D.: Mixtus: An automatic partial evaluator for full Prolog. New Generation Computing 12, 7–15 (1993)
17. Serebrenik, A., Schrijvers, T., Demoen, B.: Improving Prolog programs: Refactoring for Prolog. Theory and Practice of Logic Programming (TPLP) 8, 201–215 (2008), other version consulted `https://lirias.kuleuven.be/bitstream/123456789/164765/1/technical_note.pdf`
18. Seres, S., Spivey, M.: Higher-order transformation of logic programs. In: Lau, K.-K. (ed.) LOPSTR 2000. LNCS, vol. 2042, pp. 57–68. Springer, Heidelberg (2001)
19. Tamaki, H., Sato, T.: Unfold/fold transformations of logic programs. In: Proceedings of the 2nd International Conference on Logic Programming, ICLP 1984 (1984)
20. Tamaki, H., Sato, T.: A generalized correctness proof of the unfold/fold logic program transformation. Tech. Rep. 86-4, Ibaraki University, Japan (1986)
21. Vanhoof, W.: Searching semantically equivalent code fragments in logic programs. In: Etalle, S. (ed.) LOPSTR 2004. LNCS, vol. 3573, pp. 1–18. Springer, Heidelberg (2005)
22. Vanhoof, W., Degrave, F.: An algorithm for sophisticated code matching in logic programs. In: Garcia de la Banda, M., Pontelli, E. (eds.) ICLP 2008. LNCS, vol. 5366, pp. 785–789. Springer, Heidelberg (2008)
23. Weber Vasconcelos, W., Aragão, M.A.T.: An adaptation of dynamic slicing techniques for logic programming. In: de Oliveira, F.M. (ed.) SBIA 1998. LNCS (LNAI), vol. 1515, pp. 151–160. Springer, Heidelberg (1998)

Appendix

The following well-known program transformation rules come from the work of Tamaki and Sato [19,20] and of Pettorossi and Proietti [12,13]. Their definition are adapted to our notation.

Definition. ***[R1] Definition introduction.*** *Let* $P \cup \{\overline{C}\}$ *be a program. We define a predicate* $\overline{D} = \langle newp(\ldots) \leftarrow Body_1, \ldots, newp(\ldots) \leftarrow Body_n\rangle (1 \leq n)$, *such that (1) newp is a predicate symbol not occurring in P and (2)* $\forall i (1 \leq i \leq n)$, *all predicate symbols in the goal* $Body_i$ *occur in P. By* Definition introduction, *we derive from* $P \cup \{\overline{C}\}$ *the new program* $P' \cup \{\overline{C}\}$, *where* $P' = P \cup \{\overline{D}\}$.

Definition. ***[R2] Unfolding.*** *Let* $P \cup \{\langle C_1, \ldots, C_i, \ldots, C_k\rangle\}$ *be a program with* $C_i = H \leftarrow L, A, R$ *where A is a positive atom and L and R are (possibly empty) goals. Suppose that:*

1. $D_1, \ldots, D_n$ $(n > 0)$ *is the subsequence of all clauses of* P *such that* $\forall j(1 \leq j \leq n)$, *there is a variant of the clause* D_j, *say* D'_j, *with* $vars(C_i) \cap vars(D'_j) = \emptyset$ *and* A *is unifiable with* $hd(D'_j)$ *with most general unifier* θ_j, *and*
2. $\forall j(1 \leq j \leq n) : C'_j$ *is the clause* $(H \leftarrow L, bd(D'_j), R)\theta_j$

If we unfold C_i w.r.t. A, *we derive the clauses* $C'_1, \ldots, C'_n$ *and we get the new program* $P \cup \{\langle C_1, \ldots, C'_1, \ldots, C'_n, \ldots, C_k \rangle\}$.

Remark that the application of the unfolding rule realizes an (SLD-)resolution step to clause C_i with the selection of the positive atom A and the input clauses $D_1, \ldots, D_n$ [12].

Definition. *[R3]* ***In-situ folding.*** *Let* $P \cup \{\langle C_1, \ldots, C'_1, \ldots, C'_n, \ldots, C_k \rangle\}$ *be a program and* $D_1, \ldots, D_n$ *be a subsequence of clauses in* P. *Suppose that there exist an atom* A *and two goals* L *and* R *such that for each* $i(1 \leq i \leq n)$, *there exists a substitution* θ_i *which satisfies the following conditions:*

1. C'_i *is a variant of the clause* $H \leftarrow L, bd(D_i)\theta_i, R$,
2. $A = hd(D_i)\theta_i$,
3. *for every clause* D *of* P *not in the sequence* $D_1, \ldots, D_n$, $hd(D)$ *is not unifiable with* A, *and*
4. *for every variable* X *in the set* $vars(D_i) \setminus vars(hd(D_i))$, *we have that:*
   - X *i is a variable which does not occur in* (H, L, R) *and*
   - *the variable* $X\theta_i$ *does not occur in the term* $Y\theta_i$, *for any variable* Y *occurring in* $bd(D_i)$ *and different from* X.

If we in-situ fold $C'_1, \ldots, C'_n$ using $D_1, \ldots, D_n$, *we derive the clause* $C = H \leftarrow L, A, R$ *and we get the new program* $P \cup \{\langle C_1, \ldots, C, \ldots, C_k \rangle\}$.

Note that, in this version of the folding rule, the clauses $C'_1, \ldots, C'_n$ cannot be folded using clauses from $C'_1, \ldots, C'_n$. Moreover, the unfolding and in-situ folding rules are inverse transformation rules. Indeed, the application of the in-situ folding rule can be reversed by an application of the unfolding rule: if we derive a clause C from $C'_1, \ldots, C'_n$ using clauses $D_1, \ldots, D_n$, it is always possible to unfold C using $D_1, \ldots, D_n$ to get $C'_1, \ldots, C'_n$ [12].

Definition. *[R4]* ***Deletion of clauses with finitely failed body.*** *Let* $P \cup \{\langle C_1, \ldots, C_i, \ldots, C_k \rangle\}$ *be a program with* $C_i = H \leftarrow L, A, R$ *where* A *is a (positive or negative) atom and* L *and* R *are (possibly empty) goals. Suppose that* A *has a finitely failed SLDNF-tree in* P, *then* C *has a finite failed body. By* Deletion of Clauses with Finitely Failed Body, *we derive from* $P \cup \{\langle C_1, \ldots, C_i, \ldots, C_k \rangle\}$ *the new program* $P \cup \{\langle C_1, \ldots, C_k \rangle\}$.

Specialization with Constrained Generalization for Software Model Checking

Emanuele De Angelis[1], Fabio Fioravanti[1],
Alberto Pettorossi[2], and Maurizio Proietti[3]

[1] DEC, University 'G. D'Annunzio', Pescara, Italy
{emanuele.deangelis,fioravanti}@unich.it
[2] DICII, University of Rome Tor Vergata, Rome, Italy
pettorossi@disp.uniroma2.it
[3] IASI-CNR, Rome, Italy
maurizio.proietti@iasi.cnr.it

Abstract. We present a method for verifying properties of imperative programs by using techniques based on constraint logic programming (CLP). We consider a simple imperative language, called SIMP, extended with a nondeterministic choice operator and we address the problem of checking whether or not a *safety* property φ (that specifies that an *unsafe* configuration cannot be reached) holds for a SIMP program P. The operational semantics of the language SIMP is specified via an interpreter I written as a CLP program. The first phase of our verification method consists in specializing I with respect to P, thereby deriving a specialized interpreter I_P. Then, we specialize I_P with respect to the property φ and the input values of P, with the aim of deriving, if possible, a program whose least model is a finite set of constrained facts. To this purpose we introduce a novel generalization strategy which, during specialization, has the objecting of preserving the so called branching behaviour of the predicate definitions. We have fully automated our method and we have made its experimental evaluation on some examples taken from the literature. The evaluation shows that our method is competitive with respect to state-of-the-art software model checkers.

1 Introduction

Software model checking is a body of formal verification techniques for imperative programs that combine and extend ideas and techniques developed in the fields of static program analysis and model checking (see [19] for a recent survey).

In this paper we consider a simple imperative language SIMP acting on integer variables, with nondeterministic choice, assignment, conditional, and while-do commands (see, for instance, [29]) and we address the problem of verifying *safety* properties. Basically, a safety property states that when executing a program, an unsafe configuration cannot be reached from any initial configuration. Note that, since we consider programs that act on integer numbers, the problem of deciding whether or not an unsafe configuration is unreachable is in general undecidable.

E. Albert (Ed.): LOPSTR 2012, LNCS 7844, pp. 51–70, 2013.
© Springer-Verlag Berlin Heidelberg 2013

In order to cope with this undecidability limitation, many program analysis techniques have followed approaches based on *abstraction* [4], by which the concrete data domain is mapped to an abstract domain so that reachability is preserved, that is, if a concrete configuration is reachable, then the corresponding abstract configuration is reachable. By a suitable choice of the abstract domain one can design reachability algorithms that terminate and, whenever they prove that an abstract unsafe configuration is unreachable from an abstract initial configuration, then the program is proved to be safe (see [19] for a general abstract reachability algorithm). Notable abstractions are those based on convex polyhedra, that is, conjunctions of linear inequalities (also called *constraints* here).

Due to the use of abstraction, the reachability of an abstract unsafe configuration does not necessarily imply that the program is indeed unsafe. It may happen that the abstract reachability algorithm produces a *spurious counterexample*, that is, a sequence of configurations leading to an abstract unsafe configuration which does not correspond to any concrete computation. When a spurious counterexample is found, *counterexample-guided abstraction refinement* (CEGAR) automatically refines the abstract domain so that a new run of the abstract reachability algorithm rules out the counterexample [1,3,30]. Clearly, the CEGAR technique may not terminate because an infinite number of spurious counterexamples may be found. Thus, in order to improve the termination behaviour of that technique, several more sophisticated refinement strategies have been proposed (see, for instance, [14,16,20,32]).

In this paper in order to improve the termination of the safety verification process, we propose a technique based on the *specialization of constraint logic programs*. Constraint Logic Programming (CLP) has been shown to be very suitable for the analysis of imperative programs, because it provides a very convenient way of representing symbolic program executions and also, by using constraints, program invariants (see, for instance, [16,18,27,28]). Program specialization is a program transformation technique which, given a program P and a portion in_1 of its input data, returns a specialized program P_s that is equivalent to P in the sense that when the remaining portion in_2 of the input of P is given, then $P_s(in_2) = P(in_1, in_2)$ [12,21,22]. The specialization of CLP programs has been proposed in [27] as a pre-processing phase for program analysis. This analysis is done in various steps. First, the semantics of an imperative language is provided by means of a CLP program which defines the interpreter I of that language, and then, program I is specialized with respect to the program P whose safety property should be checked. The result of this specialization is a CLP program I_P and, since program specialization preserves semantic equivalence, we can analyze I_P for proving the properties of P.

Similarly to [27], also the technique proposed in this paper produces a specialized interpreter I_P. However, instead of applying program analysis techniques, we further specialize I_P with respect to the property characterizing the input values of P (that is, the precondition of P), thereby deriving a new program I'_P. The effect of this further specialization is the modification of the structure of the program I_P and the explicit addition of new constraints that denote invariants of the computation. Through various experiments we show that by exploiting these invariants, the construction of the least model of the program I'_P terminates in many interesting cases and, thus, it is possible to verify safety properties by simply inspecting that model.

An essential ingredient of program specialization are the *generalization steps*, which introduce new predicate definitions representing invariants of the program executions. Generalizations can be used to enforce the termination of program specialization (recall that termination occurs when no new predicate definitions are generated) and, in this respect, they are similar to the widening operators used in static program analysis [4,5]. One problem encountered with generalizations is that sometimes they introduce predicate definitions which are too general, thereby making specialization useless. In this paper we introduce a new generalization strategy, called the *constrained generalization*, whose objective is indeed to avoid the introduction of new predicate definitions that are too general.

The basic idea of the constrained generalization is related to the branching behaviour of the unfolding steps, as we now indicate. Given a sequence of unfolding steps performed during program specialization, we may consider a symbolic evaluation tree made out of clauses, such that every clause has as children the clauses which are generated by unfolding that clause. Suppose that a clause γ has n children which are generated by unfolding using clauses $\gamma_1, \ldots, \gamma_n$, and suppose that during program specialization we have to generalize clause γ. Then, we would like to perform this generalization by introducing a new predicate definition, say δ, such that by unfolding clause δ, we get again, if possible, n children and these children are due to the same clauses $\gamma_1, \ldots, \gamma_n$.

Since in this generalization the objective of preserving, if possible, the branching structure of the symbolic evaluation tree, is realized by adding extra constraints to the clause obtained after a usual generalization step (using, for instance, the widening operator [4] or the convex-hull operator [5]), we call the generalization proposed in this paper *a constrained generalization*. Similar proposals have been presented in [2,15] and in Section 7 we will briefly compare those proposals with ours.

The paper is organized as follows. In Section 2 we describe the syntax of the SIMP language and the CLP interpreter which defines the operational semantics of that language. In Section 3 we outline our software model checking approach by developing an example taken from [14]. In Sections 4 and 5 we describe our strategy for specializing CLP programs and, in particular, our novel constrained generalization technique. In Section 6 we report on some experiments we have performed by using a prototype implementation based on the MAP transformation system [26]. We also compare the results we have obtained using the MAP system with the results we have obtained using state-of-the-art software model checking systems such as ARMC [28], HSF(C) [13], and TRACER [17]. Finally, in Section 7 we discuss the related work and, in particular, we compare our method with other existing methods for software model checking.

2 A CLP Interpreter for a Simple Imperative Language

The syntax of our language SIMP, a C-like imperative language, is defined by using: (i) the set *Int* of integers, ranged over by n, (ii) the set {`true`, `false`} of booleans, and (iii) the set *Loc* of locations, ranged over by x. We have also the following derived sets: (iv) *Aexpr* of arithmetic expressions, (v) *Bexpr* of boolean expressions, (vi) *Test* of tests, and (vii) *Com* of commands. The syntax of our language is as follows.

$Aexpr \ni a ::= n \mid x \mid a_0 \; aop \; a_1$
$Bexpr \ni b ::= \texttt{true} \mid \texttt{false} \mid a_0 \; rop \; a_1 \mid \texttt{!}\, b \mid b_0 \; bop \; b_1$
$Test \;\; \ni t ::= \texttt{nd} \mid b$
$Com \;\; \ni c ::= \texttt{skip} \mid x\,\texttt{=}\,a \mid c_0\texttt{;}c_1 \mid \texttt{if (}t\texttt{) \{}c_0\texttt{\} else } c_1 \mid \texttt{while (}t\texttt{) \{}c\texttt{\}} \mid \texttt{error}$

where the arithmetic operator *aop* belongs to {`+`, `-`, `*`}, the relational operator *rop* belongs to {`<`, `<=`, `==`}, and the boolean operator *bop* belongs to {`&&`, `||`}. The constant `nd` denotes the nondeterministic choice and `error` denotes the error command. The other symbols should be understood as usual in C. We will write `if (`t`) {`c_0`}`, instead of `if (`t`) {`c_0`} else skip`.

Now we introduce a CLP program which defines the interpreter of our SIMP language. We need the following notions.

A *state* is a function from *Loc* to *Int*. It is denoted by a list of CLP terms, each of which is of the form `bn(loc(X),V)`, where `bn` is a binary constructor binding the location `X` to the value of the CLP variable `V`. We assume that the set of locations used in every command is fixed and, thus, for every command, the state has a fixed, finite length. We have two predicates operating on states: (i) `lookup(loc(X),S,V)`, which holds iff the location `X` stores the value `V` in the state `S`, and (ii) `update(loc(X),V,S1,S2)`, which holds iff the state `S2` is equal to the state `S1`, except that the location `X` stores the value `V`.

We also have the predicates `aev(A,S,V)` and `bev(B,S)`, for the evaluation of arithmetic expressions and boolean expressions, respectively. `aev(A,S,V)` holds iff the arithmetic expression `A` in the state `S` evaluates to `V`, and `bev(B,S)` holds iff the boolean expression `B` holds in the state `S`. A test `T` in a state `S` is evaluated via the predicate `tev(T,S)` defined as follows: (i) for all states `S`, *both* `tev(nd,S)` and `tev(not(nd),S)` hold (and in this sense `nd` denotes the nondeterministic choice), and (ii) for all boolean expressions `B`, `tev(B,S)` holds iff `bev(B,S)` holds.

A command c is denoted by a term built out of the following constructors: `skip` (nullary), `asgn` (binary) for the assignment, `comp` (binary) for the composition of command, `ite` (ternary) for the conditional, and `while` (binary) for the while-do. The operator '`;`' associates to the right. Thus, for instance, the command c_0`;`c_1`;`c_2 is denoted by the term `comp(c0,comp(c1,c2))`.

A *configuration* is a pair of a command and a state. A configuration is denoted by the term `cf(c,s)`, where `cf` is a binary constructor which takes as arguments the command `c` and the state `s`. The interpreter of our SIMP language, adapted from [29], is defined in terms of a *transition relation* that relates an old configuration to either a new configuration or a new state. That relation is denoted by the predicate `tr` whose clauses are given below. `tr(cf(C,S),cf(C1,S1))` holds iff the execution of the command C in the state S leads to the new configuration cf(C1,S1), and tr(cf(C,S),S1) holds iff the execution of the command C in the state S leads to the new state S1.

```
tr(cf(skip,S), S).
tr(cf(asgn(loc(X),A),S),S1) :- aev(A,S,V), update(loc(X),V,S,S1).
tr(cf(comp(C0,C1),S), cf(C1,S1)) :- tr(cf(C0,S),S1).
tr(cf(comp(C0,C1),S), cf(comp(C0',C1),S')) :- tr(cf(C0,S), cf(C0',S')).
tr(cf(ite(T,C0,C1),S), cf(C0,S)) :- tev(T,S).
tr(cf(ite(T,C0,C1),S), cf(C1,S)) :- tev(not(T),S).
tr(cf(while(T,C),S), cf(ite(T,comp(C,while(T,C)),skip),S)).
```

A state s is said to be *initial* if *initProp* holds in s. A configuration is said to be *initial* if its state is initial. A configuration is said to be *unsafe* if its command is error.

Now, we introduce a CLP program, called R, that by using a bottom-up evaluation strategy, performs in a backward way the reachability analysis over configurations. Program R checks whether or not an unsafe configuration is reachable from an initial configuration, by starting from the unsafe configurations. The semantics of program R is given by its least model, denoted $M(R)$.

Definition 1 (Reachability Program). *Given a boolean expression initProp holding in the initial states and a command com, the reachability program R is made out of the following clauses:*

```
unsafe :- initConf(X), reachable(X).
reachable(X) :- unsafeConf(X).            % unsafe configurations are reachable
reachable(X) :- tr(X,X1), reachable(X1).
initConf(cf(com,S)) :- bev(initProp,S). % initProp holds in the initial state S
unsafeConf(cf(error,S)).  % the error command defines an unsafe configuration
```

together with the clauses for the predicates tr *and* bev *and the predicates they depend upon. In the above clauses for R the terms* initProp *and* com *denote initProp and com, respectively. We will say that com is safe with respect to initProp (or com is safe, for short) iff* unsafe $\notin M(R)$.

3 Specialization-Based Software Model Checking

In this section we outline the method for software model checking we propose. By means of an example borrowed from [14], we argue that program specialization can prove program safety in some cases where the CEGAR method (as implemented in ARMC [28]) does not work.

Let the property *initProp* which characterizes the initial states be:

```
x==0 && y==0 && n>=0
```

and the SIMP command *com* be:

```
while (x<n) { x = x+1; y = y+1 };
while (x>0) { x = x-1; y = y-1 };
if (y>x) error
```

We want to prove that *com* is safe with respect to *initProp*, that is, there is no execution of *com* with input values of x, y, and n satisfying *initProp*, such that the `error` command is executed. As shown in Table 1 of Section 6, CEGAR fails to prove this safety property, because an infinite set of counterexamples is generated (see the entry '∞' for Program *rel* in the ARMC column).

By applying the specialization-based software model checking method we propose in this paper, we will be able to prove that *com* is indeed safe. As indicated in Section 2, we have to show that `unsafe` $\notin M(R)$, where R is the CLP program of Definition 1, `com` is the term:

```
comp(while(lt(loc(x),loc(n)),
   comp(asgn(loc(x),plus(loc(x),1)), asgn(loc(y),plus(loc(y),1)))),
comp(while(gt(loc(x),0),
   comp(asgn(loc(x),minus(loc(x),1)), asgn(loc(y),minus(loc(y),1)))),
ite(gt(loc(y),loc(x)),error,skip)))
```

and `initProp` is the term:

```
and(eq(loc(x),0), and(eq(loc(y),0), ge(loc(n),0)))
```

Our method consists of the three phases as we now specify.

The Software Model Checking Method

Input: A boolean expression *initProp* characterizing the initial states and a SIMP command *com*.

Output: The answer *safe* iff *com* is safe with respect to *initProp*.

Let R be the CLP program of Definition 1 defining the predicate `unsafe`.
Phase (1): $\mathit{Specialize}_{\mathtt{com}}(R, R_{\mathtt{com}})$;
Phase (2): $\mathit{Specialize}_{\mathtt{initProp}}(R_{\mathtt{com}}, R_{Sp})$;
Phase (3): $\mathit{BottomUp}(R_{Sp}, M_{Sp})$;
Return the answer *safe* iff `unsafe` $\notin M_{Sp}$.

During Phase (1), by making use of familiar transformation rules (definition introduction, unfolding, folding, removal of clauses with unsatisfiable body, and removal of subsumed clauses [7]), we 'compile away', similarly to [27], the SIMP interpreter by specializing program R with respect to `com`, thereby deriving the following program $R_{\mathtt{com}}$ which encodes the reachability relation associated with the interpreter specialized with respect to `com`:

```
1. unsafe :- X=1, Y=1,  N>=1,   new1(X,Y,N).
2. new1(X,Y,N) :- X<N,  X'=X+1, Y'=Y+1, new1(X',Y',N).
3. new1(X,Y,N) :- X>=1, X>=N,   X'=X-1, Y'=Y-1, new2(X',Y',N).
4. new1(X,Y,N) :- X=<0, X<Y,    X>=N.
5. new2(X,Y,N) :- X>=1, X'=X-1, Y'=Y-1, new2(X',Y',N).
6. new2(X,Y,N) :- X=<0, X<Y.
```

The specialization of Phase (1) is said to perform 'the removal of the interpreter'. Note that: (i) the two predicates `new1` and `new2` correspond to the two while-do commands occurring in *com*, and (ii) the assignments and the conditional occurring in *com*, do not occur in $R_{\mathtt{com}}$ because by unfolding they have been replaced by suitable constraints relating the values of `X` and `Y` (that is, the old values of the SIMP variables `x` and `y`) to the values of `X'` and `Y'` (that is, the new values of those variables `x` and `y`).

Unfortunately, the program $R_{\mathtt{com}}$ is not satisfactory for showing safety, because the bottom-up construction of the least model $M(R_{\mathtt{com}})$ does not terminate. The top-down evaluation of the `unsafe` query in $R_{\mathtt{com}}$ does not terminate either. Then, in Phase (2) we specialize program $R_{\mathtt{com}}$ with respect to the property `initProp`, thereby deriving the specialized program R_{Sp}. During this Phase (2) the constraints occurring in the definitions of `new1` and `new2` are generalized according to a suitable generalization strategy based both on widening [4,8,11] and on the novel constrained generalization strategy we propose in this paper. Suitable new predicate definitions will be introduced during this Phase (2), so that at Phase (3) we can construct the least model M_{Sp} of the derived program R_{Sp} by using a bottom-up evaluation procedure. We will show that, in our example, the construction of the least model M_{Sp} terminates and we can prove the safety of the command *com* by showing that the atom `unsafe` does not belong to that model.

Phase (2) of our method makes use of the same transformation rules used during Phase (1), but those rules are applied according to a different strategy, whose effect is the propagation of the constraints occurring in $R_{\mathtt{com}}$.

We start off by introducing the following definition:

7. `new3(X,Y,N) :- X=1, Y=1, N>=1, new1(X,Y,N).`

and then folding clause 1 by using this clause 7. We get the folded clause:

1.f `unsafe:- X=1, Y=1, N>=1, new3(X,Y,N).`

We proceed by following the usual unfold-definition-fold cycle of the specialization strategies [8,11]. Each new definition introduced during specialization determines a new node of a tree, called *DefsTree*, whose root is clause 7, which is the first definition we have introduced. (We will explain below how the tree *DefsTree* is incrementally constructed.) Then, we unfold clause 7 and we get:

8. `new3(X,Y,N) :- X=1, Y=1, N>=2, X'=2, Y'=2, new1(X',Y',N).`

9. `new3(X,Y,N) :- X=1, Y=1, N=1, X'=0, Y'=0, new2(X',Y',N).`

Now, we should fold these two clauses. Let us deal with them, one at the time, and let us first consider clause 8. In order to fold clause 8 we consider a definition, called the *candidate definition*, which is of the form:

10. `new4(X,Y,N) :- X=2, Y=2, N>=2, new1(X,Y,N).`

The body of this candidate definition is obtained by projecting the constraint in clause 8 with respect to `X'`, `Y'`, and `N`, and renaming the primed variables to unprimed variables. Since in *DefsTree* there is *an ancestor definition*, namely the root clause 7, with the predicate `new1` in the body, we apply the *widening operator*, introduced in [11], to clause 7 and clause 10, and we get the definition:

11. `new4(X,Y,N) :- X>=1, Y>=1, N>=1, new1(X,Y,N).`

(Recall that the widening operation of two clauses $c1$ and $c2$, after replacing every equality `A=B` by the equivalent conjunction `A>=B, A=<B`, keeps the atomic constraints of clause $c1$ which are implied by the constraint of clause $c2$.)

At this point, we do *not* introduce clause 11 (as we would do if we perform a usual generalization using widening alone, as indicated in [8,11]), but we apply our *constrained generalization*, which imposes the addition of some extra constraints to the body of clause 11, as we now explain.

With each predicate `newk` we associate a set of constraints, called the *regions for* `newk`, which are all the *atomic constraints* on the unprimed variables (that is, the variables in the heads of the clauses) occurring in any one of the clauses for `newk` in program $R_{\mathtt{com}}$. Then, we add to the body of the generalized definition obtained by widening, say `newp(...) :- c, newk(...)`, (clause 11, in our case), all *negated regions for* `newk` which are implied by `c`.

In our example, the regions for `new1` are: `X<N`, `X>=1`, `X>=N`, `X=<0`, `X<Y` (see clauses 2, 3, and 4) and the negated regions are, respectively: `X>=N`, `X<1`, `X<N`, `X>0`, `X>=Y`. The negated regions implied by the constraint `X=2, Y=2, N>=2`, occurring in the body of the candidate clause 10, are: `X>0` and `X>=Y`.

Thus, instead of clause 11, we introduce the following clause 12 (we wrote neither `X>0` nor `X>=1` because those constraints are implied by `X>=Y, Y>=1`):
12. `new4(X,Y,N) :- X>=Y, Y>=1, N>=1, new1(X,Y,N).`
and we say that clause 12 has been obtained by constrained generalization from clause 10. Clause 12 is placed in *DefsTree* as a child of clause 7, as clause 8 has been derived by unfolding clause 7. By folding clause 8 using clause 12 we get:
8.f `new3(X,Y,N) :- X=1, Y=1, N>=2, X'=2, Y'=2, new4(X',Y',N).`
Now, it remains to fold clause 9 and in order to do so, we consider the following candidate definition:
13. `new5(X,Y,N) :- X=0, Y=0, N=1, new2(X,Y,N).`
Clause 13 is placed in *DefsTree* as a child of clause 7, as clause 9 has been derived by unfolding clause 7. We do not make any generalization of this clause, because no definition with `new2` in its body occurs as an ancestor of clause 13 in *DefsTree*. By folding clause 9 using clause 13 we get:
9.f `new3(X,Y,N) :- X=1, Y=1, N=1, X'=0, Y'=0, new5(X',Y',N).`
Now, we consider the last two definition clauses we have introduced, that is, clauses 12 and 13. First, we deal with clause 12. Starting from that clause, we perform a sequence of unfolding-definition-folding steps similar to the sequence we have described above, when presenting the derivation of clauses 8.f and 9.f, starting from clause 7. During this sequence of steps, we introduce two predicates, `new6` and `new7` (see the definition clauses 16 and 18, respectively), for performing the required folding steps. We get the following clauses:
14.f `new4(X,Y,N):-X>=Y, X<N, Y>0, X'=X+1, Y'=Y+1, new4(X',Y',N).`
15.f `new4(X,Y,N):-X>=Y, X>=N, Y>0, N>0, X'=X-1, Y'=Y-1, new6(X',Y',N).`
17.f `new6(X,Y,N):-X>0, X>=Y, X>=N-1, Y>=0, N>0, X'=X-1, Y'=Y-1, new7(X',Y',N).`
19.f `new7(X,Y,N):-X>0, X=<Y, N>0, X'=X-1, Y'=Y-1, new7(X',Y',N).`
The tree *DefsTree* of all the definitions introduced during Phase (2), can be depicted as follows:

```
DefsTree:                 7. new3(X,Y,N):= X=1,Y=1,N>=1,new1(X,Y,N).
                                    /                \
  12. new4(X,Y,N):= X>=Y,Y>=1,N>=1,new1(X,Y,N).       \
                          |                 13. new5(X,Y,N):= X=0,Y=0,N=1,new2(X,Y,N).
16. new6(X,Y,N):= X>=Y,X+1>=N,Y>=0,N>=1,new2(X,Y,N).
                          |
   18. new7(X,Y,N):= X>=Y,N>=1,new2(X,Y,N).
```

Then, we deal with clause 13. Again, starting from that clause we perform a sequence of unfolding-definition-folding steps. By unfolding clause 13 w.r.t. `new2` we get an empty set of clauses for `new5`. Then, we delete clause 9.f because there are no clauses for `new5`.

Eventually, we get the program R_{Sp} made out of the following clauses:

1.f `unsafe :- X=1, Y=1, N>=1, new3(X,Y,N).`

7.1f `new3(X,Y,N):-X=1, Y=1, N>=2, X'=2, Y'=2, new4(X',Y',N).`

together with the clauses 14.f, 15.f, 17.f, and 19.f.

This concludes Phase (2).

Now, we can perform Phase (3) of our method. This phase terminates immediately because in R_{Sp} there are no constrained facts (that is, clauses whose bodies consist of constraints only) and $M(R_{Sp})$ is the empty set.

Thus, `unsafe` $\notin M(R_{Sp})$ and we conclude that the command *com* is safe with respect to *initProp*.

One can verify that if we were to do the generalization steps of Phase (2) using the widening technique alone (without the constrained generalization), we could not derive a program that allows us to prove safety, because during Phase (3) the execution of the *BottomUp* procedure does not terminate.

4 The Specialization Strategy

Phases (1) and (2) of our Software Model Checking method outlined in Section 3 are realized by two applications of a single, general specialization strategy for CLP programs that we now present.

This strategy is an adaptation of the specialization strategies we have presented in [8,11] and, as already mentioned in Section 3, it makes use of the following transformation rules: definition introduction, unfolding, clause removal, and folding. These rules, under suitable conditions, guarantee that the least model semantics is preserved (see, for instance, [7]).

Our general specialization strategy is realized by the following *Specialize* procedure.

Procedure *Specialize*
Input: A CLP program of the form $P \cup \{\gamma_0\}$, where γ_0 is `unsafe` $\leftarrow c, G$.
Output: A CLP program P_S such that `unsafe` $\in M(P \cup \{\gamma_0\})$ iff `unsafe` $\in M(P_S)$.

$P_S := \{\gamma_0\}$; *InDefs* $:= \{\gamma_0\}$; *Defs* $:= \emptyset$;
while there exists a clause γ in *InDefs*
do *Unfold*(γ, Γ);
  Generalize&Fold$(\mathit{Defs}, \Gamma, \mathit{NewDefs}, \Delta)$;
  $P_S := P_S \cup \Delta$; *InDefs* $:= (\mathit{InDefs} - \{\gamma\}) \cup \mathit{NewDefs}$; *Defs* $:= \mathit{Defs} \cup \mathit{NewDefs}$;
end-while

Initially, this procedure considers the clause γ_0 of the form:

$\texttt{unsafe} \leftarrow c, G$

where c is a constraint and G is a goal, and then iteratively applies the following two procedures: (i) the *Unfold* procedure, which uses the unfolding rule and the clause removal rule, and (ii) the *Generalize&Fold* procedure, which uses the definition introduction rule and the folding rule.

The *Unfold* procedure takes as input a clause γ and returns as output a set Γ of clauses derived from γ by one or more applications of the unfolding rule, which consists in: (i) replacing an atom A occurring in the body of a clause by the bodies of the clauses in P whose head is unifiable with A, and (ii) applying the unifying substitution. The first step of the *Unfold* procedure consists in unfolding γ with respect to the leftmost atom in its body. In order to guarantee the termination of the *Unfold* procedure, an atom A is selected for unfolding only if it has not been derived by unfolding a variant of A itself. More sophisticated unfolding strategies can be applied (see [22] for a survey of techniques for controlling unfolding), but our simple strategy turns out to be effective in all our examples. At the end of the *Unfold* procedure, subsumed clauses and clauses with unsatisfiable constraints are removed.

The *Generalize&Fold* procedure takes as input the set Γ of clauses produced by the *Unfold* procedure and introduces a set *NewDefs* of *definitions*, that is, clauses of the form $newp(X) \leftarrow d(X), A(X)$, where *newp* is a new predicate symbol, X is a tuple of variables, $d(X)$ is a constraint whose variables are among the ones in X, and $A(X)$ is an atom whose variables are exactly those of the tuple X. Any such definition denotes a set of states X satisfying the constraint $d(X)$. By folding the clauses in Γ using the definitions in *NewDefs* and the definitions introduced during previous iterations of the specialization procedure, the *Generalize&Fold* procedure derives a new set of specialized clauses. In particular, a clause of the form:

$newq(X) \leftarrow c(X), A(X)$

obtained by the *Unfold* procedure, is folded by using a definition of the form:

$newp(X) \leftarrow d(X), A(X)$

if for all X, $c(X)$ implies $d(X)$. This condition is also denoted by $c(X) \sqsubseteq d(X)$, where the quantification 'for all X' is silently assumed. If $c(X) \sqsubseteq d(X)$, we say that $d(X)$ is a *generalization* of $c(X)$. The result of folding is the specialized clause:

$newq(X) \leftarrow c(X), newp(X)$.

The specialization strategy proceeds by applying the *Unfold* procedure followed by the *Generalize&Fold* procedure to each clause in *NewDefs*, and terminates when no new definitions are needed for performing folding steps. Unfortunately, an uncontrolled application of the *Generalize&Fold* procedure may lead to the introduction of infinitely many new definitions, thereby causing the nontermination of the specialization procedure. In the following section we will define suitable *generalization operators* which guarantee the introduction of finitely many new definitions.

5 Constrained Generalization

In this section we define the generalization operators which are used to ensure the termination of the specialization strategy and, as mentioned in the Introduction, we also introduce *constrained* generalization operators that generalize the constraints occurring in a candidate definition and, by adding suitable extra constraints, have the objective of preventing that the set of clauses generated by unfolding the generalized definition is larger than the set of clauses generated by unfolding the candidate definition. In this sense we say the objective of constrained generalization is to preserve the branching behaviour of the candidate definitions.

Let $\mathcal{C}$ denote the set of all linear constraints. The set $\mathcal{C}$ is the minimal set of constraints which: (i) includes all atomic constraints of the form either $p_1 \leq p_2$ or $p_1 < p_2$, where p_1 and p_2 are linear polynomials with variables $X_1, \ldots, X_k$ and integer coefficients, and (ii) is closed under conjunction (which we denote by ',' and also by '$\wedge$'). An equation $p_1 = p_2$ stands for $p_1 \leq p_2 \wedge p_2 \leq p_1$. The projection of a constraint c onto a tuple X of variables, denoted $project(c, X)$, is a constraint such that $\mathcal{R} \models \forall X\ (project(c, X) \leftrightarrow \exists Y c)$, where Y is the tuple of variables occurring in c and not in X, and $\mathcal{R}$ is the structure of the real numbers.

In order to introduce the notion of a generalization operator (see also [11], where the set $\mathcal{C}$ of all linear constraints with variables $X_1, \ldots, X_k$ has been denoted Lin_k), we need the following definition [6].

Definition 2 (Well-Quasi Ordering $\precsim$). A *well-quasi ordering* (or *wqo*, for short) on a set S is a reflexive, transitive relation $\precsim$ on S such that, for every infinite sequence $e_0 e_1 \ldots$ of elements of S, there exist i and j such that $i<j$ and $e_i \precsim e_j$. Given e_1 and e_2 in S, we write $e_1 \approx e_2$ if $e_1 \precsim e_2$ and $e_2 \precsim e_1$. A wqo $\precsim$ is *thin* iff for all $e \in S$, the set $\{e' \in S \mid e \approx e'\}$ is finite.

The use of a thin wqo guarantees that during the *Specialize* procedure each definition can be generalized a finite number of times only, and thus the termination of the procedure is guaranteed.

The thin wqo *Maxcoeff*, denoted by $\precsim_M$, compares the maximum absolute values of the coefficients occurring in polynomials. It is defined as follows. For any atomic constraint a of the form $p < 0$ or $p \leq 0$, where p is $q_0 + q_1 X_1 + \ldots + q_k X_k$, we define $maxcoeff(a)$ to be $\max\{|q_0|, |q_1|, \ldots, |q_k|\}$. Given two atomic constraints a_1 of the form $p_1 < 0$ and a_2 of the form $p_2 < 0$, we have that $a_1 \precsim_M a_2$ iff $maxcoeff(a_1) \leq maxcoeff(a_2)$.

Similarly, if we are given the atomic constraints a_1 of the form $p_1 \leq 0$ and a_2 of the form $p_2 \leq 0$. Given two constraints $c_1 \equiv a_1, \ldots, a_m$, and $c_2 \equiv b_1, \ldots, b_n$, we have that $c_1 \precsim_M c_2$ iff, for $i = 1, \ldots, m$, there exists $j \in \{1, \ldots, n\}$ such that $a_i \precsim_M b_j$. For example, we have that:

(i) $(1-2X_1 < 0) \precsim_M (3+X_1 < 0)$,
(ii) $(2-2X_1+X_2 < 0) \precsim_M (1+3X_1 < 0)$, and
(iii) $(1+3X_1 < 0) \not\precsim_M (2-2X_1+X_2 < 0)$.

Definition 3 (Generalization Operator $\ominus$). Let $\precsim$ be a thin wqo on the set $\mathcal{C}$ of constraints. A function $\ominus$ from $\mathcal{C}\times\mathcal{C}$ to $\mathcal{C}$ is a *generalization operator* with respect to $\precsim$ if, for all constraints c and d, we have: (i) $d \sqsubseteq c\ominus d$, and (ii) $c\ominus d \precsim c$.

A trivial generalization operator is defined as $c\ominus d = true$, for all constraints c and d (without loss of generality we assume that $true \precsim c$ for every constraint c). This operator is used during Phase (1) of our Software Model Checking method.

Definition 3 generalizes several operators proposed in the literature, such as the widening operator [4] and the *most specific generalization* operator [23,33].

Other generalization operators defined in terms of relations and operators on constraints such as *widening* and *convex-hull*, have been defined in [11]. Some of them can be found in Appendix A.

Now we describe a method for deriving, from any given generalization operator $\ominus$, a new version of that operator, denoted $\ominus_{cns}$, which adds some extra constraints and still is a generalization operator. The operator $\ominus_{cns}$ is called the *constrained generalization operator derived from* $\ominus$. Constrained generalization operators are used during Phase (2) of our Software Model Checking method.

In order to specify the constrained generalization operator we need the following notions.

Let $P\cup\{\gamma_0\}$ be the input program of the *Specialize* procedure. For any constraint d and atom A, we define the *unfeasible clauses* for the pair (d,A), denoted $\mathit{UnfCl}(d,A)$, to be the set $\{(H_1 \leftarrow c_1, G_1),\ldots,(H_m \leftarrow c_m, G_m)\}$, of (renamed apart) clauses of $P\cup\{\gamma_0\}$ such that, for $i=1,\ldots,m$, A and H_i are unifiable via the most general unifier ϑ_i and $(d\wedge c_i)\,\vartheta_i$ is unsatisfiable.

The *head constraint* of a clause γ of the form $H \leftarrow c, A$ is the constraint $\mathit{project}(c,X)$, where X is the tuple of variables occurring in H. For any atomic constraint a, $\mathit{neg}(a)$ denotes the negation of a defined as follows: $\mathit{neg}(p<0)$ is $-p\leq 0$ and $\mathit{neg}(p\leq 0)$ is $-p<0$. Given a set C of clauses, we define the set of the *negated regions* of C, denoted $\mathit{NegReg}(C)$, as follows:

$$\mathit{NegReg}(C) = \{\mathit{neg}(a) \mid a \text{ is an atomic constraint of a head constraint of a clause in } C\}.$$

For any constraint d and atom A, we define the following constraint:

$$\mathit{cns}(d,A) = \bigwedge\{r \mid r \in \mathit{NegReg}(\mathit{UnfCl}(d,A)) \;\wedge\; d \sqsubseteq r\}.$$

We have that $d \sqsubseteq \mathit{cns}(d,A)$. Now, let $\ominus$ be a generalization operator with respect to the thin wqo $\precsim$. We define the constrained generalization operator derived from $\ominus$, as follows:

$$\ominus_{cns}(c,d,A) = (c\ominus d)\wedge \mathit{cns}(d,A).$$

Now we show that $\ominus_{cns}$ is indeed a generalization operator w.r.t. the thin wqo $\precsim_B$ we now define. Given a finite set B of (non necessarily atomic) constraints, a constraint $c_1\wedge\ldots\wedge c_n$, where $c_1,\ldots,c_n$ are atomic, and a constraint d, we define the binary relation $\precsim_B$ on constraints as follows: $c_1\wedge\ldots\wedge c_n \precsim_B d$ iff *either* (i) $(c_1\wedge\ldots\wedge c_n)\precsim d$, *or* (ii) there exists $i\in\{1,\ldots,n\}$ such that $c_i\in B$ and $(c_1\wedge\ldots\wedge c_{i-1}\wedge c_{i+1}\wedge\ldots\wedge c_n)\precsim_B d$. It can be shown that $\precsim_B$ is a thin wqo.

We observe that, for all constraints c, d, and all atoms A: (i) since $d \sqsubseteq c \ominus d$ and $d \sqsubseteq cns(d, A)$, then also $d \sqsubseteq \ominus_{cns}(c, d, A)$, and (ii) by definition of $\precsim_B$, for all constraints e, if $c \ominus d \precsim e$, then $\ominus_{cns}(c, d, A) \precsim_B e$, where $B = NegReg(P \cup \{\gamma_0\})$.
Thus, we have the following result.

Proposition 1. *For any program $P \cup \{\gamma_0\}$ given as input to the Specialize procedure, for any atom A, the operator $\ominus_{cns}(\_, \_, A)$ is a generalization operator with respect to the thin well-quasi ordering $\precsim_B$, where $B = NegReg(P \cup \{\gamma_0\})$.*

During Phase (2) in the *Specialize* procedure we use the following sub-procedure *Generalize&Fold* which is an adaptation of the one in [11].

Procedure *Generalize&Fold*
Input: (i) a set *Defs* of definitions structured as a tree of definitions, called *DefsTree*, (ii) a set Γ of clauses obtained from a clause γ by the *Unfold* procedure, and (iii) a constrained generalization operator $\ominus_{cns}$.
Output: (i) A set *NewDefs* of new definitions, and (ii) a set Δ of folded clauses.

NewDefs $:= \emptyset$; $\Delta := \Gamma$;
while in Δ there exists a clause δ: $H \leftarrow d, G_1, A, G_2$ where the predicate symbol of A occurs in the body of some clause in Γ *do*
GENERALIZE:
  Let X be the set of variables occurring in A and $d_X = project(d, X)$.
  1. *if* in *Defs* $\cup$ *NewDefs* there exists a (renamed apart) clause
     η: $newp(X) \leftarrow e, A$ such that $d_X \sqsubseteq e$ and $e \sqsubseteq cns(d_X, A)$
    then *NewDefs* := *NewDefs*
  2. *elseif* there exists a clause α in *Defs* such that:
    (i) α is of the form $newq(X) \leftarrow b, A$, and (ii) α is the most recent ancestor of γ in *DefsTree* whose body contains a variant of A
    then *NewDefs* := *NewDefs* $\cup \{newp(X) \leftarrow \ominus_{cns}(b, d_X, A), A\}$
  3. *else* *NewDefs* := *NewDefs* $\cup \{newp(X) \leftarrow d_X, A\}$
FOLD:
  $\Delta := (\Delta - \{\delta\}) \cup \{H \leftarrow d, G_1, newp(X), G_2\}$
end-while

The proof of termination of the *Specialize* procedure of Section 4 is a variant of the proof of Theorem 3 in [10]. In this variant we use Proposition 1 and we also take into account the fact that during the execution of the procedure, only a finite number of atoms are generated (modulo variants). Since the correctness of the *Specialize* procedure directly follows from the fact that the transformation rules preserve the least model semantics [7], we have the following result.

Theorem 1 (Termination and Correctness of Specialization). (i) *The Specialize procedure always terminates.* (ii) *Let program P_S be the output of the Specialize procedure. Then* `unsafe`$\in M(P)$ *iff* `unsafe`$\in M(P_S)$.

6 Experimental Evaluation

In this section we present some preliminary results obtained by applying our Software Model Checking method to some benchmark programs taken from the literature. The results show that our approach is viable and competitive with the state-of-the-art software model checkers.

Programs $ex1$, $f1a$, $f2$, and *interp* have been taken from the benchmark set of DAGGER [14]. Programs *substring* and *tracerP* are taken from [20] and [16], respectively. Programs $re1$ and *singleLoop* have been introduced to illustrate the constrained generalization strategy. Finally, *selectSort* is an encoding of the Selection sort algorithm where references to arrays have been replaced by using the nondeterministic choice operator `nd` to perform array bounds checking. The source code of all the above programs is available at `http://map.uniroma2.it/smc/`.

Our model checker uses the MAP system [26] which is a tool for transforming constraint logic programs implemented in SICStus Prolog. MAP uses the `clpr` library to operate on constraints over the reals. Our model checker consists of three modules: (i) a translator which takes a property *initProp* and a command *com* and returns their associated terms, (ii) the MAP system for CLP program specialization which performs Phases (1) and (2) of our method, and (iii) a program for computing the least models of CLP programs which performs Phase (3) of our method.

We have also run three state-of-the-art CLP-based software model checkers on the same set of programs, and we have compared their performance with that of our model checker. In particular, we have used: (i) ARMC [28], (ii) HSF(C) [13], and (iii) TRACER [17]. ARMC and HSF(C) are CLP-based software model checkers which implement the CEGAR technique. TRACER is a CLP-based model checker which uses Symbolic Execution (SE) for the verification of safety properties of sequential C programs using approximated preconditions or approximated postconditions.

Table 1. Time (in seconds) taken for performing model checking. '∞' means 'no answer within 20 minutes', and '⊥' means 'termination with error'.

Program	MAP				ARMC	HSF(C)	TRACER	
	W	W_{cns}	$CHWM$	$CHWM_{cns}$			$SPost$	$WPre$
$ex1$	1.08	1.09	1.14	1.25	0.18	0.21	∞	1.29
$f1a$	∞	∞	0.35	0.36	∞	0.20	⊥	1.30
$f2$	∞	∞	0.75	0.88	∞	0.19	∞	1.32
interp	0.29	0.29	0.32	0.44	0.13	0.18	∞	1.22
$re1$	∞	0.33	0.33	0.33	∞	0.19	∞	∞
selectSort	4.34	4.70	4.59	5.57	0.48	0.25	∞	∞
singleLoop	∞	∞	∞	0.26	∞	∞	⊥	1.28
substring	88.20	171.20	5.21	5.92	931.02	1.08	187.91	184.09
tracerP	0.11	0.12	0.11	0.12	∞	∞	1.15	1.28

Table 1 reports the results of our experimental evaluation which has been performed on an Intel Core Duo E7300 2.66Ghz processor with 4GB of memory under the GNU Linux operating system.

In Columns W and $CHWM$ we report the results obtained by the MAP system when using the procedure presented in Section 5 and the generalization operators *Widen* and *CHWidenMax* [11], respectively. In Columns W_{cns} and $CHWM_{cns}$ we report the results for the constrained versions of those generalization operators, called $Widen_{cns}$ and $CHWidenMax_{cns}$, respectively. In the remaining columns we report the results obtained by ARMC, HSF(C), and TRACER using the strongest postcondition (*SPost*) and the weakest precondition (*WPre*) options, respectively. More details on the experimental results can be found in Appendix B.

On the selected set of examples, we have that the MAP system with the $CHWidenMax_{cns}$ is able to verify 9 properties out of 9, while the other tools do not exceed 7 properties. Also the verification time is generally comparable to that of the other tools, and it is not much greater than that of the fastest tools. Note that there are two examples (*re*1 and *singleLoop*) where constrained generalization operators based on widening and convex-hull are strictly more powerful than the corresponding operators which are not constrained.

We also observe that the use of a constrained generalization operator usually causes a very small increase of the verification time with respect to the non-constrained counterparts, thus making constrained generalization a promising technique that can be used in practice for software verification.

7 Related Work and Conclusions

The specialization-based software model checking technique presented in this paper is an extension of the technique for the verification of safety properties of infinite state reactive systems, encoded as CLP programs, presented in [9,11]. The main novelties of the present paper are that here we consider imperative sequential programs and we propose a new specialization strategy which has the objective of preserving, if possible, the branching behaviour of the definitions to be generalized.

The use of constraint logic programming and program specialization for verifying properties of imperative programs has also been proposed by [27]. In that paper, the interpreter of an imperative language is encoded as a CLP program. Then the interpreter is specialized with respect to a specific imperative program to obtain a residual program on which a static analyser for CLP programs is applied. Finally, the information gathered during this process is translated back in the form of invariants of the original imperative program. Our approach does not require static analysis of CLP and, instead, we discover program invariants *during* the specialization process by means of (constrained) generalization operators.

The idea of constrained generalization which has the objective of preserving the branching behaviour of a clause, is related to the technique for preserving *characteristic trees* while applying abstraction during partial deduction [24]. Indeed, a characteristic tree provides an abstract description of the tree generated by unfolding a given goal, and abstraction corresponds to generalization. However, the partial deduction technique considered in [24] is applied to ordinary logic programs (not CLP programs) and constraints such as equations and inequations on finite terms, are only used in an intermediate phase.

In order to prove that a program satisfies a given property, software model checking methods try to automatically construct a conservative model (that is, a property-preserving model) of the program such that, if the model satisfies the given property, then also does the actual program. In constructing such a model a software model checker may follow two dual approaches: either (i) it may start from a coarse model and then progressively refine it by incorporating new facts, or (ii) it may start from a concrete model and then progressively abstract away from it some irrelevant facts.

Our verification method follows the second approach. Given a program P, we model its computational behaviour as a CLP program (Phase 1) by using the interpreter of the language in which P is written. Then, the CLP program is specialized with respect to the property to be verified, by using constrained generalization operators which have the objective of preserving, if possible, the branching behaviour of the definitions to be generalized. In this way we may avoid loss of precision, and at the same time, we enforce the termination of the specialization process (Phase 2).

In order to get a conservative model of a program, different generalization operators have been introduced in the literature. In particular, in [2] the authors introduce the *bounded widen* operator $c \nabla_B d$, defined for any given constraint c and d and any set B of constraints. This operator, which improves the precision of the *widen* operator introduced in [4], has been applied in the verification of synchronous programs and linear hybrid systems. A similar operator $c \nabla_B d$, called *widening up to* B, has been introduced in [15]. In this operator the set B of constraints is statically computed once the system to be verified is given. There is also a version of that operator, called *interpolated widen*, in which the set B is dynamically computed [14] by using the interpolants which are derived during the counterexample analysis.

Similarly to [2,5,14,15], the main objective of the constrained generalization operators introduced in this paper is the improvement of precision during program specialization. In particular, this generalization operator, similar to the bounded widen operator, limits the possible generalizations on the basis of a set of constraints defined by the CLP program obtained as output of Phase 1. Since this set of constraints which limits the generalization depends on the output of Phase 1, our generalization is more flexible than the one presented in [2]. Moreover, our generalization operator is more general than the classical widening operator introduced in [4]. Indeed, we only require that the set of constraints which have a non-empty intersection with the generalized constraint $c \ominus d$, are entailed by d.

Now let us point out some advantages of the techniques for software model checking which, like ours, use methodologies based on program specialization.

(1) First of all, the approach based on specialization of interpreters provides a parametric, and thus flexible, technique for software model checking. Indeed, by following this approach, given a program P written in the programming language L, and a property φ written in a logic M, in order to verify that φ holds for P, first (i) we specify the interpreter I_L for L and we specify the semantics S_M of M (as a proof system or a satisfaction relation) in a suitable metalanguage, then (ii) we specialize the interpreter and the semantics with respect to P and φ, and finally (iii) we analyze the derived specialized program (by possibly applying program specialization again, as done in this paper).

The metalanguage we used in this paper for Step (i) is CLP in which we have specified both the interpreter and the reachability relation (which defines the semantics of the reachability formula to be verified).

These features make program specialization a suitable framework for software model checking because it can easily adapt to the changes of the syntax and the semantics of the programming languages under consideration and also to the different logics where the properties of interest are expressed.

(2) By applying suitable generalization operators we can make sure that specialization always terminates and produces an equivalent program with respect to the property of interest. Thus, we can apply a sequence of specializations, thereby refining the analysis to the desired degree of precision.

(3) Program specialization provides a uniform framework for program analysis. Indeed, as already mentioned, abstraction operators can be regarded as particular generalization operators and, moreover, specialization can be easily combined to other program transformation techniques, such as program slicing, dead code elimination, continuation passing transformation, and loop fusion.

(4) Finally, on a more technical side, program specialization can easily accommodate polyvariant analysis [31] by introducing several specialized predicate definitions corresponding to the same point of the program to be analyzed.

Our preliminary experimental results show that our approach is viable and competitive with state-of-the-art software model checkers such as ARMC [28], HSF(C) [13] and TRACER [17].

In order to further validate of our approach, we plan in the near future to perform experiments on a larger set of examples. In particular, in order to support a larger set of input specifications, we are currently working on rewriting our translator so that it can take CIL (C Intermediate Language) programs [25]. We also plan to extend our interpreter to deal with more sophisticated features of imperative languages such as arrays, pointers, and procedure calls.

Moreover, since our specialization-based method preserves the semantics of the original specification, we also plan to explore how our techniques can be effectively used in a preprocessing step before using existing state-of-the-art software model checkers for improving both their precision and their efficiency.

References

1. Ball, T., Rajamani, S.K.: Boolean programs: a model and process for software analysis. MSR TR 2000-14, Microsoft Report (2000)
2. Bjørner, N., Browne, A., Manna, Z.: Automatic generation of invariants and intermediate assertions. In: Montanari, U., Rossi, F. (eds.) CP 1995. LNCS, vol. 976, pp. 589–623. Springer, Heidelberg (1995)
3. Clarke, E.M., Grumberg, O., Jha, S., Lu, Y., Veith, H.: Counterexample-Guided Abstraction Refinement. In: Emerson, E.A., Sistla, A.P. (eds.) CAV 2000. LNCS, vol. 1855, pp. 154–169. Springer, Heidelberg (2000)
4. Cousot, P., Cousot, R.: Abstract interpretation: A unified lattice model for static analysis of programs by construction of approximation of fixpoints. In: Proc. POPL 1977, pp. 238–252. ACM Press (1977)
5. Cousot, P., Halbwachs, N.: Automatic discovery of linear restraints among variables of a program. In: Proc. POPL 1978, pp. 84–96. ACM Press (1978)
6. Dershowitz, N.: Termination of rewriting. Journal of Symbolic Computation 3(1-2), 69–116 (1987)
7. Etalle, S., Gabbrielli, M.: Transformations of CLP modules. Theoretical Computer Science 166, 101–146 (1996)
8. Fioravanti, F., Pettorossi, A., Proietti, M.: Automated strategies for specializing constraint logic programs. In: Lau, K.-K. (ed.) LOPSTR 2000. LNCS, vol. 2042, pp. 125–146. Springer, Heidelberg (2001)
9. Fioravanti, F., Pettorossi, A., Proietti, M.: Verifying CTL properties of infinite state systems by specializing constraint logic programs. In: Proc. VCL 2001, DSSE-TR-2001-3, pp. 85–96. University of Southampton, UK (2001)
10. Fioravanti, F., Pettorossi, A., Proietti, M.: Verifying infinite state systems by specializing constraint logic programs. R. 657, IASI-CNR, Rome, Italy (2007)
11. Fioravanti, F., Pettorossi, A., Proietti, M., Senni, V.: Generalization strategies for the verification of infinite state systems. Theo. Pract. Log. Pro. 13(2), 175–199 (2013)
12. Gallagher, J.P.: Tutorial on specialisation of logic programs. In: Proc. PEPM 1993, pp. 88–98. ACM Press (1993)
13. Grebenshchikov, S., Gupta, A., Lopes, N.P., Popeea, C., Rybalchenko, A.: HSF(C): A Software Verifier based on Horn Clauses. In: Flanagan, C., König, B. (eds.) TACAS 2012. LNCS, vol. 7214, pp. 549–551. Springer, Heidelberg (2012)
14. Gulavani, B.S., Chakraborty, S., Nori, A.V., Rajamani, S.K.: Automatically Refining Abstract Interpretations. In: Ramakrishnan, C.R., Rehof, J. (eds.) TACAS 2008. LNCS, vol. 4963, pp. 443–458. Springer, Heidelberg (2008), `www.cfdvs.iitb.ac.in/~bhargav/dagger.php`
15. Halbwachs, N., Proy, Y.E., Roumanoff, P.: Verification of real-time systems using linear relation analysis. Formal Methods in System Design 11, 157–185 (1997)
16. Jaffar, J., Navas, J.A., Santosa, A.E.: Symbolic execution for verification. Computing Research Repository (2011)
17. Jaffar, J., Navas, J.A., Santosa, A.E.: TRACER: A Symbolic Execution Tool for Verification (2012), `paella.d1.comp.nus.edu.sg/tracer/`
18. Jaffar, J., Santosa, A.E., Voicu, R.: An interpolation method for CLP traversal. In: Gent, I.P. (ed.) CP 2009. LNCS, vol. 5732, pp. 454–469. Springer, Heidelberg (2009)
19. Jhala, R., Majumdar, R.: Software model checking. ACM Computing Surveys 41(4), 21:1–21:54 (2009)

20. Jhala, R., McMillan, K.L.: A Practical and Complete Approach to Predicate Refinement. In: Hermanns, H., Palsberg, J. (eds.) TACAS 2006. LNCS, vol. 3920, pp. 459–473. Springer, Heidelberg (2006)
21. Jones, N.D., Gomard, C.K., Sestoft, P.: Partial Evaluation and Automatic Program Generation. Prentice Hall (1993)
22. Leuschel, M., Bruynooghe, M.: Logic program specialisation through partial deduction: Control issues. Theo. Pract. Log. Pro. 2(4&5), 461–515 (2002)
23. Leuschel, M., Martens, B., De Schreye, D.: Controlling generalization and polyvariance in partial deduction of normal logic programs. ACM Transactions on Programming Languages and Systems 20(1), 208–258 (1998)
24. Leuschel, M., De Schreye, D.: Constrained partial deduction. In: Proc. WLP 1997, Munich, Germany, pp. 116–126 (1997)
25. Necula, G.C., McPeak, S., Rahul, S.P., Weimer, W.: CIL: Intermediate language and tools for analysis and transformation of C programs. In: Nigel Horspool, R. (ed.) CC 2002. LNCS, vol. 2304, pp. 213–228. Springer, Heidelberg (2002), `kerneis.github.com/cil/`
26. The MAP transformation system, `www.iasi.cnr.it/~proietti/system.html`
27. Peralta, J.C., Gallagher, J.P., Saglam, H.: Analysis of Imperative Programs through Analysis of Constraint Logic Programs. In: Levi, G. (ed.) SAS 1998. LNCS, vol. 1503, pp. 246–261. Springer, Heidelberg (1998)
28. Podelski, A., Rybalchenko, A.: ARMC: The Logical Choice for Software Model Checking with Abstraction Refinement. In: Hanus, M. (ed.) PADL 2007. LNCS, vol. 4354, pp. 245–259. Springer, Heidelberg (2007)
29. Reynolds, C.J.: Theories of Programming Languages. Cambridge Univ. Press (1998)
30. Saïdi, H.: Model checking guided abstraction and analysis. In: Palsberg, J. (ed.) SAS 2000. LNCS, vol. 1824, pp. 377–396. Springer, Heidelberg (2000)
31. Smith, S.F., Wang, T.: Polyvariant flow analysis with constrained types. In: Smolka, G. (ed.) ESOP 2000. LNCS, vol. 1782, pp. 382–396. Springer, Heidelberg (2000)
32. Sharygina, N., Tonetta, S., Tsitovich, A.: An abstraction refinement approach combining precise and approximated techniques. Soft. Tools Techn. Transf. 14(1), 1–14 (2012)
33. Sørensen, M.H., Glück, R.: An algorithm of generalization in positive supercompilation. In: Proc. ILPS 1995, pp. 465–479. MIT Press (1995)

Appendix A: Some Generalization Operators

Here we define some generalization operators which have been used in the experiments we have performed (see also [11]).

• (W) Given any two constraints $c \equiv a_1, \ldots, a_m$, and d, the operator *Widen*, denoted $\ominus_W$, returns the constraint $a_{i1}, \ldots, a_{ir}$, such that $\{a_{i1}, \ldots, a_{ir}\} = \{a_h \mid 1 \leq h \leq m \text{ and } d \sqsubseteq a_h\}$. Thus, *Widen* returns all atomic constraints of c that are entailed by d (see [4] for a similar widening operator used in static program analysis). The operator $\ominus_W$ is a generalization operator w.r.t. the thin wqo $\precsim_M$.

• (WM) Given any two constraints $c \equiv a_1, \ldots, a_m$, and $d \equiv b_1, \ldots, b_n$, the operator *WidenMax*, denoted $\ominus_{WM}$, returns the conjunction $a_{i1}, \ldots, a_{ir}, b_{j1}, \ldots, b_{js}$, where: (i) $\{a_{i1}, \ldots, a_{ir}\} = \{a_h \mid 1 \leq h \leq m \text{ and } d \sqsubseteq a_h\}$, and (ii) $\{b_{j1}, \ldots, b_{js}\} = \{b_k \mid 1 \leq k \leq n \text{ and } b_k \precsim_M c\}$.

The operator *WidenMax* is a generalization operator w.r.t. the thin wqo $\precsim_M$. It is similar to *Widen* but, together with the atomic constraints of c that are entailed by d, it returns also the conjunction of a subset of the atomic constraints of d.

Next we define a generalization operator by using the *convex hull* operator, which is often used in static program analysis [5].

• (CH) The *convex hull* of two constraints c and d in $\mathcal{C}$, denoted by $ch(c,d)$, is the least (w.r.t. the $\sqsubseteq$ ordering) constraint h in $\mathcal{C}$ such that $c \sqsubseteq h$ and $d \sqsubseteq h$. (Note that $ch(c,d)$ is unique up to equivalence of constraints.)

• ($CHWM$) Given any two constraints c and d, we define the operator *CHWidenMax*, denoted $\ominus_{CHWM}$, as follows: $c \ominus_{CHWM} d = c \ominus_{WM} ch(c,d)$. The operator $\ominus_{CHWM}$ is a generalization operator w.r.t. the thin wqo $\precsim_M$.

CHWidenMax returns the conjunction of a subset of the atomic constraints of c and a subset of the atomic constraints of $ch(c,d)$.

Appendix B: Detailed Experimental Results

In Table 2 we present in some more detail the time taken for proving the properties of interest by using our method for software model checking with the generalization operators *Widen* (Column W) and *CHWidenMax* (Column $CHWM$) [11], and the constrained generalization operators derived from them $Widen_{cns}$ (Column W_{cns}) and $CHWidenMax_{cns}$ (Column $CHWM_{cns}$), respectively.

Columns *Ph*1, *Ph*2, and *Ph*3 show the time required during Phases (1), (2), and (3), respectively, of our Software Model Checking method presented in Section 3. The sum of these three times for each phase is reported in Column *Tot*.

Table 2. Time (in seconds) taken for performing software model checking with the MAP system. '∞' means 'no answer within 20 minutes'. Times marked by '$\lhd$' are relative to the programs obtained after Phase (2) and have no constrained facts (thus, for those programs the times of Phase (3) are very small ($\leq 0.01\,s$)).

Program		W			W_{cns}			$CHWM$			$CHWM_{cns}$		
	*Ph*1	*Ph*2	*Ph*3	*Tot*	*Ph*2	*Ph*3	*Tot*	*Ph*2	*Ph*3	*Tot*	*Ph*2	*Ph*3	*Tot*
*ex*1	1.02	0.05	0.01	1.08	0.07 $\lhd$	0	1.09	0.11	0.01	1.14	0.23 $\lhd$	0	1.25
*f*1*a*	0.35	0.01	∞	∞	0.01	∞	∞	0 $\lhd$	0	0.35	0.01 $\lhd$	0	0.36
*f*2	0.71	0.03	∞	∞	0.13	∞	∞	0.03 $\lhd$	0.01	0.75	0.17 $\lhd$	0	0.88
interp	0.27	0.01	0.01	0.29	0.02 $\lhd$	0	0.29	0.04	0.01	0.32	0.17 $\lhd$	0	0.44
*re*1	0.31	0.01	∞	∞	0.02 $\lhd$	0	0.33	0.02 $\lhd$	0	0.33	0.02 $\lhd$	0	0.33
selectSort	4.27	0.06	0.01	4.34	0.43 $\lhd$	0	4.70	0.3	0.02	4.59	1.3 $\lhd$	0	5.57
singleLoop	0.22	0.02	∞	∞	0.02	∞	∞	0.03	∞	∞	0.04 $\lhd$	0	0.26
substring	0.24	0.01	87.95	88.20	0.02	170.94	171.2	4.96 $\lhd$	0.01	5.21	5.67 $\lhd$	0.01	5.92
tracerP	0.11	0 $\lhd$	0	0.11	0.01 $\lhd$	0	0.12	0 $\lhd$	0	0.11	0.01 $\lhd$	0	0.12

Enhancing Declarative Debugging with Loop Expansion and Tree Compression*

David Insa, Josep Silva, and César Tomás

Universitat Politècnica de València
Camino de Vera s/n, E-46022 Valencia, Spain
{dinsa,jsilva,ctomas}@dsic.upv.es

Abstract. Declarative debugging is a semi-automatic debugging technique that allows the programer to debug a program without the need to see the source code. The debugger generates questions about the results obtained in different computations and the programmer only has to answer them to find the bug. Declarative debugging uses an internal representation of programs called execution tree, whose structure highly influences its performance. In this work we introduce two techniques that optimize the execution trees structure. In particular, we expand and collapse the representation of loops allowing the debugger to find bugs with a reduced number of questions.

Keywords: Declarative Debugging, Tree Compression, Loop Expansion.

1 Introduction

Debugging is one of the most difficult and less automated tasks of software engineering. This is due to the fact that bugs are usually hidden under complex conditions that only happen after particular interactions of software components. Programmers cannot consider all possible computations of their pieces of software, and those unconsidered computations usually produce a bug. In words of Brian Kernighan, the difficulty of debugging is explained as follows:

> *"Everyone knows that debugging is twice as hard as writing a program in the first place. So if you're as clever as you can be when you write it, how will you ever debug it?"*
>
> The Elements of Programming Style, 2nd edition

The problems caused by bugs are highly expensive. Sometimes more than the product development price. For instance, the NIST report [14] calculated that

* This work has been partially supported by the Spanish *Ministerio de Economía y Competitividad (Secretaría de Estado de Investigación, Desarrollo e Innovación)* under grant TIN2008-06622-C03-02 and by the *Generalitat Valenciana* under grant PROMETEO/2011/052. David Insa was partially supported by the Spanish *Ministerio de Eduación* under FPU grant AP2010-4415.

© Springer-Verlag Berlin Heidelberg 2013

undetected software bugs produce a cost to the USA economy of $59 billion per year.

There have been many attempts to define automatic techniques for debugging, but in general with poor results. One notable exception is Declarative Debugging [16,17]. In this work we present a technique to improve the performance of Declarative Debugging reducing the debugging time.

Declarative debugging is a semi-automatic debugging technique that automatically generates questions about the results obtained in subcomputations. Then, the programmer answers the questions and with this information the debugger is able to precisely identify the bug in the source code. Roughly speaking, the debugger discards parts of the source code associated with correct computations until it isolates a small part of the code (usually a function or procedure). One interesting property of this technique is that programmers do not need to see the source code during debugging. They only need to know the actual and intended results produced by a computation with given inputs. Therefore, if we have available a formal specification of the pieces of software that is able to answer the questions, then the technique is fully automatic.

In declarative debugging, a data structure called *Execution Tree* (ET) represents a program execution, where each node is associated with a particular method execution.[1] Moreover, declarative debugging uses a navigation strategy to select ET nodes, and to ask the programmer about their validity. Each node contains a particular method execution with its inputs and outputs. If the programmer marks a node as wrong, then the bug must be in its subtree; and it must be in the rest of the tree if it is marked as correct. We find a *buggy node* when the programmer marks a node as wrong, and all its children are marked as correct. Hence, the debugger reports the associated method as buggy.

Let us explain the technique with an example. Consider the Java program shown in Fig. 1. This program initializes the elements of a matrix to 1, and then traverses the matrix to sum all of them. The ET associated with this example is shown in Fig. 2 (left). Because it is not relevant in our technique, we omit the information of the nodes in the ETs of this paper, and instead we label them (inside) with the number of the method associated with them. Observe that `main` (0) calls once to the constructor `Matrix` (1), and then calls nine times to `position` (2) inside two nested loops. In the ET, every node has a number (outside) that indicates the number of questions needed to find the bug when it is the buggy node. To compute this number, we have considered the navigation strategy Divide and Query [16], that always selects the ET node that minimizes the difference between the numbers of descendants and non-descendants (i.e., better divides the ET by the half).

Declarative debugging can produce long series of questions making the debugging session too long. Moreover, it works at the level of methods, thus the

[1] Nodes represent computations. Hence, depending on the underlying paradigm, they can represent methods, functions, procedures, clauses, etc. Our technique can be applied to any paradigm, but, for the sake of concreteness, we will center the discussion on the object-oriented paradigm and our examples on Java.

```
     public class Matrix {
       private int numRows;
       private int numColumns;
       private int[][] matrix;

(1)    public Matrix(int numRows, int numColumns) {
         this.numRows = numRows;
         this.numColumns = numColumns;
         this.matrix = new int[numRows][numColumns];
         for (int i = 0; i < numRows; i++)
           for (int j = 0; j < numColumns; j++)
             this.matrix[i][j] = 1;
       }

(2)    public int position(int numRow, int numColumn) {
         return matrix[numRow][numColumn];
       }
     }

     public class SumMatrix {
(0)    public static void main(String[] args) {
         int result = 0;
         int numRows = 3;
         int numColumns = 3;
         Matrix m = new Matrix(numRows, numColumns);
         for (int i = 0; i < numRows; i++)
           for (int j = 0; j < numColumns; j++)
             result += m.position(i, j);
         System.out.println(result);
       }
     }
```

Fig. 1. Iterative Java program

granularity level of the bug found is a method. In this work we propose two new techniques that reduce (i) the number of questions needed to detect a bug, (ii) the complexity of the questions, and also (iii) the granularity level of the bug found. The first technique improves a previous technique called *Tree Compression* [4], and the second one is a new technique named *Loop Expanssion*. While the first technique is based on a transformation of the ET, the second one is based on a transformation of the source code. In both cases, the produced ET can be debugged more efficiently using the standard algorithms. Therefore, the techniques are conservative with respect to previous implementations and they can be integrated in any debugger as a preprocessing stage. The cost of the transformations is low compared with the cost of generating the whole ET. In fact, they are efficient enough as to be always used in all declarative debuggers before the ET exploration phase. According to our experiments (summarized in Table 1), as an average, they reduce the number of questions in 27.65 % with a temporal cost of 274 ms for loop expansion and 123 ms for tree compression.

The rest of the paper has been organized as follows. In Section 2 we discuss the related work. Section 3 introduces some preliminary definitions that will be used in the rest of the paper. In Section 4 we explain our techniques and their main applications, and we introduce the algorithms that improve the structure of the ET. Then, in Section 6 we provide details about our implementation and some experiments carried out with real Java programs. Finally, Section 7 concludes.

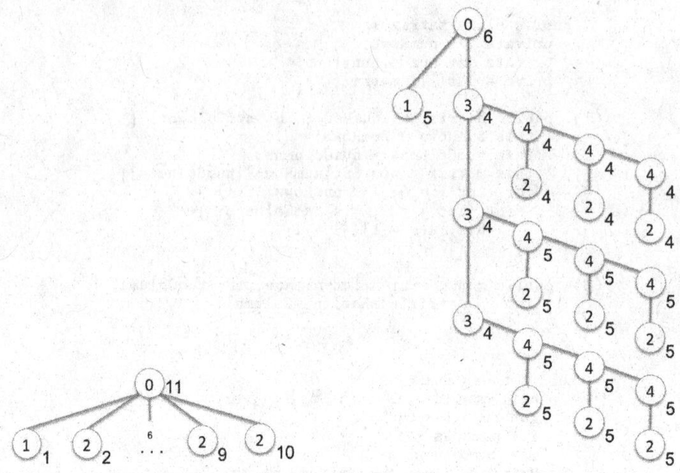

Fig. 2. ETs of the examples in Fig. 1 (left) and 8 (right)

2 Related Work

Reducing the number of questions asked by declarative debuggers is a well-known objective in the field, and there exist several works devoted to achieve this goal. Some of them face the problem by defining different ET transformations that modify the structure of the ET to explore it more efficiently.

For instance, in [15], authors improve the declarative debugging of Maude balancing the ET by introducing nodes. These nodes represent transitivity inferences done by their inference system. Although these nodes could be omitted, if they are kept, the ET becomes more balanced. Balanced ETs are very convenient for strategies such as Divide and Query [17], because it is possible to prune almost half of the tree after every answer, thus obtaining a logarithmic number of questions with respect to the number of nodes in the ET. This approach is related to our technique, but it has some drawbacks: it can only be applied where transitivity inferences took place while creating the ET, and thus most of the parts of the tree cannot be balanced, and even in these cases the balancing only affects two nodes. Our techniques, in contrast, balance loops, that usually contain many nodes.

Our first technique is based on Tree Compression (TC) introduced by Davie and Chitil [4]. In particular, we define an algorithm to decide when TC must be applied (or partially applied) in an ET. TC is a conservative approach that transforms an ET into an equivalent (smaller) ET where we can detect the same bugs. The objective of this technique is essentially different to previous ones: it tries to reduce the size of the ET by removing redundant nodes, and it is only applicable to recursive calls. For each recursive call, TC removes the child node associated with the recursive call and all its children become children of the parent node. Let us explain it with an example.

Example 1. Consider the ET in Fig. 3. Here, TC removes six nodes, thus statically reducing the size of the tree. Observe that the average number of questions has been reduced ($\frac{72}{17}$ vs $\frac{42}{11}$) thanks to the use of TC.

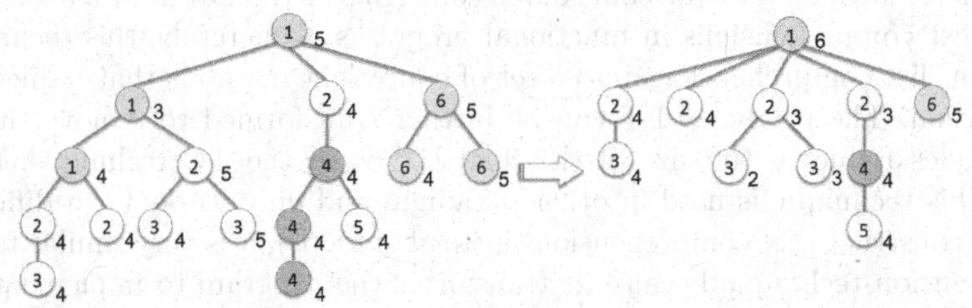

Fig. 3. Example of Tree Compression

Unfortunately, TC does not always produce good results. Sometimes reducing the number of nodes causes a worse ET structure that is more difficult to debug and thus the number of questions is increased, producing the contrary effect to the intended one.

Example 2. Consider the ET in Fig. 4 (top). In this ET, the average number of questions needed to find the bug is $\frac{33}{9}$. Nevertheless, after compressing the recursive calls (the dark nodes), the average number of questions is augmented to $\frac{28}{7}$ (see the ET at the left). The reason is that in the new compressed ET we cannot prune any node because its structure is completely flat. The previous structure allowed us to prune some nodes because deep trees are more convenient for declarative debugging. However, if we only compress one of the two recursive calls, the number of questions is reduced to $\frac{27}{8}$ (see the ET at the right).

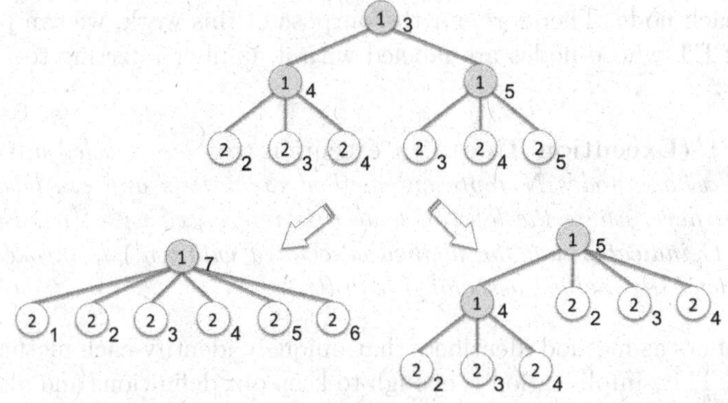

Fig. 4. Negative and positive effects of Tree Compression

Example 2 clearly shows that TC should not be always applied. From the best of our knowledge, there does not exist an algorithm to decide when to apply TC, and current implementations always compress all recursive calls [3,7]. Our new technique solves this problem with an analysis to decide when to compress them.

A similar approach to TC is declarative source debugging [2], that instead of modifying the tree prevents the debugger from selecting questions related to nodes generated by recursive calls. Another related approach was presented in [13]. Here, authors introduced a source code (rather than an ET) transformation for list comprehensions in functional programs. Concretely, this technique transforms list comprehensions into a set of equivalent methods that implement the iteration. The produced ET can be further transformed to remove the internal nodes reducing the size of the final ET as in the TC technique. Even though this technique is used in other paradigm and only works for a different program construct (list comprehensions instead of loops), it is very similar to our loop expansion technique because it transforms the program to implement the list comprehension iterations with recursive functions. This is somehow equivalent to our transformation for `for-each` loops. However, the objective of their technique is different. Their objective is to divide a question related to a list comprehension in different (probably easier) questions, while our objective is to balance the tree, and thus they are optimized in a different way. Of course, their transformation is orthogonal to our technique and it can be applied before.

Even though the techniques discussed can be applied to any language, they only focus on recursion. This means that they cannot improve ETs that use loops, avoiding their use in the imperative or the object-oriented paradigm where loops predominate. Our second technique is based on an automatic transformation of loops into recursive methods. Hence, it allows the previously discussed transformations to work in presence of iteration.

3 Preliminaries

Our ET transformations are based on its structure and the signature of the method in each node. Therefore, for the purpose of this work, we can provide a definition of ET whose nodes are labeled with a number referring to a specific method.

Definition 1 (Execution Tree). *An* execution tree *is a labeled directed tree $T = (N, E)$ whose nodes N represent method executions and are labeled with method identifiers, where the label of node n is referenced with $l(n)$. Each edge $(n \rightarrow n') \in E$ indicates that the method associated with $l(n')$ is invoked during the execution of the method associated with $l(n)$.*

We use numbers as method identifiers that uniquely identify each method in the source code. This simplification is enough to keep our definitions and algorithms precise and simple. Given an ET, *simple recursion* is represented with a branch of chained nodes with the same identifier. *Nested recursion* happens when a recursive branch is descendant of another recursive branch. *Multiple recursion* happens when a node labeled with an identifier n has two or more children labeled with n.

The weight of a node is the number of nodes contained in the tree rooted at this node. We refer to the weight of node n as w_n. In the following, we will refer to the two most used navigation strategies for declarative debugging: Top-Down [12] and Divide and Query (D&Q) [16]. In both cases, we will always implicitly refer to the most efficient version of both strategies, respectively named, (i) *Heaviest First* [1], which always traverses the ET from the root to the leaves selecting always the heaviest node; and (ii) *Hirunkitti's Divide and Query* [6], which always selects the node in the ET that better divides the number of nodes in the ET by the half. A comparative study of these techniques can be found in [17].

For the comparison of strategies we use function $Questions(T, s)$ that computes the number of questions needed (as an average) to find the bug in an ET T using the navigation strategy s.

4 Execution Trees Optimization

In this section we present two new techniques for the optimization of ETs: Tree Compression (TC) and Loop Expansion (LE). TC was defined and described in [4]. Here, we introduce an algorithm to compress a recursive branch of the ET in any case (i.e., simple recursion, nested recursion, and multiple recursion). We also discuss how navigation strategies are affected by TC. The other technique introduced is Loop Expansion that essentially transforms a loop into an equivalent recursive method. Then, TC adequately balances the iterations to obtain a new ET as optimized as possible. We explain each technique separately and propose algorithms for them that can work independently.

4.1 When to Apply Tree Compression

Tree compression was proposed as a general technique for declarative debugging. However, it was defined in the context of a functional language (Haskell) and with the use of a particular strategy (Hat-Delta). The own authors realized that TC can produce wide trees that are difficult to debug and, for this reason, defined strategies that avoid asking about the same method repeatedly. These strategies do not prevent to apply TC. They just assume that the ET has been totally compressed and they follow a top-down traversal of the ET that can jump to any node when the probability of this node to contain the bug is high. This way of proceeding somehow partially mitigates the bad structure of the produced ET when it is totally compressed. Our approach is radically different: We do not create a new strategy to avoid the bad ET structure; but we transform the ET to ensure a good structure.

Even though TC can produce bad ETs (as shown in Example 4), its authors did not study how this technique works with other (more extended) strategies such as Top-Down or D&Q. So it is not clear at all when to use it. To study when to use TC, we can consider the most general case of a simple recursion in an ET. It is shown in Fig. 5 where clouds represent possibly empty sets of subtrees and the dark nodes are the recursion branch with a length of $n \geq 2$ calls.

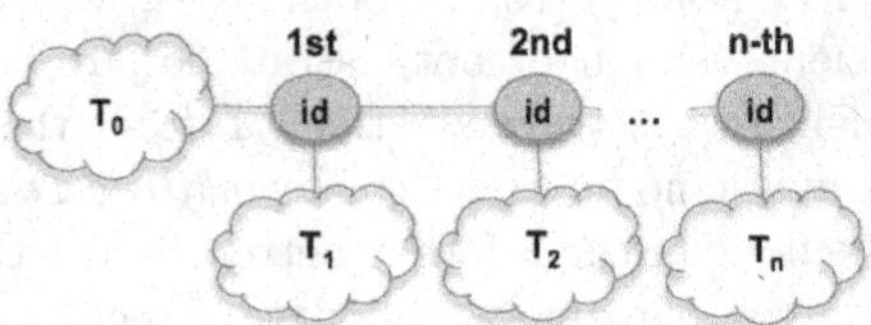

Fig. 5. Recursion branch in an ET

It should be clear that the recursion branch can be useful to prune nodes of the tree. For instance, in the figure, if we ask for the node $n/2$, we prune $n/2$ subtrees. Therefore, in the case that the subtrees $T_i, 1 \leq i \leq n$, are empty, then no pruning is possible. In that case, only the nodes in the recursion branch could be buggy; but, because they form a recursive chain, all of them have the same label. Thus no matter which one is buggy because all of them refer to the same method. Hence, TC must be used to reduce the recursive branch to a single node avoiding navigation strategies to explore this branch. Hence, we can conclude that *we must compress every node whose only child is a recursive call.* This result can be formally stated for Top-Down as follows.

Theorem 1. *Let T be an ET with a recursion branch $R = n_1 \to n_2 \to \ldots \to n_m$ where the only child of a node n_i, $1 \leq i \leq m-1$, is n_{i+1}. And let T' be an ET equivalent to T except that nodes n_i and n_{i+1} have been compressed. Then, Questions(T',Top-Down) $<$ Questions(T,Top-Down).*

The associated proof can be found in [10].

This theorem shows us that, in some situations, TC must be used to statically improve the ET structure. But TC is not the panacea, and we need to identify in what cases it should be used. D&Q is a good example of strategy where TC has a negative effect.

***Tree Compression* for D&Q**

In general, when debugging an ET with the strategy D&Q, TC should only be applied in the case described by Theorem 1 ($T_i, 1 \leq i \leq n$, are empty). The reason is that D&Q can jump to any node of the ET without following a predefined path. This allows D&Q to ask about any node of the recursion branch without asking about the previous nodes in the branch. Note that this does not happen in other strategies such as Top-Down. Therefore, D&Q has the ability to use the recursion branch as a mean to prune half of the iterations.

Observe in Fig. 6 that, except for very small recursion branches (e.g., $n \leq 3$), D&Q can take advantage of the recursion branch to prune half of the iterations. The greater is n the more nodes pruned. Observe that D&Q can prune nodes even in the case when there is only one child for every node in the recursion branch (e.g., $T_i, 1 \leq i \leq n$, is a single node). Therefore, if we add more nodes to the subtrees Ti, then more nodes can be pruned and D&Q will behave even better.

Fig. 6. TC applied to a recursive method

Tree Compression for Top-Down

In the case of Top-Down-based strategies, it is not trivial at all to decide when to apply TC. Considering again the ET in Fig. 5, there are two factors that must be considered: (i) the length n of the recursive branch, and (ii) the size of the trees $T_i, 1 \leq i \leq n$. In order to decide when TC should be used, we provide Algorithm 1 that takes an ET and compresses all recursion branches whenever it improves the ET structure.

Essentially, Algorithm 1 analyzes for each recursion what is the effect of applying TC, and it is finally applied only when it produces an improvement. This analysis is done little by little, analyzing each pair of parent-child (recursive) nodes in the sequence separately. Thus, it is possible that the final result is to only compress one (or several) parts of one recursion branch. For this, variable `recs` initially contains all nodes of the ET with a recursive child. Each of these nodes is processed with the loop in line 1 bottom-up (lines 2-3). That is, the nodes closer to the leaves are processed first. In order to also consider multiple recursion, the algorithm uses the loops in lines 5 and 8. These loops store in the variable `improvement` the improvement achieved when compressing each recursive branch. In addition to the functions (`Cost` and `Compress`) shown here, the algorithm uses three more functions whose code has not been included because they are trivial: function `Children` computes the set of children of a node in the ET (i.e., `Children(m)` = $\{$`n` $|$ (`m` $\rightarrow$ `n`) $\in$ `E`$\}$); function `Sort` takes a set of nodes and produces an ordered sequence where nodes have been decreasingly ordered by their weights; and function `Pos` takes a node and a sequence of nodes and returns the position of the node in the sequence.

Given two nodes `parent` and `child` candidates to make a TC, the algorithm first sorts the children of both the `parent` and the `child` (lines 9-10) in the order in which Top-Down would ask them (sorted by their weight). Then, it combines the children of both nodes simulating a TC (line 11). Finally, it compares the average number of questions when compressing or not the nodes (line 12). The equation that appears in line 12 is one of the main contributions of the algorithm, because this equation determines when to perform TC between two nodes in

Algorithm 1. Optimized Tree Compression

```
Input: An ET T = (N, E)
Output: An ET T'
Inicialization: T' = T and recs = {n | n, n' ∈ N ∧ (n → n') ∈ E ∧ l(n) = l(n')}

begin
1) while (recs ≠ ∅)
2)      take n ∈ recs such that ∄n' ∈ recs with (n → n') ∈ E⁺
3)      recs = recs\{n}
4)      parent = n
5)      do
6)           maxImprovement = 0
7)           children = {c | (n → c) ∈ E ∧ l(n) = l(c)}
8)           for each child ∈ children
9)                pchildren = Sort(Children(parent))
10)               cchildren = Sort(Children(child))
11)               comb = Sort((pchildren ∪ cchildren)\{child})
12)               improvement = (Cost(pchildren)+Cost(cchildren)) / w_parent − Cost(comb) / (w_parent − 1)
13)               if (improvement > maxImprovement)
14)                    maxImprovement = improvement
15)                    bestNode = child
16)          end for each
17)          if (maxImprovement ≠ 0)
18)               T' = Compress(T', parent, bestNode)
19)     while (maxImprovement ≠ 0)
20) end while
end
return T'

function Cost(sequence)
begin
21) return ∑{Pos(node, sequence) ∗ w_node | node ∈ sequence} + |sequence|
end

function Compress(T = (N, E), parent, child)
begin
22) nodes = Children(child)
23) E' = E\{(child → n) ∈ E | n ∈ nodes}
24) E' = E' ∪ {(parent → n) | n ∈ nodes}
25) N' = N\{child}
end
return T' = (N', E')
```

a branch with the strategy Top-Down. This equation depends in turn on the formula (line 21 in function Cost) used to compute the average cost of exploring an ET with Top-Down.

If we analyze Algorithm 1, we can easily realize that its asymptotic cost is quadratic with the number of recursive calls $\mathcal{O}(N^2)$ because in the worst case, all recursive calls would be compared between them. Note also that the algorithm could be used with incomplete ETs [8] (this is useful when we try to debug a program while the ET is being generated). In this case, the algorithm can still be applied locally, i.e., to those subtrees of the ET that are totally generated.

4.2 Loop Expansion

Recursive calls group the iterations in different subtrees whose roots belong to the recursion branch. This is very convenient because it allows the debugger to prune different iterations. Therefore, recursion is beneficial for declarative debugging except in the cases discussed in the previous section. Contrarily, loops produce very wide trees where all iterations are represented as trees with a common root. In this structure, it is impossible to prune more than one iteration at a time, being the debugging of these trees very expensive.

To solve this problem, in this section we present a technique for declarative debugging that transforms loops into equivalent recursive methods. Because iteration is more efficient than recursion, there exist many approaches to transform recursive methods into equivalent loops (e.g., [5,11]). However, there exist few approaches to transform loops into equivalent recursive methods. An exception is the one presented in [18] to improve performance in multi-level memory hierarchies. Nevertheless, we are not aware of any algorithm of this kind proposed for Java or for any other object-oriented language. Hence, we had to implement this algorithm as a Java library and made it public for the community: `http://users.dsic.upv.es/~jsilva/loops2recursion/`. Moreover, it has been also integrated into a declarative debugger [7]. Due to lack of space we cannot describe here the algorithm, but we made a technical report with a detailed description [9]. This algorithm has an asymptotic cost linear with the number of loops in the program and is the basis of LE. Basically, it transforms each loop into an equivalent recursive method. The transformation is slightly different for each kind of loop (`while`, `do`, `for` or `foreach`). In the case of `for`-loops, it can be explained with the code in Fig. 7 where A, B, C and D represent blocks of code. If we observe the transformed ET we see that each iteration is represented with a different node of the recursive branch $r(1) \to r(2) \to \ldots \to r(10)$, thus it is possible to detect a bug in a single iteration. This means that, in the case that function f had a bug, thanks to the transformation, the debugger could detect that a bug exists in the code in B + C or in A + D. Note that this is not possible in the original ET where the debugger would report that A + B + C + D has a bug. This is a very important result because it augments the granularity level of the reported bugs, detecting bugs inside loops and not only inside methods.

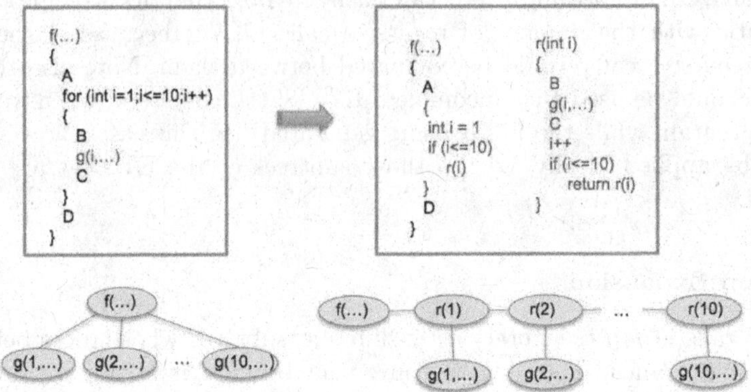

Fig. 7. ET transformation from a loop to a recursive method

Nested recursion augments the possibilities of pruning. For instance, class Matrix in Fig. 1 can be automatically transformed[2] to the code in Fig. 8. The ET associated with the transformed program is shown in Fig. 2 (right). Observe that there is a recursion branch for each executed loop, and thus, we have recursive branches (those labelled with 4) inside a recursive branch (labelled with 3). Hence, the new nodes added by the transformation are used to represent each single iteration; and thanks to them, now it is possible to prune loops, iterations, or single calls inside an iteration.

Our implementation combines the use of TC and LE as follows: (i) Expand all loops of the source code with LE, (ii) generate the ET with the transformed code, and (iii) use Algorithm 1 to compress the ET with the current strategy—observe that if the strategy is later changed, all nodes removed by the compression can be introduced again, and the ET compressed for the new strategy—. In this way, we produce an ideal representation of loops where each individual loop is partially or totally expanded to produce an optimal debugging session.

5 Correctness

In this section we prove that after our transformations, all bugs that could be detected in the original ET can still be detected in the transformed one. An even more interesting result is that the transformed ET can contain more buggy nodes than the original one, and thus, we can detect bugs that before were undetectable. Regarding TC, its correctness has been proved in [4].

[2] For the sake of clarity, in the figure we replaced the names of the generated recursive methods by `sumRows` and `sumColumns`. In the implementation, if a loop has a Java label in the original source code, the transformation uses this label to name the recursive method. If this label does not exist, then the name of the loop is the name of the method that contains this loop followed by "_loopN", where N is an autonumeric. While debugging, the user can see the source code of the loop, and she can change its name if she wants to do it.

```
    public class SumMatrix {
(0)   public static void main(String[] args) {
        int result = 0;
        int numRows = 3;
        int numColumns = 3;
        Matrix m = new Matrix(numRows, numColumns);
        // For loop
        { // Init for loop
           int i = 0;
           // First iteration
           if (i < numRows) {
              Object[] res = SumMatrix.sumRows(m, i, numRows, numColumns, result);
              result = (Integer)res[0];
           }
        }
        System.out.println(result);
      }
(3)   private static Object[] sumRows(Matrix m, int i, int numRows, int numColumns, int result) {
        // For loop
        { // Init for loop
           int j = 0;
           // First iteration
           if (j < numColumns) {
              Object[] res = SumMatrix.sumColumns(m, i, j, numColumns, result);
              result = (Integer)res[0];
           }
        }
        // Update for loop
        i++;
        // Next iteration
        if (i < numRows)
           return SumMatrix.sumRows(m, i, numRows, numColumns, result);
        return new Object[]{result};
      }
(4)   private static Object[] sumColumns(Matrix m, int i, int j, int numColumns, int result) {
        result += m.position(i, j);
        // Update for loop
        j++;
        // Next iteration
        if (j < numColumns)
           return SumMatrix.sumColumns(m, i, j, numColumns, result);
        return new Object[]{result};
      }
    }
```

Fig. 8. Recursive version of the program in Fig. 1

Our algorithm does not influence this correctness property because it only decides what nodes should be compressed, but the TC algorithm is the standard one. The correctness of LE is stated in the following. The associated proofs can be found in [10].

Theorem 2. *[Completeness] Let $\mathcal{P}$ be a program, let T be the ET associated with $\mathcal{P}$, and let T' be the ET obtained by applying loop expansion to T. For each buggy node in T, there is at least one buggy node in T'.*

Theorem 3. *[Soundness] Let $\mathcal{P}$ be a program, let T be the ET associated with $\mathcal{P}$, and let T' be the ET obtained by applying loop expansion to T. If T' contains a buggy node associated with code $f \subseteq \mathcal{P}$, then T contains a buggy node associated with code $g \subseteq \mathcal{P}$ and $f \subseteq g$.*

From Theorem 2 and 3 we have a very interesting corollary that reveals that the transformed tree can find more bugs than the original ET.

Corollary 1. *Let $\mathcal{P}$ be a program, let T be the ET associated with $\mathcal{P}$, and let T' be the ET obtained by applying loop expansion to T. If T contains n buggy nodes, then T' contains n' buggy nodes with $n \leq n'$.*

Table 1. Summary of the experiments

Benchmark	Nodes				LE	Time		Questions				%	
	ET	LE	TC_{ori}	TC_{opt}		LE	TC	ET	LE	TC_{ori}	TC_{opt}	LETC	TC
Factoricer	55	331	51	51	5	151	105	11.62	8.50	7.35	7.35	63.25	100.0
Classifier	25	57	22	24	3	184	4	8.64	6.19	6.46	6.29	72.80	97.36
LegendGame	87	243	87	87	10	259	31	12.81	8.28	11.84	11.84	92.43	100.0
Romanic	121	171	112	113	3	191	12	16.24	7.74	10.75	9.42	58.00	87.62
FibRecursive	5378	6192	98	101	12	251	953	15.64	12.91	9.21	8.00	51.15	86.86
FactTrans	197	212	24	26	3	181	26	10.75	7.88	6.42	5.08	47.26	79.13
BinaryArrays	141	203	100	100	5	172	79	12.17	7.76	7.89	7.89	64.83	100.0
FibFactAna	178	261	44	49	7	202	33	7.90	8.29	8.50	6.06	76.71	71.29
RegresionTest	13	121	15	15	5	237	4	4.77	7.17	4.20	4.20	88.05	100.0
BoubleFibArrays	16	164	10	10	10	213	27	9.31	8.79	4.90	4.90	52.63	100.0
StatsMeanFib	19	50	23	23	6	195	21	7.79	8.12	6.78	6.48	83.18	95.58
Integral	5	8	8	8	3	152	2	6.80	5.75	7.88	5.88	86.47	74.62
TestMath	3	5	3	3	3	195	2	7.67	6.00	9.00	7.67	100.0	85.22
TestMath2	92	2493	13	13	3	211	607	14.70	11.54	15.77	12.77	86.87	80.98
Figures	2	10	10	10	24	597	13	9.00	7.20	6.60	6.60	73.33	100.0
FactCalc	128	179	75	75	3	206	46	8.45	7.60	7.96	7.96	94.20	100.0
SpaceLimits	95	133	98	100	15	786	10	36.26	12.29	18.46	14.04	38.72	76.06

6 Implementation

We have implemented the original TC algorithm and the optimized version presented in this paper; and also the LE algorithm in such a way that they all can work together. This implementation has been integrated into the Declarative Debugger for Java DDJ [7]. The experiments, the source code of the tool, the benchmarks, and other materials can be found at `http://www.dsic.upv.es/~jsilva/DDJ/`.

All the implementation has been done in Java. The optimized TC algorithm contains around 90 LOC, and the LE algorithm contains around 1700 LOC. We conducted a series of experiments in order to measure the influence of both techniques in the performance of the debugger. Table 1 summarizes the obtained results.

The first column in Table 1 shows the name of the benchmarks. For each benchmark, column `nodes` shows the number of nodes descendant of a loop[3] in the original ET (`ET`), in the ET after applying LE (`LE`), in the ET after applying LE first and then the original version of TC—compressing all nodes—(TC_{ori}), and in the ET after applying LE first and then the optimized version of TC—Algorithm 1—(TC_{opt}); column `LE` shows the number of loops expanded; column `Time` shows the time (in milliseconds) needed to apply LE and TE; column `Questions` shows the average number of questions asked with each of the previously described ETs. Each benchmark has been analyzed assuming that the bug could be in any node of its associated ET. This means that each value in column

[3] We consider these nodes because the part of the ET that is not descendant of a loop remains unchanged after applying our technique, and thus the number of questions needed to find the bug is the same before and after the transformations.

`Questions` represents the average of a set of experiments. For instance, in order to obtain the information associated with `Factoricer`, this benchmark has been debugged 55 times with the original ET, 331 with the ET after applying loop expansion, etc. In total, `Factoricer` was debugged 55+331+51+51=488 times, considering all ET transformations and assuming each time that the bug was a different node (and computing the average of all tests for each ET); finally, column (`%`) shows, on the one hand, the percentage of questions asked after applying our transformations (LE and TC) with respect to the original ET (`LETC`); and, on the other hand, the percentage of questions asked using Algorithm 1 to decide when to apply TC, with respect to always applying TC (`TC`). From the table we can conclude that our transformations produce a reduction of 27.65 % in the number of questions asked by the debugger. Moreover, the use of Algorithm 1 to decide when to apply TC also produces an important reduction in the number of questions with an average of 9.72 %.

7 Conclusions

Declarative debugging can generate too many questions to find a bug, and once it is found, the debugger reports a whole method as the buggy code. In this work we make a step forward to solve these problems. We introduce techniques that reduce the number of questions by improving the structure of the ET. This is done with two transformations called tree compression and loop expansion. Moreover, loop expansion also faces the second problem, and it allows us to detect bugs in loops and not only in methods, augmenting in this way the granularity level of the bug found. As a side effect, being able to ask about the correctness of loops allows us to reduce the complexity of questions: the programmer can answer about a part of a method (a single loop), and not only to the whole method mixing the effects of different loops. We think that this is an interesting result that opens new possibilities for future work related to the complexity of questions. The idea of transforming loops into recursive methods in an ET is novel, and it allows us to apply all previous techniques based on recursion in the imperative and object-oriented paradigms.

Acknowledgements. We want to thank Rafael Caballero and Adrián Riesco for productive comments and discussions at the initial stages of this work. We also thank the anonymous referees of LOPSTR'12 for useful feedback and constructive criticism which has improved this work.

References

1. Binks, D.: Declarative Debugging in Gödel. PhD thesis, University of Bristol (1995)
2. Calejo, M.: A Framework for Declarative Prolog Debugging. PhD thesis, New University of Lisbon (1992)

3. Davie, T., Chitil, O.: Hat-delta: One Right Does Make a Wrong. In: Butterfield, A. (ed.) Draft Proceedings of the 17th International Workshop on Implementation and Application of Functional Languages, IFL 2005, p. 11. Tech. Report No: TCD-CS-2005-60, University of Dublin, Ireland (September 2005)
4. Davie, T., Chitil, O.: Hat-delta: One Right Does Make a Wrong. In: Seventh Symposium on Trends in Functional Programming, TFP 2006 (April 2006)
5. Harrison, P.G., Khoshnevisan, H.: A new approach to recursion removal. Theor. Comput. Sci. 93(1), 91–113 (1992)
6. Hirunkitti, V., Hogger, C.J.: A Generalised Query Minimisation for Program Debugging. In: Adsul, B. (ed.) AADEBUG 1993. LNCS, vol. 749, pp. 153–170. Springer, Heidelberg (1993)
7. Insa, D., Silva, J.: An Algorithmic Debugger for Java. In: Proc. of the 26th IEEE International Conference on Software Maintenance, pp. 1–6 (2010)
8. Insa, D., Silva, J.: Debugging with Incomplete and Dynamically Generated Execution Trees. In: Proc. of the 20th International Symposium on Logic-based Program Synthesis and Transformation, LOPSTR 2010, Austria (2010)
9. Insa, D., Silva, J.: A Transformation of Iterative Loops into Recursive Loops. Technical Report DSIC/05/12, Universidad Politécnica de Valencia (2012), `http://www.dsic.upv.es/~jsilva/research.htm#techs`
10. Insa, D., Silva, J., Tomás, C.: Enhancing Declarative Debugging with Loop Expansion and Tree Compression. Technical Report DSIC/11/12, Universidad Politécnica de Valencia (2012), `http://www.dsic.upv.es/~jsilva/research.htm#techs`
11. Liu, Y.A., Stoller, S.D.: From recursion to iteration: what are the optimizations? In: Proceedings of the 2000 ACM SIGPLAN Workshop on Partial Evaluation and Semantics-Based Program Manipulation, PEPM 2000, pp. 73–82. ACM, New York (2000)
12. Lloyd, J.W.: Declarative error diagnosis. New Gen. Comput. 5(2), 133–154 (1987)
13. Nilsson, H.: Declarative Debugging for Lazy Functional Languages. PhD thesis, Linköping, Sweden (May 1998)
14. NIST: The Economic Impacts of Inadequate Infrastructure for Software Testing. USA National Institute of Standards and Technology, NIST Planning Report 02-3 (May 2002)
15. Riesco, A., Verdejo, A., Martí-Oliet, N., Caballero, R.: Declarative Debugging of Rewriting Logic Specifications. Journal of Logic and Algebraic Programming (September 2011)
16. Shapiro, E.Y.: Algorithmic Program Debugging. MIT Press (1982)
17. Silva, J.: A Survey on Algorithmic Debugging Strategies. Advances in Engineering Software 42(11), 976–991 (2011)
18. Yi, Q., Adve, V., Kennedy, K.: Transforming loops to recursion for multi-level memory hierarchies. In: Proceedings of the SIGPLAN 2000 Conference on Programming Language Design and Implementation, pp. 169–181 (2000)

Appendix 1: Proofs of Technical Results

This section presents the proofs of Theorems 1, 2, 3 and Corollary 1.

Theorem 1. *Let T be an ET with a recursion branch $R = n_1 \to n_2 \to \ldots \to n_m$ where the only child of a node n_i, $1 \leq i \leq m-1$, is n_{i+1}. And let T' be an ET equivalent to T except that nodes n_i and n_{i+1} have been compressed. Then, $Questions(T', Top\text{-}Down) < Questions(T, Top\text{-}Down)$.*

Proof. Let us consider the two nodes that form the sub-branch to be compressed. For the proof we can call them $n_1 \to n_2$. Firstly, the number of questions needed to find a bug in any ancestor of n_1 is exactly the same if we compress or not the sub-branch. Therefore, it is enough to prove that $Questions(T_{1c},\text{Top-Down}) < Questions(T_1,\text{Top-Down})$ where T_1 is the subtree whose root is n_1 and T_{1c} is the subtree whose root is n_1 after tree compression.
Let us assume that n_2 has j children. Thus we call T_2 the subtree whose root is n_2, and T_{2_i} the subtree whose root is the i-th child of n_2. Then,

$$Questions(T_2,\text{Top-Down}) = \frac{(j+1) + \sum_{i=1}^{j} |T_{2_i}| * (i + Questions(T_{2_i},\text{Top-Down}))}{|T_2|}$$

Here, $(j+1)$ are the questions needed to find a bug in n_2. To reach the children of n_2, the own n_2 and the previous $i-1$ children must be asked first, and this is why we need to add i to $Questions(T_{2_i},\text{Top-Down})$, Finally, $|T_x|$ represents the number of nodes in the (sub)tree T_x.
Therefore, $Questions(T_{1c},\text{Top-Down}) = Questions(T_2,\text{Top-Down})$
and $Questions(T_1,\text{Top-Down}) = \frac{2+|T_2|*(i+Questions(T_2,\text{Top-Down}))}{|T_2|+1}$.
Clearly, $Questions(T_{1c},\text{Top-Down}) < Questions(T_1,\text{Top-Down})$, and thus the claim follows.

Theorem 2. *Let $\mathcal{P}$ be a program, let T be the ET associated with $\mathcal{P}$, and let T' be the ET obtained by applying loop expansion to T. For each buggy node in T, there is at least one buggy node in T'.*

Proof. Let us prove the theorem for an arbitrary buggy node n in T associated with a function f. Firstly, because loop expansion only transforms iterative loops into recursive loops, all functions executed in T are also executed in T'. This means that every node in T has a counterpart (equivalent) node in T' that represents the same (sub)computation. Therefore, we can call n' to the node that represents in T' the same execution than n in T. Because n is buggy, then n is wrong, and all the children of n (if any) are correct. Hence, n' is also wrong. Moreover, if f does not contain a loop, then loop expansion has no effect on the code of f and thus n and n' will have exactly the same children, and thus, trivially, n' is also buggy in T'. If we assume the existence of a loop in f, then we will have a situation as the one shown in the ETs of Figure 7. We can consider for the proof that the ET at the left is the subtree of n and the ET at the right is the subtree of n'. Then, because n is buggy, all nodes labeled with g are correct (in both ETs) and, thus, either n' or one of the nodes labeled with r are buggy.

Theorem 3. *Let $\mathcal{P}$ be a program, let T be the ET associated with $\mathcal{P}$, and let T' be the ET obtained by applying loop expansion to T. If T' contains a buggy node associated with code $f \subseteq \mathcal{P}$, then, T contains a buggy node associated with code $g \subseteq \mathcal{P}$ and $f \subseteq g$.*

Proof. According to the proof of Theorem 2, every node in T has a counterpart (equivalent) node in T'. Hence, let n be the buggy node in T and let n' be the associated buggy node in T'. If f does not have a loop, then both n and n' point to the same function (f) and thus the theorem holds trivially. If f contains a loop that has been expanded, then, as stated in the proof of Theorem 2, either n' or one of its descendants (say n'') that represent the iterations of the loop are buggy. But we know that the code of n'' is the code of the loop that is included in the code of f. Therefore, in all cases $f \subseteq g$.

Corollary 1. *Let $\mathcal{P}$ be a program, let T be the ET associated with $\mathcal{P}$, and let T' be the ET obtained by applying loop expansion to T. If T contains n buggy nodes, then T' contains n' buggy nodes with $n \leq n'$.*

Proof. Trivial from Theorems 2 and 3. On the one hand, equality is ensured with Theorem 2 because for each buggy node in T, there is at least one buggy node in T'. On the other hand, if a node in T is associated with a function whose code contains more than one loop that has been expanded, then T' can contain more than one new buggy node not present in T.

XACML 3.0 in Answer Set Programming

Carroline Dewi Puspa Kencana Ramli,
Hanne Riis Nielson, and Flemming Nielson

Department of Applied Mathematics and Computer Science
Danmarks Tekniske Universitet
Lyngby, Denmark
{cdpu,hrni,fnie}dtu.dk

Abstract. We present a systematic technique for transforming XACML 3.0 policies in Answer Set Programming (ASP). We show that the resulting logic program has a unique answer set that directly corresponds to our formalisation of the standard semantics of XACML 3.0 from [9]. We demonstrate how our results make it possible to use off-the-shelf ASP solvers to formally verify properties of access control policies represented in XACML, such as checking the completeness of a set of access control policies and verifying policy properties.

Keywords: XACML, access control, policy language, Answer Set Programming.

1 Background

XACML (eXtensible Access Control Markup Language) is a prominent access control language that is widely adopted both in industry and academia. XACML is an international standard in the field of information security and in February 2005, XACML version 2.0 was ratified by OASIS.[1] XACML represents a shift from a more static security approach as exemplified by ACLs (Access Control Lists) towards a dynamic approach, based on Attribute Based Access Control (ABAC) systems. These dynamic security concepts are more difficult to understand, audit and interpret in real-world implications. The use of XACML requires not only the right tools but also well-founded concepts for policy creation and management.

The problem with XACML is that its specification is described in natural language (c.f. [8,11]) and manual analysis of the overall effect and consequences of a large XACML policy set is a very daunting and time-consuming task. How can a policy developer be certain that the represented policies capture all possible requests? Can they lead to conflicting decisions for some request? Do the policies satisfy all required properties? These complex problems cannot be solved easily without some automatised support.

[1] The Organization for the Advancement of Structured Information Standards (OASIS) is a global consortium that drives the development, convergence, and adoption of e-business and web service standards.

E. Albert (Ed.): LOPSTR 2012, LNCS 7844, pp. 89–105, 2013.
© Springer-Verlag Berlin Heidelberg 2013

To address this problem we propose a logic-based XACML analysis framework using Answer Set Programming (ASP). With ASP we model an XACML Policy Decision Point (PDP) that loads XACML policies and evaluates XACML requests against these policies. The expressivity of ASP and the existence of efficient implementations of the answer set semantics, such as `clasp`[2] and `DLV`[3], provide the means for declarative specification and verification of properties of XACML policies.

Our work is depicted in Figure 1. There are two main modules, viz. the PDP simulation module and the access control (AC) security property verification module. In the first module, we transform an XACML query and XACML policies from the original format in XML syntax into abstract syntax, which is more compact than the original. Subsequently we generate a query program Π_Q and XACML policies program Π_{XACML} that correspond to the XACML query and the XACML policies, respectively. We show that the corresponding answer set of $\Pi_Q \cup \Pi_{\text{XACML}}$ is unique and it coincides with the semantics of original XACML policy evaluation. In the second module, we demonstrate how our results make it possible to use off-the-shelf ASP solvers to formally verify properties of AC policies represented in XACML. First we encode the AC security property and a generator for each possible domain of XACML policies into logic programs $\Pi_{AC\_property}$ and $\Pi_{generator}$, respectively. The encoding of AC property is in the negated formula in order to show at a later stage that each answer set corresponds to a counter example that violates the AC property. Together with the combination of $\Pi_{\text{XACML}} \cup \Pi_{AC\_property} \cup \Pi_{generator}$ we show that the XACML policies satisfy the AC property when there is no available answer set.

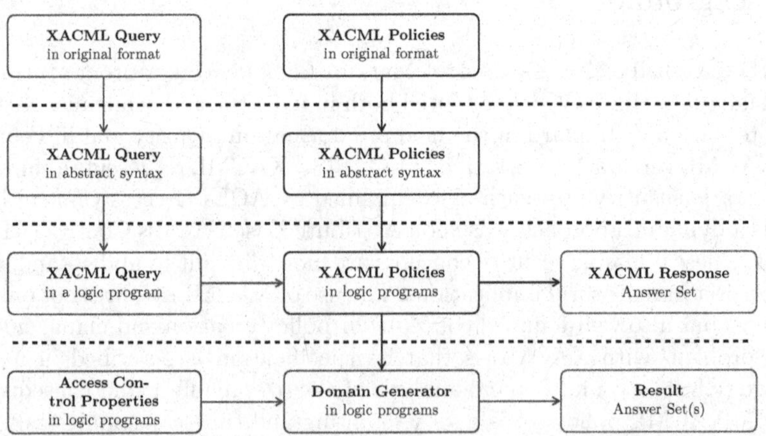

Fig. 1. Translation Process from Original XACML to XACML-ASP

Outline. We consider the current version, XACML 3.0, Committee Specification 01, 10 August 2010. In Section 2 we explain the abstract syntax and semantics of

[2] `http://www.cs.uni-potsdam.de/clasp/`
[3] `http://www.dlvsystem.com/`

XACML 3.0. Then we describe the transformation of XACML 3.0 components into logic programs in Section 3. We show the relation between XACML 3.0 semantics and ASP semantics in Section 4. Next, in Section 5, we show how to verify AC properties, such as checking the completeness of a set of policies. In Section 6 we discuss the related work. We end the paper with conclusions and future work.

2 XACML 3.0

In order to avoid superfluous syntax of XACML 3.0, first we present the abstract syntax of XACML 3.0 which only shows the important components of XACML 3.0. We continue the explanation by presenting the semantics of XACML 3.0 components' evaluation based on XACML 3.0 Committee Specification [11]. We take the work of Ramli *et. al* [9] as our reference.

2.1 Abstract Syntax of XACML 3.0

Table 1 shows the abstract syntax of XACML 3.0. We use bold font for non-terminal **symbols**, typewriter font for terminal `symbols` and *identifiers* and *values* are written in italic font. A symbol followed by the star symbol (*) indicates that there are zero or more occurrences of that symbol. Similarly, a symbol followed by the plus symbol ($^+$) indicates that there are one or more occurrences of that symbol. We consider that each policy has a unique identifier (ID). We use initial capital letter for XACML components such as PolicySet, Policy, Rule, etc., and small letters for English terminology.

Table 1. Abstraction of XACML 3.0 Components

		XACML Policy Components
PolicySet	$\mathcal{PS}$	$::= \mathcal{PS}_{id} = [\mathcal{T}, \langle(\mathcal{PS}_{id} \mid \mathcal{P}_{id})^*\rangle, \textbf{CombID}]$
Policy	$\mathcal{P}$	$::= \mathcal{P}_{id} = [\mathcal{T}, \langle \mathcal{R}_{id}{}^+\rangle, \textbf{CombID}]$
Rule	$\mathcal{R}$	$::= \mathcal{R}_{id} = [\textbf{Effect}, \mathcal{T}, \mathcal{C}]$
Condition	$\mathcal{C}$	$::=$ `true` $\mid f^{\text{bool}}(a_1, \ldots, a_n)$
Target	$\mathcal{T}$	$::=$ `null` $\mid \bigwedge \mathcal{E}^+$
AnyOf	$\mathcal{E}$	$::= \bigvee \mathcal{A}^+$
AllOf	$\mathcal{A}$	$::= \bigwedge \mathcal{M}^+$
Match	$\mathcal{M}$	$::= \boldsymbol{Attr}$
	CombID	$::=$ `po` | `do` | `fa` | `ooa`
	Effect	$::=$ `p` | `d`
Attribute	$\boldsymbol{Attr}$	$::=$ *category*(*attribute_value*)
		XACML Request Component
Request	$\mathcal{Q}$	$::= \{\, (\boldsymbol{Attr} \mid$ `error`$(\boldsymbol{Attr}))^+ \,\}$

There are three levels of policies in XACML, namely PolicySet, Policy and Rule. PolicySet or Policy can act as the root of a set of access control policies, while Rule is a single entity that describes one particular access control policy. Throughout this paper we consider that PolicySet is the root of the set of access control policies.

Both PolicySet and Policy are containers for a sequence of PolicySet, Policy or Rule. A PolicySet contains either a sequence of PolicySet elements or a sequence of Policy elements, while a Policy can only contain a sequence of Rule elements. Every sequence of PolicySet, Policy or Rule elements has an associated *combining algorithm.* There are four common combining algorithms defined in XACML 3.0, namely *permit-overrides* (po), *deny-overrides* (do), *first-applicable* (fa) and *only-one-applicable* (ooa).

A Rule describes an individual access control policy. It regulates whether an access should be *permitted* (p) or *denied* (d). All PolicySet, Policy and Rule are applicable whenever their Target matches with the Request. When the Rule's Target matches the Request, then the applicability of the Rule is refined by its Condition.

A Target element identifies the set of decision requests that the parent element is intended to evaluate. The Target element must appear as a child of a PolicySet and Policy element and may appear as a child of a Rule element. The empty Target for Rule element is indicated by `null` attribute. The Target element contains a conjunctive sequence of AnyOf elements. The AnyOf element contains a disjunctive sequence of AllOf elements, while the AllOf element contains a conjunctive sequence of Match elements. Each Match element specifies an attribute that a Request should match.

A Condition is a Boolean function over attributes or functions of attributes. In this abstraction, the user is free to define the Condition as long as its expression returns a Boolean value, i.e., either true or false. Empty Condition is always associated to true.

A Request contains a set of attribute values for a particular access request and the error messages that occurred during the evaluation of attribute values.

2.2 XACML 3.0 Formal Semantics

The evaluation of XACML policies starts from the evaluation of Match elements and continues bottom-up until the evaluation of the root of the XACML element, i.e., the evaluation of PolicySet. For each XACML element X we denote by $[\![X]\!]$ a semantic function associated to X. To each Request element, this function assigns a value from a set of values that depends on the particular type of the XACML element X. For example, the semantic function $[\![X]\!]$, where X is a Match element, ranges over the set $\{\mathsf{m}, \mathsf{nm}, \mathsf{idt}\}$, while its range is the set $\{\mathsf{t}, \mathsf{f}, \mathsf{idt}\}$ when X is a Condition element. A further explanation will be given below. An XACML component returns an indeterminate value whenever the decision cannot be made. This happens when there is an error during the evaluation process. See [9] for further explanation of the semantics of XACML 3.0.

Evaluation of Match, AllOf, AnyOf and Target Components. Let X be either a Match, an AllOf, an AnyOf or a Target component and let $\mathbf{Q}$ be a set of all possible Requests. A *Match semantic function* is a mapping $[\![X]\!] : \mathbf{Q} \to \{\,\mathsf{m}, \mathsf{nm}, \mathsf{idt}\,\}$, where m, nm and idt denote *match*, *no-match* and *indeterminate*, respectively.

Our evaluation of Match element is based on equality function.[4] We check whether there are any attribute values in Request element that match the Match attribute value.

Let $\mathcal{Q}$ be a Request element and let $\mathcal{M}$ be a Match element. The evaluation of Match $\mathcal{M}$ is as follows

$$[\![\mathcal{M}]\!](\mathcal{Q}) = \begin{cases} \mathsf{m} & \text{if } \mathcal{M} \in \mathcal{Q} \text{ and } \mathtt{error}(\mathcal{M}) \notin \mathcal{Q} \\ \mathsf{nm} & \text{if } \mathcal{M} \notin \mathcal{Q} \text{ and } \mathtt{error}(\mathcal{M}) \notin \mathcal{Q} \\ \mathsf{idt} & \text{if } \mathtt{error}(\mathcal{M}) \in \mathcal{Q} \end{cases} \tag{1}$$

The evaluation of AllOf is a conjunction of a sequence of Match elements. The value of m, nm and idt corresponds to true, false and undefined in 3-valued logic, respectively.

Given a Request $\mathcal{Q}$, the evaluation of AllOf, $\mathcal{A} = \bigwedge_{i=1}^{n} \mathcal{M}_i$, is as follows

$$[\![\mathcal{A}]\!](Q) = \begin{cases} \mathsf{m} & \text{if } \forall i : [\![\mathcal{M}_i]\!](Q) = \mathsf{m} \\ \mathsf{nm} & \text{if } \exists i : [\![\mathcal{M}_i]\!](Q) = \mathsf{nm} \\ \mathsf{idt} & \text{otherwise} \end{cases} \tag{2}$$

where each $\mathcal{M}_i$ is a Match element.

The evaluation of AnyOf element is a disjunction of a sequence of AllOf elements. Given a Request $\mathcal{Q}$, the evaluation of AnyOf, $\mathcal{E} = \bigvee_{i=1}^{n} \mathcal{A}_i$, is as follows

$$[\![\mathcal{E}]\!](\mathcal{Q}) = \begin{cases} \mathsf{m} & \text{if } \exists i : [\![\mathcal{A}_i]\!](\mathcal{Q}) = \mathsf{m} \\ \mathsf{nm} & \text{if } \forall i : [\![\mathcal{A}_i]\!](\mathcal{Q}) = \mathsf{nm} \\ \mathsf{idt} & \text{otherwise} \end{cases} \tag{3}$$

where each $\mathcal{A}_i$ is an AllOf element.

The evaluation of Target element is a conjunction of a sequence of AnyOf elements. An empty Target, indicated by `null` attribute, is always evaluated to m. Given a Request $\mathcal{Q}$, the evaluation of Target, $\mathcal{T} = \bigwedge_{i=1}^{n} \mathcal{E}_i$, is as follows

$$[\![\mathcal{T}]\!](\mathcal{Q}) = \begin{cases} \mathsf{m} & \text{if } \forall i : [\![\mathcal{E}_i]\!](Q) = \mathsf{m} \text{ or } \mathcal{T} = \mathtt{null} \\ \mathsf{nm} & \text{if } \exists i : [\![\mathcal{E}_i]\!](Q) = \mathsf{nm} \\ \mathsf{idt} & \text{otherwise} \end{cases} \tag{4}$$

where each $\mathcal{E}_i$ is an AnyOf element.

Evaluation of Condition. Let X be a Condition component and let $\mathbf{Q}$ be a set of all possible Requests. A *Condition semantic function* is a mapping $[\![X]\!] : \mathbf{Q} \to \{\,\mathsf{t}, \mathsf{f}, \mathsf{idt}\,\}$, where t, f and idt denote *true*, *false* and *indeterminate*, respectively.

The evaluation of Condition element is based on the evaluation of its Boolean function as described in its element. To keep it abstract, we do not specify specific functions; however, we use an unspecified function, eval, that returns $\{\,\mathsf{t}, \mathsf{f}, \mathsf{idt}\,\}$.

[4] Our Match evaluation is a simplification compared with [11].

Given a Request $\mathcal{Q}$, the evaluation of Condition $\mathcal{C}$ is as follows

$$[\![\mathcal{C}]\!](\mathcal{Q}) = \mathsf{eval}(\mathcal{C}, \mathcal{Q}) \tag{5}$$

Evaluation of Rule. Let X be a Rule component and let $\mathbf{Q}$ be a set of possible Requests. A *Rule semantic function* is a mapping $[\![X]\!] : \mathcal{Q} \to \{\, \mathsf{p}, \mathsf{d}, \mathsf{i_p}, \mathsf{i_d}, \mathsf{na} \,\}$, where $\mathsf{p}, \mathsf{d}, \mathsf{i_p}, \mathsf{i_d}$ and na correspond to *permit, deny, indeterminate permit, indeterminate deny* and *not – applicable*, respectively.

Given a Request $\mathcal{Q}$, the evaluation of Rule $\mathcal{R}_{id} = [E, \mathcal{T}, \mathcal{C}]$ is as follows

$$[\![\mathcal{R}_{id}]\!](\mathcal{Q}) = \begin{cases} E & \text{if } [\![\mathcal{T}]\!](\mathcal{Q}) = \mathsf{m} \text{ and } [\![\mathcal{C}]\!](\mathcal{Q}) = \mathsf{t} \\ \mathsf{na} & \text{if } ([\![\mathcal{T}]\!](\mathcal{Q}) = \mathsf{m} \text{ and } [\![\mathcal{C}]\!](\mathcal{Q}) = \mathsf{f}) \text{ or } ([\![\mathcal{T}]\!](\mathcal{Q}) = \mathsf{nm}) \\ \mathsf{i}_E & \text{otherwise} \end{cases} \tag{6}$$

where E is an effect, $E \in \{\, \mathsf{p}, \mathsf{d} \,\}$, $\mathcal{T}$ is a Target element and $\mathcal{C}$ is a Condition element.

Evaluation of Policy and PolicySet. Let X be either a Policy or a PolicySet component and let $\mathbf{Q}$ be a set of all possible Requests. A *Policy semantic function* is a mapping $[\![X]\!] : \mathcal{Q} \to \{\, \mathsf{p}, \mathsf{d}, \mathsf{i_p}, \mathsf{i_d}, \mathsf{i_{dp}}, \mathsf{na} \,\}$, where $\mathsf{p}, \mathsf{d}, \mathsf{i_p}, \mathsf{i_d}, \mathsf{i_{dp}}$ and na correspond to *permit, deny, indeterminate permit, indeterminate deny, indeterminate deny permit* and *not – applicable*, respectively.

Given a Request $\mathcal{Q}$, the evaluation of Policy $\mathcal{P}_{id} = [T, \langle \mathcal{R}_1, \ldots, \mathcal{R}_n \rangle, \mathsf{CombID}]$ is as follows

$$[\![\mathcal{P}_{id}]\!](\mathcal{Q}) = \begin{cases} \mathsf{i_d} & \text{if } [\![\mathcal{T}]\!](\mathcal{Q}) = \mathsf{idt} \text{ and } \bigoplus_{\mathsf{CombID}}(\mathbf{R}) = \mathsf{d} \\ \mathsf{i_p} & \text{if } [\![\mathcal{T}]\!](\mathcal{Q}) = \mathsf{idt} \text{ and } \bigoplus_{\mathsf{CombID}}(\mathbf{R}) = \mathsf{p} \\ \mathsf{na} & \text{if } [\![T]\!](Q) = \mathsf{nm} \text{ or } \forall i : [\![R_i]\!](Q) = \mathsf{na} \\ \bigoplus_{\mathsf{CombID}}(\mathbf{R}) & \text{otherwise} \end{cases} \tag{7}$$

where $\mathcal{T}$ is a Target element, and each $\mathcal{R}_i$ is a Rule element. We use $\mathbf{R}$ to denote $\langle [\![\mathcal{R}_1]\!](\mathcal{Q}), \ldots, [\![\mathcal{R}_n]\!](\mathcal{Q}) \rangle$.

Note: The combining algorithm denoted by $\bigoplus_{\mathsf{CombID}}$ will be explained in Section 2.3.

The evaluation of PolicySet is exactly like the evaluation of Policy except that it differs in terms of input parameter. While in Policy we use a sequence of Rule elements as an input, in the evaluation of PolicySet we use a sequence of Policy or PolicySet elements.

2.3 XACML Combining Algorithms

There are four common combining algorithms defined in XACML 3.0, namely permit-overrides (po), deny-overrides (do), first-applicable (fa) and only-one-applicable (ooa). In this paper, we do not consider the deny-overrides combining algorithm since it is the mirror of the permit-overrides combining algorithm.

Permit-Overrides (po) Combining Algorithm. The permit-overrides combining algorithm is intended for use if a permit decision should have priority over a deny decision. This algorithm has the following behaviour [11].

1. If any decision is "permit", then the result is "permit".
2. Otherwise, if any decision is "indeterminate deny permit", then the result is "indeterminate deny permit".
3. Otherwise, if any decision is "indeterminate permit" and another decision is "indeterminate deny" or "deny", then the result is "indeterminate deny permit".
4. Otherwise, if any decision is "indeterminate permit", then the result is "indeterminate permit".
5. Otherwise, if decision is "deny", then the result is "deny".
6. Otherwise, if any decision is "indeterminate deny", then the result is "indeterminate deny".
7. Otherwise, the result is "not applicable".

Let $\langle s_1, \ldots, s_n \rangle$ be a sequence of element of $\{\, \mathsf{p}, \mathsf{d}, \mathsf{i_p}, \mathsf{i_d}, \mathsf{i_{dp}}, \mathsf{na} \,\}$. The *permit-overrides combining operator* is defined as follows

$$\bigoplus_{\mathsf{po}}(\langle s_1, \ldots, s_n \rangle) = \begin{cases} \mathsf{p} & \text{if } \exists i : s_i = \mathsf{p} \\ \mathsf{i_{dp}} & \text{if } \forall i : s_i \neq \mathsf{p} \text{ and} \\ & \quad (\exists j : s_j = \mathsf{i_{dp}} \\ & \quad\quad \text{or } (\exists j, j' : s_j = \mathsf{i_p} \text{ and } (s_{j'} = \mathsf{i_d} \text{ or } s_{j'} = \mathsf{d})) \\ \mathsf{i_p} & \text{if } \exists i : s_i = \mathsf{i_p} \text{ and } \forall j : s_j \neq \mathsf{i_p} \Rightarrow s_j = \mathsf{na} \\ \mathsf{d} & \text{if } \exists i : s_i = \mathsf{d} \text{ and } \forall j : s_j \neq \mathsf{d} \Rightarrow (s_j = \mathsf{i_d} \text{ or } s_j = \mathsf{na}) \\ \mathsf{i_d} & \text{if } \exists i : s_i = \mathsf{i_d} \text{ and } \forall j : s_j \neq \mathsf{i_d} \Rightarrow s_j = \mathsf{na} \\ \mathsf{na} & \text{otherwise} \end{cases} \tag{8}$$

First-Applicable (fa) Combining Algorithm. Each Rule must be evaluated in the order in which it is listed in the Policy. If a particular Rule is applicable, then the result of first-applicable combining algorithm must be the result of evaluating the Rule. If the Rule is "not applicable" then the next Rule in the order must be evaluated. If no further Rule in the order exists, then the first-applicable combining algorithm must return "not applicable".

Let $\langle s_1, \ldots, s_n \rangle$ be a sequence of element of $\{\, \mathsf{p}, \mathsf{d}, \mathsf{i_p}, \mathsf{i_d}, \mathsf{i_{dp}}, \mathsf{na} \,\}$. The *first-applicable combining operator* is defined as follows:

$$\bigoplus_{\mathsf{fa}}(\langle s_1, \ldots, s_n \rangle) = \begin{cases} s_i & \text{if } \exists i : s_i \neq \mathsf{na} \text{ and } \forall j : (j < i) \Rightarrow (s_j = \mathsf{na}) \\ \mathsf{na} & \text{otherwise} \end{cases} \tag{9}$$

Only-One-Applicable (ooa) Combining Algorithm. If only one Policy is considered applicable by evaluation of its Target, then the result of the only-one-applicable combining algorithm must be the result of evaluating the Policy. If in the entire sequence of Policy elements in the PolicySet, there is no Policy that is applicable, then the result of the only-one-applicable combining algorithm must be "not applicable". If more than one Policy is considered applicable, then the result of the only-one-applicable combining algorithm must be "indeterminate". We follow [9] for the equation of only-one-applicable combining algorithm.

Let $\langle s_1, \ldots, s_n \rangle$ be a sequence of element of $\{ \mathsf{p}, \mathsf{d}, \mathsf{i_p}, \mathsf{i_d}, \mathsf{i_{dp}}, \mathsf{na} \}$. The *only-one-applicable combining operator* is defined as follows:

$$\bigoplus_{\mathsf{ooa}}(\langle s_1, \ldots, s_n \rangle) = \begin{cases} \mathsf{i_{dp}} & \text{if } (\exists i : s_i = \mathsf{i_{dp}}) \text{ or} \\ & \quad (\exists i, j : i \neq j \text{ and } s_i = (\mathsf{d} \text{ or } \mathsf{i_d}) \wedge s_j = (\mathsf{p} \text{ or } \mathsf{i_p})) \\ \mathsf{i_d} & \text{if } (\forall i : s_i \neq (\mathsf{p} \text{ or } \mathsf{i_p} \text{ or } \mathsf{i_{dp}})) \text{ and} \\ & \quad ((\exists j : s_j = \mathsf{i_d}) \text{ or } (\exists j, k : j \neq k \text{ and } s_j = s_k = \mathsf{d})) \\ \mathsf{i_p} & \text{if } (\forall i : s_i \neq (\mathsf{d} \text{ or } \mathsf{i_d} \text{ or } \mathsf{i_{dp}})) \text{ and} \\ & \quad ((\exists j : s_j = \mathsf{i_p}) \text{ or } (\exists j, k : j \neq k \text{ and } s_j = s_k = \mathsf{p})) \\ s_i & \text{if } \exists i : s_i \neq \mathsf{na} \text{ and } \forall j : j \neq i \Rightarrow s_j = \mathsf{na} \\ \mathsf{na} & \text{otherwise} \end{cases} \tag{10}$$

3 Transforming XACML Components into Logic Programs

In this section we show, step by step, how to transform XACML 3.0 components into logic programs. We begin by introducing the syntax of logic programs (LPs). Then we show the transformation of XACML component into LPs starting from Request element to PolicySet element. We also present transformations for combining algorithms. The transformation of each XACML element is based on its formal semantics explained in Section 2.2 and Section 2.3.

3.1 Preliminaries

We recall basic notation and terminology that we use in the remainder of this paper.

First-Order Language. We consider an *alphabet* consisting of (finite or countably infinite) disjoint sets of variables, constants, function symbols, predicate symbols, connectives $\{ \textbf{not}, \wedge, \leftarrow \}$, punctuation symbols { "(", ",", ")", "." } and special symbols $\{ \top, \bot \}$. We use upper case letters to denote variables and lower case letters to denote constants, function and predicate symbols. Terms, atoms, literals and formulae are defined as usual. The *language* given by an alphabet consists of the set of all formulae constructed from the symbols occurring in the alphabet.

Logic Programs. A *rule* is an expression of the form

$$A \leftarrow B_1 \wedge \cdots \wedge B_m \wedge \textbf{not } B_{m+1} \wedge \cdots \wedge \textbf{not } B_n. \tag{11}$$

where A is either an atom or $\bot$ and each B_i, $1 \leq i \leq n$, is an atom or $\top$. $\top$ is a valid formula. We usually write $B_1 \wedge \cdots \wedge B_m \wedge \textbf{not } B_{m+1} \wedge \cdots \wedge \textbf{not } B_n$ simply as $B_1, \ldots, B_m, \textbf{not } B_{m+1}, \ldots, \textbf{not } B_n$. We call the rule as a *constraint* when $A = \bot$. One should observe that the body of a rule must not be empty. A *fact* is a rule of the form $A \leftarrow \top$.

A *logic program* is a finite set of rules. We denote *ground*(Π) for the set of all ground instances of rules in the program Π.

3.2 XACML Components Transformation into Logic Programs

The transformation of XACML components is based on the semantics of each component explained in Section 2.2. Please note that the calligraphic font in each transformation indicates the XACML component's name, that is, it does not represent a variable in LP.

3.2.1 Request Transformation

XACML Syntax: Let $\mathcal{Q} = \{ \mathcal{A}t_1, \ldots, \mathcal{A}t_n \}$ be a Request component. We transform all members of Request element into facts. The transformation of Request, $\mathcal{Q}$, into LP $\Pi_{\mathcal{Q}}$ is as follows

$$\mathcal{A}t_i \leftarrow \top. \;\; 1 \leq i \leq n$$

3.2.2 XACML Policy Components Transformation

We use a two-place function val to indicate the semantics of XACML components where the first argument is the name of XACML component and the second argument is its value.

Transformation of Match, AnyOf, AllOf and Target Components. Given a semantic equation of the form $[\![X]\!](\mathcal{Q}) = v$ if $cond_1$ and ... and $cond_n$, we produce a rule of the form $\mathsf{val}(X, v) \leftarrow cond_1, \ldots, cond_n$. Given a semantic equation of the form $[\![X]\!](\mathcal{Q}) = v$ if $cond_1$ or ... or $cond_n$, we produce a rule of the form $\mathsf{val}(X, v) \leftarrow cond_i$. $1 \leq i \leq n$. For example, the Match evaluation $[\![\mathcal{M}]\!](\mathcal{Q}) = \mathsf{m}$ if $\mathcal{M} \in \mathcal{Q}$ and $\mathtt{error}(\mathcal{M}) \notin \mathcal{Q}$ is transformed into a rule: $\mathsf{val}(\mathcal{M}, \mathsf{m}) \leftarrow \mathcal{M}, \mathbf{not}\ \mathtt{error}(\mathcal{M})$. The truth value of $\mathcal{M}$ depends on whether $\mathcal{M} \leftarrow \top$ is in $\Pi_{\mathcal{Q}}$ and the same is the case also for the truth value of $\mathtt{error}(\mathcal{M})$.

Let $\mathcal{M}$ be a Match component. The transformation of Match $\mathcal{M}$ into LP $\Pi_{\mathcal{M}}$ is as follows (see (1) for Match evaluation)

$$\begin{array}{l} \mathsf{val}(\mathcal{M}, \mathsf{m}) \leftarrow \mathcal{M}, \mathbf{not}\ \mathtt{error}(\mathcal{M}). \\ \mathsf{val}(\mathcal{M}, \mathsf{nm}) \leftarrow \mathbf{not}\ cat(a), \mathbf{not}\ \mathtt{error}(\mathcal{M}). \\ \mathsf{val}(\mathcal{M}, \mathsf{idt}) \leftarrow \mathtt{error}(\mathcal{M}). \end{array}$$

Let $\mathcal{A} = \bigwedge_{i=1}^{n} \mathcal{M}_i$ be an AllOf component where each $\mathcal{M}_i$ is a Match component. The transformation of AllOf $\mathcal{A}$ into LP $\Pi_{\mathcal{A}}$ is as follows (see (2) for AllOf evaluation)

$$\begin{array}{l} \mathsf{val}(\mathcal{A}, \mathsf{m}) \leftarrow \mathsf{val}(\mathcal{M}_1, \mathsf{m}), \ldots, \mathsf{val}(\mathcal{M}_n, \mathsf{m}). \\ \mathsf{val}(\mathcal{A}, \mathsf{nm}) \leftarrow \mathsf{val}(\mathcal{M}_i, \mathsf{nm}). \; (1 \leq i \leq n) \\ \mathsf{val}(\mathcal{A}, \mathsf{idt}) \leftarrow \mathbf{not}\ \mathsf{val}(\mathcal{A}, \mathsf{m}), \mathbf{not}\ \mathsf{val}(\mathcal{A}, \mathsf{nm}). \end{array}$$

Let $\mathcal{E} = \bigvee_{i=1}^{n} \mathcal{A}_i$ be an AnyOf component where each $\mathcal{A}_i$ is an AllOf component. The transformation of AnyOf $\mathcal{E}$ into LP $\Pi_{\mathcal{E}}$ is as follows (see (3) for AnyOf evaluation)

$$\begin{array}{l} \mathsf{val}(\mathcal{E}, \mathsf{m}) \leftarrow \mathsf{val}(\mathcal{A}_i, \mathsf{m}). \; (1 \leq i \leq n) \\ \mathsf{val}(\mathcal{E}, \mathsf{nm}) \leftarrow \mathsf{val}(\mathcal{A}_1, \mathsf{nm}), \ldots, \mathsf{val}(\mathcal{A}_n, \mathsf{nm}). \\ \mathsf{val}(\mathcal{E}, \mathsf{idt}) \leftarrow \mathbf{not}\ \mathsf{val}(\mathcal{A}, \mathsf{m}), \mathbf{not}\ \mathsf{val}(\mathcal{E}, \mathsf{nm}). \end{array}$$

Let $\mathcal{T} = \bigwedge_{i=1}^{n} \mathcal{T}_i$ be a Target component where each $\mathcal{E}_i$ is an AnyOf component. The transformation of Target $\mathcal{T}$ into LP $\Pi_{\mathcal{T}}$ is as follows (see (4) for Target evaluation)

$$\begin{aligned}
&\mathsf{val}(\mathsf{null}, \mathsf{m}) \leftarrow \top. \\
&\mathsf{val}(\mathcal{T}, \mathsf{m}) \quad \leftarrow \mathsf{val}(\mathcal{E}_1, \mathsf{m}), \ldots, \mathsf{val}(\mathcal{E}_n, \mathsf{m}). \\
&\mathsf{val}(\mathcal{T}, \mathsf{nm}) \leftarrow \mathsf{val}(\mathcal{E}_i, \mathsf{nm}).\ (1 \leq i \leq n) \\
&\mathsf{val}(\mathcal{T}, \mathsf{idt}) \ \ \leftarrow \textbf{not}\ \mathsf{val}(\mathcal{T}, \mathsf{m}), \textbf{not}\ \mathsf{val}(\mathcal{T}, \mathsf{nm}).
\end{aligned}$$

Transformation of Condition Component. The transformation of Condition $\mathcal{C}$ into LP $\Pi_{\mathcal{C}}$ is as follows

$$\mathsf{val}(\mathcal{C}, V) \leftarrow \mathsf{eval}(\mathcal{C}, V).$$

Moreover, the transformation of Condition also depends on the transformation of eval function into LP. Since we do not describe specific eval functions, we leave this transformation to the user.

Example 1. A possible eval function for "rule r1: patient only can see his or her patient record" is

$$\begin{aligned}
&\Pi_{cond(r1)} : \\
&\mathsf{val}(cond(r1), V) \quad \leftarrow \mathsf{eval}(cond(r1), V). \\
&\mathsf{eval}(cond(r1), \mathsf{t}) \quad \leftarrow patient_id(X), patient_record_id(X), \\
&\qquad\qquad \textbf{not}\ \mathtt{error}(patient_id(X)), \textbf{not}\ \mathtt{error}(patient_record_id(X)). \\
&\mathsf{eval}(cond(r1), \mathsf{f}) \quad \leftarrow patient_id(X), patient_record_id(Y), X \neq Y, \\
&\qquad\qquad \textbf{not}\ \mathtt{error}(patient_id(X)), \textbf{not}\ \mathtt{error}(patient_record_id(Y)). \\
&\mathsf{eval}(cond(r1), \mathsf{idt}) \leftarrow \textbf{not}\ \mathsf{eval}(cond(r1), \mathsf{t}), \textbf{not}\ \mathsf{eval}(cond(r1), \mathsf{f}).
\end{aligned}$$

The $\mathtt{error}(patient\_id(X))$ and $\mathtt{error}(patient\_record\_id(X))$ indicate possible errors that might occur, e.g., the system could not connect to the database so that the system does not know the identifier of the patient. □

Transformation of Rule Component. The general step of the transformation of Rule component is similar to the transformation of Match component.

Let $\mathcal{R} = [e, \mathcal{T}, \mathcal{C}]$ be a Rule component where $e \in \{\mathsf{p}, \mathsf{d}\}$, $\mathcal{T}$ is a Target and $\mathcal{C}$ is a Condition. The transformation of Rule $\mathcal{R}$ into LP $\Pi_{\mathcal{R}}$ is as follows (see (6) for Rule evaluation)

$$\begin{aligned}
&\mathsf{val}(\mathcal{R}, e) \quad \leftarrow \mathsf{val}(\mathcal{T}, \mathsf{m}), \mathsf{val}(\mathcal{C}, \mathsf{t}). \\
&\mathsf{val}(\mathcal{R}, \mathsf{na}) \leftarrow \mathsf{val}(\mathcal{T}, \mathsf{m}), \mathsf{val}(\mathcal{C}, \mathsf{f}). \\
&\mathsf{val}(\mathcal{R}, \mathsf{na}) \leftarrow \mathsf{val}(\mathcal{T}, \mathsf{nm}). \\
&\mathsf{val}(\mathcal{R}, \mathsf{i}_e) \quad \leftarrow \textbf{not}\ \mathsf{val}(\mathcal{R}, e), \textbf{not}\ \mathsf{val}(\mathcal{R}, \mathsf{na}).
\end{aligned}$$

Transformation of Policy and PolicySet Components. Given a Policy component $\mathcal{P}_{id} = [\mathcal{T}, \langle \mathcal{R}_1, \ldots, \mathcal{R}_n \rangle, \mathsf{CombID}]$ where $\mathcal{T}$ is a Target, $\langle \mathcal{R}_1, \ldots, \mathcal{R}_n \rangle$ is a sequence of Rule elements and CombID is a combining algorithm identifier. In order to indicate that the Policy contains Rule $\mathcal{R}_i$, for every Rule $\mathcal{R}_i \in \langle \mathcal{R}_1, \ldots, \mathcal{R}_n \rangle$, $\Pi_{\mathcal{P}_{id}}$ contains:

$$\mathsf{decision\_of}(\mathcal{P}_{id}, \mathcal{R}_i, V) \leftarrow \mathsf{val}(\mathcal{R}_i, V).\ (1 \leq i \leq n)$$

The transformation for Policy Π into LP $\Pi_{\mathcal{P}_{id}}$ is as follows (see (7) for Policy evaluation)

$\mathsf{val}(\mathcal{P}_{id}, \mathsf{i_d}) \leftarrow \mathsf{val}(\mathcal{T}, \mathsf{idt}), \mathsf{algo}(\mathsf{CombID}, \mathcal{P}_{id}, \mathsf{d}).$
$\mathsf{val}(\mathcal{P}_{id}, \mathsf{i_p}) \leftarrow \mathsf{val}(\mathcal{T}, \mathsf{idt}), \mathsf{algo}(\mathsf{CombID}, \mathcal{P}_{id}, \mathsf{p}).$
$\mathsf{val}(\mathcal{P}_{id}, \mathsf{na}) \leftarrow \mathsf{val}(\mathcal{T}, \mathsf{nm}).$
$\mathsf{val}(\mathcal{P}_{id}, \mathsf{na}) \leftarrow \mathsf{val}(\mathcal{R}_1, \mathsf{na}), \ldots, \mathsf{val}(\mathcal{R}_n, \mathsf{na}).$
$\mathsf{val}(\mathcal{P}_{id}, V') \leftarrow \mathsf{val}(\mathcal{T}, \mathsf{m}), \mathsf{decision\_of}(\mathcal{P}_{id}, \mathcal{R}, V), V \neq \mathsf{na}, \mathsf{algo}(\mathsf{CombID}, \mathcal{P}_{id}, V').$
$\mathsf{val}(\mathcal{P}_{id}, V') \leftarrow \mathsf{val}(\mathcal{T}, \mathsf{idt}), \mathsf{decision\_of}(\mathcal{P}_{id}, \mathcal{R}, V), V \neq \mathsf{na}, \mathsf{algo}(\mathsf{CombID}, \mathcal{P}_{id}, V'), V' \neq \mathsf{p}.$
$\mathsf{val}(\mathcal{P}_{id}, V') \leftarrow \mathsf{val}(\mathcal{T}, \mathsf{idt}), \mathsf{decision\_of}(\mathcal{P}_{id}, \mathcal{R}, V), V \neq \mathsf{na}, \mathsf{algo}(\mathsf{CombID}, \mathcal{P}_{id}, V'), V' \neq \mathsf{d}.$

We write a formula $\mathsf{decision\_of}(\mathcal{P}_{id}, \mathcal{R}, V), V \neq \mathsf{na}$ to make sure that there is a Rule in the Policy that is not evaluated to na. We do this to avoid a return value from a combining algorithm that is not na, even tough all of the Rule elements are evaluated to na. The transformation of PolicySet is similar to the transformation of Policy component.

3.3 Combining Algorithm Transformation

We define generic LPs for permit-overrides combining algorithm and only-one-applicable combining algorithm. Therefore, we use a variable P to indicate a variable over Policy identifier and R, R_1 and R_2 to indicate variables over Rule identifiers. In case the evaluation of PolicySet, the input P is for PolicySet identifier, R, R_1 and R_2 are for Policy (or PolicySet) identifiers.

Permit-Overrides Transformation. Let Π_{po} be a LP obtained by permit-overrides combining algorithm transformation (see (8) for the permit-overrides combining algorithm semantics). Π_{po} contains:

$\mathsf{algo}(\mathsf{po}, P, \mathsf{p}) \leftarrow \mathsf{decision\_of}(P, R, \mathsf{p}).$
$\mathsf{algo}(\mathsf{po}, P, \mathsf{i_{dp}}) \leftarrow \textbf{not } \mathsf{algo}(\mathsf{po}, P, \mathsf{p}), \mathsf{decision\_of}(P, R, \mathsf{i_{dp}}).$
$\mathsf{algo}(\mathsf{po}, P, \mathsf{i_{dp}}) \leftarrow \textbf{not } \mathsf{algo}(\mathsf{po}, P, \mathsf{p}), \mathsf{decision\_of}(P, R_1, \mathsf{i_p}), \mathsf{decision\_of}(P, R_2, \mathsf{d}).$
$\mathsf{algo}(\mathsf{po}, P, \mathsf{i_{dp}}) \leftarrow \textbf{not } \mathsf{algo}(\mathsf{po}, P, \mathsf{p}), \mathsf{decision\_of}(P, R_1, \mathsf{i_p}), \mathsf{decision\_of}(P, R_2, \mathsf{i_d}).$
$\mathsf{algo}(\mathsf{po}, P, \mathsf{i_p}) \leftarrow \textbf{not } \mathsf{algo}(\mathsf{po}, P, \mathsf{p}), \textbf{not } \mathsf{algo}(\mathsf{po}, P, \mathsf{i_{dp}}), \mathsf{decision\_of}(P, R, \mathsf{i_p}).$
$\mathsf{algo}(\mathsf{po}, P, \mathsf{d}) \leftarrow \textbf{not } \mathsf{algo}(\mathsf{po}, P, \mathsf{p}), \textbf{not } \mathsf{algo}(\mathsf{po}, P, \mathsf{i_{dp}}), \textbf{not } \mathsf{algo}(\mathsf{po}, P, \mathsf{i_p}), \mathsf{decision\_of}(P, R, \mathsf{d}).$
$\mathsf{algo}(\mathsf{po}, P, \mathsf{i_d}) \leftarrow \textbf{not } \mathsf{algo}(\mathsf{po}, P, \mathsf{p}), \textbf{not } \mathsf{algo}(\mathsf{po}, P, \mathsf{i_{dp}}), \textbf{not } \mathsf{algo}(\mathsf{po}, P, \mathsf{i_p}), \textbf{not } \mathsf{algo}(\mathsf{po}, P, \mathsf{d}), \mathsf{decision\_of}(P, R, \mathsf{i_d}).$
$\mathsf{algo}(\mathsf{po}, P, \mathsf{na}) \leftarrow \textbf{not } \mathsf{algo}(\mathsf{po}, P, \mathsf{p}), \textbf{not } \mathsf{algo}(\mathsf{po}, P, \mathsf{i_{dp}}), \textbf{not } \mathsf{algo}(\mathsf{po}, P, \mathsf{i_p}), \textbf{not } \mathsf{algo}(\mathsf{po}, P, \mathsf{d}), \textbf{not } \mathsf{algo}(\mathsf{po}, P, \mathsf{i_d}).$

First-Applicable Transformation. Let Π_{fa} be a LP obtained by first-applicable combining algorithm transformation (see (9) for the first-applicable combining algorithm semantics). For each Policy (or PolicySet) that uses this combining algorithm, $\mathcal{P}_{id} = [\mathcal{T}, \langle \mathcal{R}_1, \ldots, \mathcal{R}_n \rangle, \mathsf{fa}]$, $\Pi_{\mathcal{P}_{id}}$ contains:

$\mathsf{algo}(\mathsf{fa}, \mathcal{P}_{id}, E) \leftarrow \mathsf{decision\_of}(\mathcal{P}_{id}, \mathcal{R}_1, V), V \neq \mathsf{na}.$
$\mathsf{algo}(\mathsf{fa}, \mathcal{P}_{id}, E) \leftarrow \mathsf{decision\_of}(\mathcal{P}_{id}, \mathcal{R}_1, \mathsf{na}), \mathsf{decision\_of}(\mathcal{P}_{id}, \mathcal{R}_2, E), E \neq \mathsf{na}.$

$\vdots$

$\mathsf{algo}(\mathsf{fa}, \mathcal{P}_{id}, E) \leftarrow \mathsf{decision\_of}(\mathcal{P}_{id}, \mathcal{R}_1, \mathsf{na}), \ldots, \mathsf{decision\_of}(\mathcal{P}_{id}, \mathcal{R}_{n-1}, \mathsf{na}), \mathsf{decision\_of}(P, R_n, E).$

Only-One-Applicable Transformation. Let Π_{ooa} be a LP obtained by only-one-applicable combining algorithm transformation (see (10) for the only-one-applicable combining algorithm semantics). Π_{ooa} contains:

$$
\begin{array}{ll}
\mathsf{algo}(\mathsf{ooa}, P, \mathsf{i_{dp}}) & \leftarrow \mathsf{decision\_of}(P, R, \mathsf{i_{dp}}).\\
\mathsf{algo}(\mathsf{ooa}, P, \mathsf{i_{dp}}) & \leftarrow \mathsf{decision\_of}(P, R_1, \mathsf{i_d}), \mathsf{decision\_of}(P, R_2, \mathsf{i_p}), R_1 \neq R_2.\\
\mathsf{algo}(\mathsf{ooa}, P, \mathsf{i_{dp}}) & \leftarrow \mathsf{decision\_of}(P, R_1, \mathsf{i_d}), \mathsf{decision\_of}(P, R_2, \mathsf{p}), R_1 \neq R_2.\\
\mathsf{algo}(\mathsf{ooa}, P, \mathsf{i_{dp}}) & \leftarrow \mathsf{decision\_of}(P, R_1, \mathsf{d}), \mathsf{decision\_of}(P, R_2, \mathsf{i_p}), R_1 \neq R_2.\\
\mathsf{algo}(\mathsf{ooa}, P, \mathsf{i_{dp}}) & \leftarrow \mathsf{decision\_of}(P, R_1, \mathsf{d}), \mathsf{decision\_of}(P, R_2, \mathsf{p}), R_1 \neq R_2.\\
\mathsf{algo}(\mathsf{ooa}, P, \mathsf{i_p}) & \leftarrow \mathbf{not}\ \mathsf{algo}(\mathsf{ooa}, P, \mathsf{i_{dp}}), \mathsf{decision\_of}(P, R, \mathsf{i_p}).\\
\mathsf{algo}(\mathsf{ooa}, P, \mathsf{i_p}) & \leftarrow \mathbf{not}\ \mathsf{algo}(\mathsf{ooa}, P, \mathsf{i_{dp}}), \mathsf{decision\_of}(P, R_1, \mathsf{p}),\\
& \quad\ \mathsf{decision\_of}(P, R_2, \mathsf{p}), R_1 \neq R_2.\\
\mathsf{algo}(\mathsf{ooa}, P, \mathsf{i_d}) & \leftarrow \mathbf{not}\ \mathsf{algo}(\mathsf{ooa}, P, \mathsf{i_{dp}}), \mathsf{decision\_of}(P, R, \mathsf{i_d}).\\
\mathsf{algo}(\mathsf{ooa}, P, \mathsf{i_d}) & \leftarrow \mathbf{not}\ \mathsf{algo}(\mathsf{ooa}, P, \mathsf{i_{dp}}), \mathsf{decision\_of}(P, R_1, \mathsf{d}),\\
& \quad\ \mathsf{decision\_of}(P, R_2, \mathsf{d}), R_1 \neq R_2.\\
\mathsf{algo}(\mathsf{ooa}, P, \mathsf{p}) & \leftarrow \mathbf{not}\ \mathsf{algo}(\mathsf{ooa}, P, \mathsf{i_{dp}}),\\
& \quad\ \mathbf{not}\ (\mathsf{ooa}, P, \mathsf{i_p}), \mathsf{decision\_of}(P, R, \mathsf{p}).\\
\mathsf{algo}(\mathsf{ooa}, P, \mathsf{d}) & \leftarrow \mathbf{not}\ \mathsf{algo}(\mathsf{ooa}, P, \mathsf{i_{dp}}), \mathbf{not}\ (\mathsf{ooa}, P, \mathsf{i_d}), \mathbf{not}\ (\mathsf{ooa}, P, \mathsf{i_p}),\\
& \quad\ \mathsf{decision\_of}(P, R, \mathsf{d}).\\
\mathsf{algo}(\mathsf{ooa}, P, \mathsf{na}) & \leftarrow \mathbf{not}\ \mathsf{algo}(\mathsf{ooa}, P, \mathsf{i_{dp}}), \mathbf{not}\ (\mathsf{ooa}, P, \mathsf{i_d}), \mathbf{not}\ (\mathsf{ooa}, P, \mathsf{i_p}),\\
& \quad\ \mathbf{not}\ \mathsf{decision\_of}(P, R, \mathsf{d}), \mathbf{not}\ \mathsf{decision\_of}(P, R, \mathsf{p}).
\end{array}
$$

4 Relation between XACML-ASP and XACML 3.0 Semantics

In this section we discuss the relationship between the ASP semantics and XACML 3.0 semantics. First, we recall the semantics of logic programs based on their answer sets. Then we show that the program obtained from transforming XACML components into LPs (Π_{XACML}) merges with the query program (Π_Q) has a unique answer set and its unique answer set corresponds to the semantics of XACML 3.0.

4.1 ASP Semantics

The declarative semantics of a logic program is given by a model-theoretic semantics of formulae in the underlying language. The formal definition of answer set semantics can be found in much literature such as [3,6].

The answer set semantics of logic program Π assigns to Π a collection of *answer sets* – interpretations of $ground(\Pi)$. An interpretation I of $ground(\Pi)$ is an answer set for Π if I is minimal (w.r.t. set inclusion) among the interpretations satisfying the rules of

$$
\Pi^I = \{A \leftarrow B_1, \ldots, B_m \mid A \leftarrow B_1, \ldots, B_m, \mathbf{not}\ B_{m+1}, \ldots, \mathbf{not}\ B_n \in \Pi \text{ and } I(\mathbf{not}\ B_{m+1}, \ldots, \mathbf{not}\ B_n) = true\}
$$

A logic program can have a single unique answer set, many or no answer set(s). Therefore, we show that programs with a particular characteristic are guaranteed to have a unique answer set.

Acyclic Programs. We say that a program is *acyclic* when there is no cycle in the program.The acyclicity in the program is guaranteed by the existence of a certain fixed assignment of natural numbers to atoms that is called a *level mapping.*

A *level mapping* for a program Π is a function

$$l : \mathcal{B}_\Pi \to \mathbf{N}$$

where $\mathbf{N}$ is the set of natural numbers and $\mathcal{B}_\Pi$ is the Herbrand base for Π. We extend the definition of level mapping to a mapping from ground literals to natural numbers by setting $l(\mathbf{not}\ A) = l(A)$.

Let Π be a logic program and l be a level mapping for Π. Π is *acyclic with respect to* l if for every clause $A \leftarrow B_1, \ldots, B_m, \mathbf{not}\ B_{m+1}, \ldots, \mathbf{not}\ B_n$ in $ground(\Pi)$ we find

$$l(A) > l(B_i) \quad \text{for all } i \text{ with } 1 \le i \le n$$

Π is *acyclic* if it is acyclic with respect to some degree of level mapping. Acyclic programs are guaranteed to have a unique answer set [3].

4.2 XACML Semantics Based on ASP Semantics

We can see from Section 3 that all of the XACML 3.0 transformation programs are acyclic. Thus, it is guaranteed that Π_{XACML} has a unique answer set.

Proposition 1. *Let Π_{XACML} be a program obtained from XACML 3.0 element transformations and let $\Pi_{\mathcal{Q}}$ be a program transformation of Request $\mathcal{Q}$. Let M be the answer set of $\Pi_{XACML} \cup \Pi_{\mathcal{Q}}$. Then the following equation holds*

$$[\![X]\!](\mathcal{Q}) = V \qquad \textit{if and only if} \qquad \mathsf{val}(X, V) \in M\ .$$

where X is an XACML component.

Note: We can see that there is no cycle in all of the program transformations. Thus, there is a guarantee that the answer set of $\Pi_{\text{XACML}} \cup \Pi_{\mathcal{Q}}$ is unique. The transformation of each component into a logic program is based on exactly the definition of its XACML evaluation. The proof of this proposition can be seen in the extended version in [10].

5 Analysis XACML Policies Using Answer Set Programming

In this section we show how to use ASP for analysing access control security properties through Π_{XACML}. In most cases, ASP solver can solve combinatorial problems efficiently. There are several combinatorial problems in analysis access control policies, e.g., gap-free property and conflict-free property [14,5]. In this section we look at gap-free analysis since in XACML 3.0 conflicts never occur.[5] We also present a mechanism for the verification of security properties against a set of access control policies.

5.1 Query Generator

In order to analyse access control property, sometimes we need to analyse all possible queries that might occur. We use *cardinality constraint* (see [15,16]) to generate all possible values restored in the database for each attribute. For example, we have the following generator:

$$
\begin{array}{lll}
\mathcal{P}_{generator}: & & \\
(1) & 1\{subject(X) : subject\_db(X)\}1 & \leftarrow \top. \\
(2) & 1\{action(X) : action\_db(X)\}1 & \leftarrow \top. \\
(3) & 1\{resource(X) : resource\_db(X)\}1 & \leftarrow \top. \\
(4) & 1\{environment(X) : environment\_db(X)\}1 & \leftarrow \top.
\end{array}
$$

The first line of the encoding means that we only consider one and only one *subject* attribute value obtained from the subject database. The rest of the encoding means the same as the *subject* attribute.

5.2 Gap-Free Analysis

A set of policies is *gap-free* if there is no access request for which there is an absence of decision. XACML defines that there is one PolicySet as the root of a set of policies. Hence, we say that there is a gap whenever we can find a request that makes the semantics of the $\mathcal{PS}_{root}$ is assigned to na. We force ASP solver to find the gap by the following encoding.

$$
\begin{array}{ll}
\Pi_{gap}: & \\
gap & \leftarrow \mathsf{val}(\mathcal{PS}_{root}, \mathsf{na}). \\
\bot & \leftarrow \textbf{not } gap.
\end{array}
$$

In order to make sure that a set of policies is gap-free we should generate all possible requests and test whether at least one request is not captured by the set of policies. Thus, the answer sets of program $\mathcal{P} = \Pi_{\text{XACML}} \cup \Pi_{generator} \cup \Pi_{gap}$ are witnesses that the set of policies encoded in Π_{XACML} is incomplete. When there is no model that satisfies the program then we are sure that the set of policies captures all of possible cases.

5.3 Property Analysis

The problem of verifying a security property Φ on XACML policies is not only to show that the property Φ holds on Π_{XACML} but also that we want to see the

[5] A conflict decision never occurs when we strictly use the standard combining algorithms defined in XACML 3.0, since every combining algorithm always returns one value.

witnesses whenever the property Φ does not hold in order to help the policy developer refine the policies. Thus, we can see this problem as finding models for $\Pi_{\mathrm{XACML}} \cup \Pi_{generator} \cup \Pi_{\neg\Phi}$. The founded model is the witness that the XACML policies cannot satisfy the property Φ.

Example 2. Suppose we have a security property:

Φ: An anonymous person **cannot** read any patient records.

Thus, the negation of property Φ is as follows

$\neg\Phi$: An anonymous person **can** read any patient records.

We define that anonymous persons are those who are neither patients, nor guardians, nor doctors, nor nurses. We encode $\mathcal{P}_{\neg\Phi}$ as follows

(1) $anonymous \leftarrow \textbf{not}\ subject(patient), \textbf{not}\ subject(guardian), \textbf{not}\ subject(doctor), \textbf{not}\ subject(nurse).$
(2) $\bot \leftarrow \textbf{not}\ anonymous.$
(3) $action(read) \leftarrow \top.$
(4) $resource(patient\_record) \leftarrow \top.$
(5) $\bot \leftarrow \textbf{not}\ \mathsf{val}(PS_{root}, \mathsf{p}).$

We list all of the requirements (lines 1 – 4). We force the program to find an anonymous person (line 2). Later we force that the returned decision should be to permit (line 5). When the program $\Pi_{\mathrm{XACML}} \cup \Pi_{generator} \cup \Pi_{\neg\Phi}$ returns models, we conclude that the property Φ does not hold and the returned models are the flaws in the policies. On the other hand, we conclude that the property Φ is satisfied if no model is found.

6 Related Work

There are some approaches to defining AC policies in LPs, for instance, Barker *et al.* use constraint logic program to define role-based access control in [4], while Jajodia *et al.* adopt the FAM / CAM language [7] – a logical language that uses a fixed set of predicates. However, their approaches are based on their own access control policy language whereas our approach is to define a well-known access control policy language, XACML.

Our approach is inspired by the work of Ahn *et al.* [1,2]. There are three main differences between our approach and the work of Ahn *et al.*

First, while they consider XACML version 2.0 [8], we address the newer version, XACML 3.0. The main difference between XACML 3.0 and XACML 2.0 is the treatment of indeterminate values. As a consequence, the combining algorithms in XACML 3.0 are more complex than the ones in XACML 2.0. XACML 2.0 only has a single indeterminate value while XACML 3.0 distinguishes between the following three types of indeterminate values:

i. *Indeterminate permit* ($\mathsf{i_p}$) – an indeterminate value arising from a policy which could have been evaluated to permit but not deny;

ii. *Indeterminate deny* ($\mathsf{i_d}$) – an indeterminate value arising from a policy which could have been evaluated to deny but not permit;
iii. *Indeterminate deny permit* ($\mathsf{i_{dp}}$) – an indeterminate value arising from a policy which could have been evaluated as both deny and permit.

Second, Ahn *et al.* produce a monolithic logic program that can be used for the analysis of XACML policies while we take a more modular approach by first modelling an XACML Policy Decision Point as a logic program and then using this encoding within a larger program for property analysis. While Ahn, *et al.* only emphasize the "indeterminate" value in the combining algorithms, we deal with the "indeterminate" value in all XACML components, i.e., in Match, AnyOf, AllOf, Target, Condition, Rule, Policy and PolicySet components.

Finally, Ahn *et al.* translate the XACML specification directly into logic programming, so the ambiguities in the natural language specification of XACML are also reflected in their encodings. To avoid this, we base our encodings on our formalisation of XACML from [9].

7 Conclusion and Future Work

We have modelled the XACML Policy Decision Point in a declarative way using the ASP technique by transforming XACML 3.0 elements into logic programs. Our transformation of XACML 3.0 elements is directly based on XACML 3.0 semantics [11] and we have shown that the answer set of each program transformation is unique and that it agrees with the semantics of XACML 3.0. Moreover, we can help policy developers analyse their access control policies such as checking policies' completeness and verifying policy properties by inspecting the answer set of $\Pi_{\text{XACML}} \cup \Pi_{generator} \cup \Pi_{configuration}$ – the program obtained by transforming XACML 3.0 elements into logic programs joined with a query generator program and a configuration program.

For future work, we can extend our work to handle role-based access control in XACML 3.0 [13] and to handle delegation in XACML 3.0 [12]. Also, we can extend our work for checking reachability of policies. A policy is reachable if we can find a request such that this policy is applicable. Thus, by removing unreachable policies we will not change the behaviour of the whole set of policies.

References

1. Ahn, G.-J., Hu, H., Lee, J., Meng, Y.: Reasoning about XACML policy descriptions in answer set programming (preliminary report). In: NMR 2010 (2010)
2. Ahn, G.-J., Hu, H., Lee, J., Meng, Y.: Representing and reasoning about web access control policies. In: COMPSAC. IEEE Computer Society (2010)
3. Baral, C.: Knowledge Representation, Reasoning and Declarative Problem Solving. Cambridge University Press (2003)
4. Barker, S., Stuckey, P.J.: Flexible access control policy specification with constraint logic programming. TISSEC 6 (2003)

5. Bruns, G., Huth, M.: Access-control via Belnap logic: Effective and efficient composition and analysis. In: 21st IEEE Computer Security Foundations Symposium (2008)
6. Gelfond, M.: Handbook of knowledge representation. In: Porter, B., van Harmelen, F., Lifschitz, V. (eds.) Foundations of Artificial Intelligence, vol. 3, ch. Answer Sets, pp. 285–316. Elsevier (2007)
7. Jajodia, S., Samarati, P., Subrahmanian, V.S., Bertino, E.: A unified framework for enforcing multiple access control policies. In: Proceedings of ACM SIGMOD International Conference on Management of Data (1997)
8. Moses, T.: eXtensible Access Control Markup Language (XACML) version 2.0. Technical report, OASIS (August 2010), `http://docs.oasis-open.org/xacml/2.0/access_control-xacml-2.0-core-spec-os.pdf`
9. Kencana Ramli, C.D.P., Nielson, H.R., Nielson, F.: The logic of XACML. In: Arbab, F., Ölveczky, P.C. (eds.) FACS 2011. LNCS, vol. 7253, pp. 205–222. Springer, Heidelberg (2012)
10. Ramli, C.D.P.K., Nielson, H.R., Nielson, F.: XACML 3.0 in answer set programming – extended version. Technical report, arXiv.org. (February 2013)
11. Rissanen, E.: eXtensible Access Control Markup Language (XACML) version 3.0 (committe specification 01). Technical report, OASIS (August 2010), `http://docs.oasis-open.org/xacml/3.0/xacml-3.0-core-spec-cs-01-en.pdf`
12. Rissanen, E.: XACML v3.0 administration and delegation profile version 1.0 (committe specification 01). Technical report, OASIS (August 2010), `http://docs.oasis-open.org/xacml/3.0/xacml-3.0-administration-v1-spec-cs-01-en.pdf`
13. Rissanen, E.: XACML v3.0 core and hierarchical role based access control (rbac) profile version 1.0 (committe specification 01). Technical report, OASIS (August 2010), `http://docs.oasis-open.org/xacml/3.0/xacml-3.0-rbac-v1-spec-cs-01-en.pdf`
14. Samarati, P., de Capitani di Vimercati, S.: Access control: Policies, models, and mechanisms. In: Focardi, R., Gorrieri, R. (eds.) FOSAD 2000. LNCS, vol. 2171, pp. 137–196. Springer, Heidelberg (2001)
15. Simons, P., Niemelá, I., Soininen, T.: Extending and implementing the stable model semantics. Artificial Intelligence 138, 181–234 (2002)
16. Syrjänen, T.: Lparse 1.0 User's Manual

Types vs. PDGs in Information Flow Analysis

Heiko Mantel and Henning Sudbrock

Computer Science Department, TU Darmstadt, Germany
{mantel,sudbrock}@mais.informatik.tu-darmstadt.de

Abstract. Type-based and PDG-based information flow analysis techniques are currently developed independently in a competing manner, with different strengths regarding coverage of language features and security policies. In this article, we study the relationship between these two approaches. One key insight is that a type-based information flow analysis need not be less precise than a PDG-based analysis. For proving this result we establish a formal connection between the two approaches which can also be used to transfer concepts from one tradition of information flow analysis to the other. The adoption of rely-guarantee-style reasoning from security type systems, for instance, enabled us to develop a PDG-based information flow analysis for multi-threaded programs.

Keywords: Information flow security, Security type system, Program dependency graph.

1 Introduction

When giving a program access to confidential data one wants to be sure that the program does not leak any secrets to untrusted sinks, like, e.g., to untrusted servers on the Internet. Such confidentiality requirements can be characterized by information flow properties. For verifying that a program satisfies an information flow property, a variety of program analysis techniques can be employed.

The probably most popular approach to information flow analysis is the use of security type systems. Starting with [24], type-based information flow analyses were developed for programs with various language features comprising procedures (e.g., [25]), concurrency (e.g., [23]), and objects (e.g., [16]). Security type systems were proposed for certifying a variety of information flow properties, including timing-sensitive and timing-insensitive properties (e.g., [22] and [2]) and properties supporting declassification (e.g., [14]).

Besides type systems, one can also employ other program analysis techniques for certifying information flow security. For instance, it was proposed in [10] to use program dependency graphs (PDGs) for information flow analysis. A PDG [4] is a graph-based program representation that captures dependencies caused by the data flow and the control flow of a program. PDG-based information flow analyses recently received new attention, resulting in, e.g., a PDG-based information flow analysis for object-oriented programs and a PDG-based information flow analysis supporting declassification [8,7].

E. Albert (Ed.): LOPSTR 2012, LNCS 7844, pp. 106–121, 2013.
© Springer-Verlag Berlin Heidelberg 2013

Type-based and PDG-based information flow analyses are currently developed independently. The two sub-communities both see potential in their approach, but the pros and cons of the two techniques have not been compared in detail. In this article, we compare type-based and PDG-based information flow analyses with respect to their precision. Outside the realm of information flow security there already exist results that compare the precision of data-flow oriented and type-based analyses, for instance, for safety properties [18,17]. Here, we clarify the relation between type-based and PDG-based analyses in the context of information flow security. We investigate whether (a) one approach has superior precision, (b) both have pros and cons, or (c) both are equally precise. To be able to establish a precise relation, we consider two prominent analyses that are both fully formalized for a simple while language, namely the type-based analysis from Hunt and Sands [11] and the PDG-based analysis from Wasserrab, Lohner, and Snelting [27].

Our main result is that the two analyses have exactly the same precision. This result was surprising for us, because one motivation for using PDGs in an information flow analysis was their precision [8]. We derive our main result based on a formal connection between the two kinds of security analyses, which we introduce in this article. It turned out that this connection is also interesting in its own right, because it can be used for transferring ideas from type-based to PDG-based information flow analyses and vice versa. In this article, we illustrate this possibility in one direction, showing how to derive a novel PDG-based information flow analysis that is suitable for multi-threaded programs by exploiting our recently proposed solution for rely-guarantee-style reasoning in a type-based security analysis [15]. The resulting analysis is compositional and, thereby, enables a modular security analysis. This is an improvement over the analysis from [5], the only provably sound PDG-based information flow analysis for multi-threaded programs developed so far. Moreover, in contrast to [5] our novel analysis supports programs with nondeterministic public output.

In summary, the main contributions of this article are

1. the formal comparison of the precision of a type-based and a PDG-based information flow analysis, showing that they have the same precision;
2. the demonstration that our formal connection between the type-based and the PDG-based analysis can be used to transfer concepts from one approach to the other (by transferring rely-guarantee-style reasoning as mentioned above); and
3. a provably sound PDG-based information flow analysis for multi-threaded programs that is compositional with respect to the parallel composition of threads and compatible with nondeterministic public output.

We believe that the connection between type- and PDG-based information flow analysis can serve as a basis for further mutual improvements of the analysis techniques. Such a transfer is desirable because there are other relevant aspects than an analysis' precision like, e.g., efficiency and availability of tools. Moreover, we hope that the connection between the two approaches to information flow analysis fosters mutual understanding and interaction between the two communities.

2 Type-Based Information Flow Analyses

If one grants a program access to secret information, one wants to be sure that the program does not leak secrets to untrusted sinks like, e.g., untrusted servers in a network. A secure program should not only refrain from directly copying secrets to untrusted sinks (as, e.g., with an assignment "*sink*:=*secret*"), but also should not reveal secrets indirectly (as, e.g., by executing "if (*secret* > 0) then *sink*:=1").

It is popular to formalize information flow security by the property that values written to public sinks do not depend on secrets, the probably best known such property being *Noninterference* [6,13]. In the following, we define information flow security for programs by such a property, and we present a security type system for certifying programs with respect to this property.

2.1 Execution Model and Security Property

We consider a set of *commands Com* that is defined by the grammar

$$c ::= \mathsf{skip} \mid x{:=}e \mid c; c \mid \mathsf{if}\ (e)\ \mathsf{then}\ c\ \mathsf{else}\ c\ \mathsf{fi} \mid \mathsf{while}\ (e)\ \mathsf{do}\ c\ \mathsf{od},$$

where $x \in Var$ is a variable and $e \in Exp$ is an expression. *Expressions* are terms built from variables and from operators that we do not specify further. The set of *free variables* in expression $e \in Exp$ is denoted with $fv(e)$. A *memory* is a function $mem : Var \to Val$ that models a snapshot of a program's memory, where Val is a set of *values* and $mem(x)$ is the *value of* x. Judgments of the form $\langle c, mem\rangle \Downarrow mem'$ model program execution, with the interpretation that command c, if executed with initial memory mem, terminates with memory mem'. The rules for deriving the judgments are as usual for big-step semantics.[1]

To define the security property, we consider a *security lattice* $\mathcal{D} = \{l, h\}$ with two *security domains* where $l \sqsubseteq h$ and $h \not\sqsubseteq l$. This models the requirement that no information flows from domain h to domain l. This is the simplest policy capturing information flow security.[2] A *domain assignment* is a function $dom : Var \to \mathcal{D}$ that associates a security domain with each program variable. We say that variables in the set $L = \{x \in Var \mid dom(x) = l\}$ are public or *low*, and that variables in the set $H = \{x \in Var \mid dom(x) = h\}$ are secret or *high*. The resulting security requirement is that the final values of low variables do not depend on the initial values of high variables. This requirement captures security with respect to an attacker who sees the initial and final values of low variables, but cannot access values of high variables (i.e., access control works correctly).

Definition 1. *Two memories mem and mem′ are* low-equal *(written $mem =_L mem'$) if and only if $mem(x) = mem'(x)$ for all $x \in L$.*

A command c is noninterferent *if whenever $mem_1 =_L mem_2$ and $\langle c, mem_1\rangle \Downarrow mem'_1$ and $\langle c, mem_2\rangle \Downarrow mem'_2$ are derivable then $mem'_1 =_L mem'_2$.*

[1] The rules and detailed proofs of theorems in this article are available on the authors' website (`http://www.mais.informatik.tu-darmstadt.de/Publications`).

[2] The results in this article can be lifted to other security lattices.

$$[\mathsf{exp}]\,\frac{}{\Gamma \vdash e : \bigsqcup_{x \in fv(e)} \Gamma(x)} \qquad [\mathsf{if}]\,\frac{\Gamma \vdash e : t \quad pc \sqcup t \vdash \Gamma\,\{c_1\}\,\Gamma_1' \quad pc \sqcup t \vdash \Gamma\,\{c_2\}\,\Gamma_2'}{pc \vdash \Gamma\,\{\mathsf{if}\ (e)\ \mathsf{then}\ c_1\ \mathsf{else}\ c_2\ \mathsf{fi}\}\,\Gamma_1' \sqcup \Gamma_2'}$$

$$[\mathsf{assign}]\,\frac{\Gamma \vdash e : t}{pc \vdash \Gamma\,\{x{:=}e\}\,\Gamma[x \mapsto pc \sqcup t]} \qquad [\mathsf{seq}]\,\frac{pc \vdash \Gamma\,\{c_1\}\,\Gamma' \quad pc \vdash \Gamma'\,\{c_2\}\,\Gamma''}{pc \vdash \Gamma\,\{c_1; c_2\}\,\Gamma''}$$

$$[\mathsf{skip}]\,\frac{}{pc \vdash \Gamma\,\{\mathsf{skip}\}\,\Gamma} \qquad [\mathsf{while}]\,\frac{\begin{array}{c}\Gamma_i' \vdash e : t_i \quad pc \sqcup t_i \vdash \Gamma_i'\,\{c\}\,\Gamma_i'' \quad 0 \le i \le k \\ \Gamma_0' = \Gamma \quad \Gamma_{i+1}' = \Gamma_i'' \sqcup \Gamma \quad \Gamma_{k+1}' = \Gamma_k'\end{array}}{pc \vdash \Gamma\,\{\mathsf{while}\ (e)\ \mathsf{do}\ c\ \mathsf{od}\}\,\Gamma_k'}$$

Fig. 1. Type system from [11]

Example 1. Consider command c to the right and assume that $dom(x) = dom(y) = l$ and $dom(z) = h$. Consider furthermore low-equal memories mem_1 and mem_2 with $mem_1(x) = mem_2(x) = mem_1(y) = mem_2(y) = 1$, $mem_1(z) = -1$, and $mem_2(z) = 1$.

```
if (z < 0) then
  while (y > 0) do
    y:=y + z od else
  skip fi;
x:=y
```

Then $\langle c, mem_1 \rangle \Downarrow mem_1'$ and $\langle c, mem_2 \rangle \Downarrow mem_2'$ are derivable with $mem_1'(x) = 0$ and $mem_2'(x) = 1$. Since $mem_1' \neq_L mem_2'$ command c is not noninterferent.

2.2 The Type-Based Information Flow Analysis by Hunt and Sands

A type-based analysis uses a collection of typing rules to inductively define a subset of programs. The intention is that every program in the subset satisfies an information flow property like, e.g., the one from Definition 1. Starting with [24], many type-based information flow analyses were developed (see [21] for an overview). Here, we recall the type system from Hunt and Sands [11] which is, unlike many other type-based security analyses, flow-sensitive (i.e., it takes the order of program statements into account to improve precision).

In [11], typing judgments have the form $pc \vdash \Gamma\,\{c\}\,\Gamma'$, where c is a command, $\Gamma, \Gamma' : Var \to \mathcal{D}$ are *environments*, and pc is a security domain. The interpretation of the judgment is as follows: For each variable $y \in Var$, $\Gamma'(y)$ is a valid upper bound on the security level of the value of y after command c has been run if (a) for each $x \in Var$, $\Gamma(x)$ is an upper bound on the security level of the value of x before c has been run and (b) pc is an upper bound on the security level of all information on which it might depend whether c is run.

The typing rules from [11] are displayed in Figure 1, where $\sqcup$ denotes the least upper bound operator on $\mathcal{D}$, which is extended to environments by $(\Gamma \sqcup \Gamma')(x) := \Gamma(x) \sqcup \Gamma'(x)$. The typing rules ensure that for any given c, Γ, and pc there is an environment Γ' such that $pc \vdash \Gamma\,\{c\}\,\Gamma'$ is derivable. Moreover, this environment is uniquely determined by c, Γ, and pc [11, Theorem 4.1].

Definition 2. *Let $c \in Com$, $\Gamma(x) = dom(x)$ for all $x \in Var$, and Γ' be the unique environment such that $l \vdash \Gamma\,\{c\}\,\Gamma'$ is derivable. Command c is* accepted by the type-based analysis *if $\Gamma'(x) \sqsubseteq dom(x)$ for all $x \in Var$.*

Example 2. Consider command c and domain assignment dom from Example 1. Let $\Gamma(x) = dom(x)$ for all $x \in Var$. Then the judgment $l \vdash \Gamma\,\{c\}\,\Gamma'$ is derivable if and only if $\Gamma'(x) = \Gamma'(y) = \Gamma'(z) = h$ and $\Gamma'(x') = \Gamma(x')$ for all other $x' \in Var$. Since $\Gamma'(x) \not\sqsubseteq dom(x)$ command c is not accepted by the type-based analysis.

Theorem 1. *Commands accepted by the type-based analysis are noninterferent.*

The theorem follows from Theorem 3.3 in [11].

3 PDG-Based Information Flow Analyses

PDG-based information flow analyses, firstly proposed in [10], exploit that the absence of certain paths in a *program dependency graph* (PDG) [4] is a sufficient condition for the information flow security of a program. In this section, we recall the PDG-based analysis from [27] which is sound with respect to the property from Definition 1. In order to make the article self-contained, we recall the construction of control flow graphs (CFGs) and PDGs in Sections 3.1 and 3.2.

3.1 Control Flow Graphs

Definition 3. *A* directed graph *is a pair* (N, E) *where* N *is a set of nodes and* $E \subseteq N \times N$ *is a set of edges. A* path p from node n_1 to node n_k *is a non-empty sequence of nodes* $\langle n_1, \ldots, n_k \rangle \in N^+$ *where* $(n_i, n_{i+1}) \in E$ *for all* $i \in \{1, \ldots, k-1\}$. *We call a path* trivial *if it is of the form* $\langle n \rangle$ *(i.e., a sequence of length* 1*), and* non-trivial *otherwise. Moreover, we say that node* n *is* on the path $\langle n_1, \ldots, n_k \rangle$ *if* $n = n_i$ *for some* $i \in \{1, \ldots, k\}$.

Definition 4. *A* control flow graph with def and use sets *is a tuple* (N, E, def, use) *where* (N, E) *is a directed graph,* N *contains two distinguished nodes* $start$ *and* $stop$, *and* $def, use : N \to \mathcal{P}(Var)$ *are functions returning the def and use set, respectively, for a node. (The set* $\mathcal{P}(Var)$ *denotes the powerset of the set* Var.)

Nodes *start* and *stop* represent program start and termination, respectively, and the remaining nodes represent program statements and control conditions. An edge $(n, n') \in E$ models that n' might immediately follow n in a program run. Finally, the sets $def(n)$ and $use(n)$ contain all variables that are defined and used, respectively, at a node n. In the remainder of this article we simply write "CFG" instead of "CFG with def and use sets."

We recall the construction of the CFG for a command following [26], where statements and control conditions are represented by numbered nodes.

Definition 5. *We denote with* $|c|$ *the number of statements and control conditions of* $c \in Com$, *and define* $|c|$ *recursively by* $|\mathsf{skip}| = 1$, $|x{:=}e| = 1$, $|c_1; c_2| = |c_1| + |c_2|$, $|\mathsf{if}\ (e)\ \mathsf{then}\ c_1\ \mathsf{else}\ c_2\ \mathsf{fi}| = 1 + |c_1| + |c_2|$, *and* $|\mathsf{while}\ (e)\ \mathsf{do}\ c\ \mathsf{od}| = 1 + |c|$.

Definition 6. *For $c \in Com$ and $1 \leq i \leq |c|$ we denote with $c[i]$ the ith statement or control condition in c, which we define recursively as follows: If $c = \mathsf{skip}$ or $c = x{:=}e$ then $c[1] = c$. If $c = c_1; c_2$ then $c[i] = c_1[i]$ for $1 \leq i \leq |c_1|$ and $c[i] = c_2[i - |c_1|]$ for $|c_1| < i \leq |c|$. If $c = \mathsf{if}\ (e)\ \mathsf{then}\ c_1\ \mathsf{else}\ c_2\ \mathsf{fi}$ then $c[1] = e$, $c[i] = c_1[i-1]$ for $1 < i \leq 1 + |c_1|$, and $c[i] = c_2[i-1-|c_1|]$ for $1 + |c_1| < i \leq |c|$. If $c = \mathsf{while}\ (e)\ \mathsf{do}\ c_1\ \mathsf{od}$ then $c[1] = e$ and $c[i] = c_1[i-1]$ for $1 < i \leq |c|$.*

Note that the ith statement or control condition, i.e., $c[i]$, is either an expression, an assignment, or a skip-statement.

Definition 7. *For $c \in Com$, $N_c = \{1, \ldots, |c|\} \cup \{start, stop\}$.*

We define an operator $\ominus : (\mathbb{N} \cup \{start, stop\}) \times \mathbb{N} \to \mathbb{Z} \cup \{start, stop\}$ by $n \ominus z = n - z$ if $n \in \mathbb{N}$ and $n \ominus z = n$ if $n \in \{start, stop\}$.

Definition 8. *For $c \in Com$ the set $E_c \subseteq N_c \times N_c$ is defined recursively by:*
- $E_{\mathsf{skip}} = E_{x:=e} = \{(start, 1), (1, stop), (start, stop)\}$,
- $E_{\mathsf{if}\ (e)\ \mathsf{then}\ c_1\ \mathsf{else}\ c_2\ \mathsf{fi}} = \{(start, 1), (start, stop)\} \cup$
  $\{(1, n') \mid (start, n' \ominus 1) \in E_{c_1} \wedge n' \neq stop\} \cup$
  $\{(1, n') \mid (start, n' \ominus (1+|c_1|)) \in E_{c_2} \wedge n' \neq stop\} \cup$
  $\{(n, n') \mid (n \ominus 1, n' \ominus 1) \in E_{c_1} \wedge n \neq start\} \cup$
  $\{(n, n') \mid (n \ominus (1+|c_1|), n' \ominus (1+|c_1|)) \in E_{c_2} \wedge n \neq start\}$,
- $E_{c_1;c_2} = \{(start, stop)\} \cup$
  $\{(n, n') \mid (n, n') \in E_{c_1} \wedge n' \neq stop\} \cup$
  $\{(n, n') \mid (n \ominus |c_1|, n' \ominus |c_1|) \in E_{c_2} \wedge n \neq start\} \cup$
  $\{(n, n') \mid (n, stop) \in E_{c_1} \wedge (start, n' \ominus |c_1|) \in E_{c_2} \wedge n \neq start \wedge n' \neq stop\}$, and
- $E_{\mathsf{while}\ (e)\ \mathsf{do}\ c\ \mathsf{od}} = \{(start, 1), (start, stop)\} \cup$
  $\{(1, n') \mid (start, n' \ominus 1) \in E_c\} \cup$
  $\{(n', 1) \mid (n' \ominus 1, stop) \in E_c\} \cup$
  $\{(n, n') \mid (n \ominus 1, n' \ominus 1) \in E_c \wedge n \neq start \wedge n' \neq stop\}$.

Definition 9. *For $c \in Com$ we define $def_c : N_c \to \mathcal{P}(Var)$ by $def_c(n) = \{x\}$ if $n \in \{1, \ldots, |c|\}$ and $c[n] = x{:=}e$, and by $def_c(n) = \{\}$ otherwise. Moreover, we define $use_c : N_c \to \mathcal{P}(Var)$ by $use_c(n) = fv(e)$ if $n \in \{1, \ldots, |c|\}$ and $c[n] = x{:=}e$ or $c[n] = e$, and by $use_c(n) = \{\}$ otherwise.*

Definition 10. *The* control flow graph of c *is* $CFG_c = (N_c, E_c, def_c, use_c)$.

Note that, by definition, an edge from *start* to *stop* is contained in E_c. This edge models the possibility that c is not executed.

We now augment CFGs with def and use sets by two nodes *in* and *out* to capture the program's interaction with its environment. Two sets of variables $I, O \subseteq Var$, respectively, specify which variables may be initialized by the environment before program execution and which variables may be read by the environment after program execution. This results in the following variant of CFGs:

Definition 11. *Let $CFG = (N, E, def, use)$ and $I, O \subseteq Var$. Then $CFG^{I,O} = (N', E', def', use')$ where $N' = N \cup \{in, out\}$, $E' = \{(start, stop), (start, in),$*

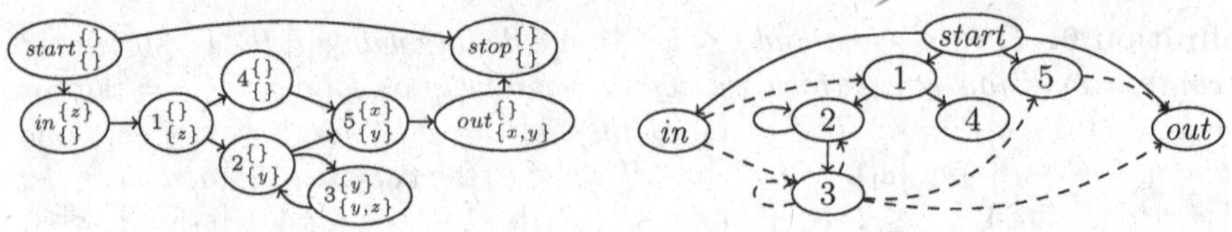

Fig. 2. The CFG and the PDG for the command from Example 1

$(out, stop)\}\cup\{(in, n') \mid (start, n')\in E\wedge n' \neq stop\}\cup\{(n, out) \mid (n, stop)\in E\wedge n \neq start\} \cup \{(n, n')\in E \mid n \notin \{start, stop\} \wedge n' \notin \{start, stop\}\}$, $def'(in) = I$, $use'(in) = def'(out) = \{\}$, $use'(out) = O$, *and* $def'(n) = def(n)$ *and* $use'(n) = use(n)$ *for* $n \in N$.

Definitions 10 and 11 both augment the usual notion of control flow graphs (see Definition 4). In the remainder of this article, we use the abbreviation CFG for arbitrary control flow graphs (including those that satisfy Definition 10 or 11).

We use a graphical representation for displaying CFGs where we depict nodes with ellipses and edges with solid arrows. For each node n we label the corresponding ellipse with n_Y^X where $X = def(n)$ and $Y = use(n)$.

Example 3. Command c in Example 1 contains three statements and two control conditions (i.e., $|c|=5$). Hence, $N_c = \{1,\ldots,5, start, stop\}$. Nodes 1–5 represent the statements and control conditions in Lines 1–5 of the program, respectively. The control flow graph $CFG_c^{\{z\},\{x,y\}}$ is displayed at the left hand side of Figure 2.

3.2 The PDG-Based Information Flow Analysis by Wasserrab et al

PDGs are directed graphs that represent dependencies in imperative programs [4]. PDGs were extended to programs with various languages features like procedures (e.g., [9]), concurrency (e.g., [3]), and objects (e.g., [8]). We recall the construction of PDGs from CFGs for the language from Section 2 based on the following notions of data dependency and control dependency.

Definition 12. *Let* (N, E, def, use) *be a CFG and* $n, n'\in N$. *If* $x\in def(n)$ *we say that* the definition of x at n reaches n' *if there is a non-trivial path* p *from* n *to* n' *such that* $x \notin def(n'')$ *for every node* n'' *on* p *with* $n''\neq n$ *and* $n''\neq n'$.

Node n' *is* data dependent *on node* n *if there exists* $x \in Var$ *such that* $x \in def(n)$, $x \in use(n')$, *and the definition of* x *at* n *reaches* n'.

Intuitively, a node n' is data dependent on a node n if n' uses a variable that has not been overwritten since being defined at n.

Example 4. Consider the CFG on the left hand side of Figure 2. The definition of y at Node 3 reaches Node 5 because $\langle 3, 2, 5\rangle$ is a non-trivial path and $y \notin def(2)$. Hence, Node 5 is data dependent on Node 3 because $y \in def(3)$, $y \notin def(2)$, and $y \in use(5)$. Note that Node 2 is also data dependent on Node 3, and that Node 3 is data dependent on itself.

Definition 13. *Let (N, E, def, use) be a CFG. Node n'* postdominates *node n if $n \neq n'$ and every path from n to stop contains n'.*

Node n' is control dependent *on node n if there is a non-trivial path p from n to n' such that n' postdominates all nodes $n'' \notin \{n, n'\}$ on p and n' does not postdominate n.*

Intuitively, a node n' is control dependent on a node n if n represents the innermost control condition that guards the execution of n'.

Example 5. Consider again the CFG in Figure 2. Node 5 postdominates Node 1 because Node 5 is on all paths from Node 1 to Node *stop*. Hence, Node 5 is not control dependent on Node 1. Nodes 2, 3, and 4 do not postdominate Node 1. Node 3 is not control dependent on Node 1 because Node 3 does not postdominate Node 2 and all paths from Node 1 to Node 3 contain Node 2. However, Node 3 is control dependent on Node 2. Moreover, Nodes 2 and 4 are control dependent on Node 1 because $\langle 1, 2\rangle$ and $\langle 1, 4\rangle$ are non-trivial paths in the CFG.

Definition 14. *Let $CFG = (N, E, def, use)$ be a control flow graph. The directed graph (N', E') is the* PDG *of CFG (denoted with $PDG(CFG)$) if $N' = N$ and $(n, n') \in E'$ if and only if n' is data dependent or control dependent on n in CFG.*

We use the usual graphical representation for displaying PDGs, depicting Node n with an ellipse labeled with n, edges that reflect control dependency with solid arrows, and edges that reflect data dependency with dashed arrows. Moreover, we do not display nodes that have neither in- nor outgoing edges.

Example 6. The PDG of the CFG at the left of Figure 2 is displayed right of the CFG. Node *stop* is not displayed because it has neither in- nor outgoing edges.

The PDG-based information flow analysis from Wasserrab et al [27] for a command c is based on the PDG of $CFG_c^{H,L}$ (cf. Definition 11).

Definition 15. *The command $c \in Com$ is* accepted by the PDG-based analysis *if and only if there is no path from in to out in $PDG(CFG_c^{H,L})$.*

Example 7. For command c and domain assignment *dom* from Example 1 the graph $PDG(CFG_c^{H,L})$ is displayed at the right of Figure 2. It contains a path from Node *in* to Node *out*, (e.g., the path $\langle in, 3, out\rangle$). In consequence, c is not accepted by the PDG-based analysis.

Theorem 2. *Commands accepted by the PDG-based analysis are noninterferent.*

The theorem follows from [27, Theorem 8].

4 Comparing the Type- and the PDG-Based Analysis

While both the type-based analysis from Section 2 and the PDG-based analysis from Section 3 are sound, both analyses are also incomplete. I.e., for both analyses there are programs that are noninterferent (according to Definition 1), but that are not accepted by the analysis. A complete analysis is impossible, because the noninterference property is undecidable (this can be proved in a standard way by showing that the decidability of the property would imply the decidability of the halting problem for the language under consideration [21]). This raises the question if one of the two analyses is more precise than the other. In this section, we answer this question. As an intermediate step, we establish a relation between the two analyses:

Lemma 1. *Let $c \in Com$, $y \in Var$, and Γ be an environment. Let Γ' be the unique environment such that $l \vdash \Gamma\,\{c\}\,\Gamma'$ is derivable in the type system from Section 2. Moreover, let X be the set of all $x \in Var$ such that there exists a path from in to out in $PDG(CFG_c^{\{x\},\{y\}})$. Then $\Gamma'(y) = \bigsqcup_{x \in X} \Gamma(x)$ holds.*

Proof sketch. We argue that the following more general statement holds: If the judgment $pc \vdash \Gamma\,\{c\}\,\Gamma'$ is derivable, then the equality $\Gamma'(y) = \bigsqcup_{x \in X} \Gamma(x)$ holds if there is no path from *start* to *out* in $PDG(CFG_c^{\{\},\{y\}})$ that contains a node $n \notin \{start, out\}$, and the equality $\Gamma'(y) = pc \sqcup \left(\bigsqcup_{x \in X} \Gamma(x)\right)$ holds if there is such a path. Intuitively, the absence of a path from *start* to *out* with more than two nodes guarantees that y is not changed during any execution of c, while y might be changed during an execution of c if a path from *start* to *out* with more than two nodes exists. Hence, the security domain pc (determined by the type-based analysis as an upper bound on the security level of all information on which it depends whether c is executed) is included in the formula for $\Gamma'(y)$ only if such a path exists in the PDG. Formally, the more general statement is proven by induction on the structure of the command c. A detailed proof is available on the authors' website. Lemma 1 follows from this more general statement by instantiating pc with the security level l. □

Lemma 1 is the key to establishing the following theorem that relates the precision of the type-based analysis to the precision of the PDG-based analysis, showing that the analyses have exactly the same precision.

Theorem 3. *A command $c \in Com$ is accepted by the type-based analysis if and only if it is accepted by the PDG-based analysis.*

Proof. Our proof is by contraposition. Let $\Gamma(x) = dom(x)$ for all $x \in Var$, and let Γ' be the unique environment such that $l \vdash \Gamma\,\{c\}\,\Gamma'$ is derivable in the type system from Section 2. If c is not accepted by the type-based analysis then $dom(y) = l$ and $\Gamma'(y) = h$ for some $y \in Var$. Hence, by Lemma 1 there exists $x \in Var$ with $dom(x) = h$ and a path $\langle in, \ldots, out\rangle$ in $PDG(CFG_c^{\{x\},\{y\}})$. Hence, there is a path $\langle in, \ldots, out\rangle$ in $PDG(CFG_c^{H,L})$. Thus, c is not accepted by the PDG-based analysis. If c is not accepted by the PDG-based analysis then there is a path $\langle in, \ldots, out\rangle$ in $PDG(CFG_c^{H,L})$. But then there exist variables x, y with $dom(x) = h$ and $dom(y) = l$ such that there is a path $\langle in, \ldots, out\rangle$ in $PDG(CFG_c^{\{x\},\{y\}})$. Hence, by Lemma 1, $\Gamma'(y) = h$. Since $dom(y) = l$ it follows that $\Gamma'(y) \not\sqsubseteq dom(y)$. Thus, c is not accepted by the type-based analysis. □

Theorem 3 shows that the information flow analyses from [11] and [27] have exactly the same precision. More generally this means that, despite their conceptual simplicity, type-based information flow analyses need not be less precise than PDG-based information flow analyses.

Given that both analyses have equal precision, the choice of an information flow analysis should be motivated by other aspects. For instance, if a program's environment is subject to modifications, one might desire a compositional analysis, and, hence, choose a type-based analysis. On the other hand, if a program is not accepted by the analyses one could use the PDG-based analysis to localize the source of potential information leakage by inspecting the path in the PDG that leads to the rejection of the program.

Beyond clarifying the connection between type-based and PDG-based information flow analyses, Theorem 3 also provides a bridge that can be used to transfer concepts from the one tradition of information flow analysis to the other. In the following section, we exploit this bridge to transfer the concept of rely-guarantee-style reasoning for the analysis of multi-threaded programs from type-based to PDG-based information flow analysis.

5 Information Flow Analysis of Multi-threaded Programs

Multi-threaded programs may exhibit subtle information leaks that do not occur in single-threaded programs. Such a leak is illustrated by the following example.

Example 8. Consider two threads with shared memory that execute commands $c_1 = \mathsf{if}\ (x)\ \mathsf{then\ skip; skip\ else\ skip\ fi}; y{:=}\mathsf{True}$ and $c_2 = \mathsf{skip}; \mathsf{skip}; y{:=}\mathsf{False}$, respectively, and that are run under a Round-Robin scheduler that selects them alternately starting with the first thread and rescheduling after each execution step. If initially $x = \mathsf{True}$ then c_1 assigns True to y after c_2 assigns False to y. Otherwise, c_1 assigns True to y prior to the assignment to y in c_2. I.e., the initial value of x is copied into y. Such leaks are also known as *internal timing leaks*.

Many type-based analyses detect such leaks (for instance, [23,22,28,15]). Regarding PDG-based analyses, this is only the case for a recently proposed analysis [5]. However, this analysis has serious limitations: It forbids publicly observable nondeterminism, and it is not compositional (cf. Section 6 for a more detailed comparison). This motivated us to choose this domain for illustrating how the connection between type-based and PDG-based information flow analysis (from Section 4) can be exploited to transfer ideas from the one analysis style to the other. More concretely, we show how rely-guarantee-style reasoning can be transferred from a type-based to a PDG-based information flow analysis. The outcome is a sound PDG-based information flow analysis for multi-threaded programs that is superior to the one in [5] in the sense that it supports publicly observable nondeterminism.

5.1 A Type-Based Analysis for Multi-threaded Programs

We consider multi-threaded programs executing a fixed number of threads that interact via shared memory, i.e., configurations have the form $\langle (c_1, \ldots, c_k), mem \rangle$ where the commands c_i model the threads and mem models the shared memory.

In the following, we recall the type-based analysis from [15] that exploits rely-guarantee-style reasoning, where typing rules for single threads exploit assumptions about when and how variables might be accessed by other threads. Assumptions are modeled by *modes* in the set $Mod = \{asm\text{-}noread, asm\text{-}nowrite\}$, where *asm-noread* and *asm-nowrite* are interpreted as the assumption that no other thread reads and writes a given variable, respectively.[3] The language for commands from Section 2 is extended as follows with a notation for specifying when one starts and stops making assumptions for a thread, respectively:

$$ann ::= \mathsf{acq}(m, x) \mid \mathsf{rel}(m, x) \qquad c ::= \ldots \mid /\!\!/ ann /\!\!/ \ c,$$

where $m \in Mod$ and $x \in Var$. The annotations $/\!\!/\mathsf{acq}(m, x)/\!\!/$ and $/\!\!/\mathsf{rel}(m, x)/\!\!/$, respectively, indicate that an assumption for x is acquired or released.

Typing judgments for commands have the form $\vdash \Lambda\{c\}\Lambda'$ where $\Lambda, \Lambda' : Var \rightharpoonup \mathcal{D}$ are *partial environments.* Partial environments provide an upper bound on the security level only for low variables for which a no-read and for high variables for which a no-write assumption is made, respectively. For other variables, the typing rules ensure that $dom(x)$ is an upper bound on the security level of the value of x. We write $\Lambda\langle x \rangle$ for the resulting upper bound (defined by $\Lambda\langle x \rangle = \Lambda(x)$ if Λ is defined for x and by $\Lambda\langle x \rangle = dom(x)$ otherwise). This reflects that (a) low variables that might be read by other threads must not store secrets because the secrets might be leaked in other threads (i.e., the upper bound for low variables without no-read assumption must be l), and that (b) other threads might write secrets into high variables without no-write assumption (and, hence, the upper bound for high variables without no-write assumption cannot be l).

The security type system contains two typing rules for assignments:

$$\frac{\Lambda(x) \text{ is defined} \quad \Lambda' = \Lambda[x \mapsto (\bigsqcup_{y \in fv(e)} \Lambda\langle y \rangle)]}{\vdash \Lambda\,\{x{:=}e\}\,\Lambda'} \qquad \frac{(\bigsqcup_{y \in fv(e)} \Lambda\langle y \rangle) \sqsubseteq dom(x) \quad \Lambda(x) \text{ is not defined} \quad \Lambda' = \Lambda}{\vdash \Lambda\,\{x{:=}e\}\,\Lambda'}$$

The left typing rule is like in the type system from Section 2. The rule applies if Λ is defined for the assigned variable. The right typing rule reflects that if Λ is not defined for the assigned variable x then $dom(x)$ must remain an upper bound on the security level of the value of x, and, hence, only expressions that do not contain secret information may be assigned to x if $dom(x) = l$.

A slightly simplified[4] variant of the typing rule for conditionals is as follows:

$$\frac{\vdash \Lambda\,\{c_1\}\,\Lambda' \quad \vdash \Lambda\,\{c_2\}\,\Lambda' \quad l = \bigsqcup_{x \in fv(e)} \Lambda\langle x \rangle}{\vdash \Lambda\,\{\mathsf{if}\ (e)\ \mathsf{then}\ c_1\ \mathsf{else}\ c_2\ \mathsf{fi}\}\,\Lambda'}$$

[3] We omit the modes *guar-noread* and *guar-nowrite* representing guarantees from [15], because they are irrelevant for the security type system.

[4] The original rule from [15] permits that guards depend on secrets (i.e., $h = \bigsqcup_{x \in fv(e)} \Lambda\langle x \rangle$) if the branches are in a certain sense indistinguishable.

In contrast to the corresponding typing rule in Section 2, the guard is required to be low. This ensures that programs with leaks like in Example 8 are not typable.

For the complete set of typing rules we refer to [15].

Remark 1. In contrast to the type system in Section 2 the security level *pc* is not considered here, because the typing rules ensure that control flow does not depend on secrets (permitting some exceptions as indicated in Footnote 4).

Definition 16. *A multi-threaded program consisting of commands* $c_1, \ldots, c_k$ *is* accepted by the type-based analysis for multi-threaded programs *if the judgment* $\vdash \Lambda_0 \{c_i\} \Lambda'_i$ *is derivable for each* $i \in \{1, \ldots, k\}$ *for some* Λ'_i *(where* Λ_0 *is undefined for all* $x \in Var$*) and the assumptions made are valid for the program.*

Validity of assumptions is formalized in [15] by *sound usage of modes*, a notion capturing that the assumptions made for single threads are satisfied in any execution of the multi-threaded program. Theorem 6 in [15] ensures that the type-based analysis for multi-threaded programs is sound with respect to *SIFUM-security*, an information flow security property for multi-threaded programs. We refer the interested reader to [15] for the definitions of sound usage of modes and of SIFUM-security.

5.2 A Novel PDG-Based Analysis for Multi-threaded Programs

We define a PDG-based analysis for multi-threaded programs by transferring rely-guarantee-style reasoning from the type-based analysis (Definition 16) to PDGs. To this end, we augment the set of edges of the program dependency graph $PDG(CFG_c^{H,L})$, obtaining a novel program dependency graph $PDG^{\|}(CFG_c^{H,L})$. Using this graph, the resulting analysis for multi-threaded programs is as follows:

Definition 17. *A multi-threaded program consisting of commands* $c_1, \ldots, c_k$ *is* accepted by the PDG-based analysis for multi-threaded programs *if there is no path from in to out in* $PDG^{\|}(CFG_{c_i}^{H,L})$ *for each* $i \in \{1, \ldots, k\}$ *and the assumptions made are valid for the program.*

It follows from Definition 17 that the analysis is compositional with respect to the parallel composition of threads.

We now define the graph $PDG^{\|}(CFG_c^{H,L})$, where the additional edges in $PDG^{\|}(CFG_c^{H,L})$ model dependencies for nodes *in* and *out* that result from the concurrent execution of threads that respect the assumptions made for *c*.

Definition 18. *If command c is not of the form* $/\!/ ann /\!/ c'$ *we say that c* does not acquire $m \in Mod$ for $x \in Var$ *and that c* does not release $m \in Mod$ for $x \in Var$. *Moreover, if an arbitrary command c does not release m for x then the command* $/\!/\mathsf{acq}(x,m)/\!/ c$ acquires m for x, *and if c does not acquire m for x then the command* $/\!/\mathsf{rel}(x,m)/\!/ c$ releases m for x.

For $c \in Com$ *we define the function* $modes_c : (N_c \times Mod) \to \mathcal{P}(Var)$ *by* $x \in modes_c(n,m)$ *if and only if for all paths* $p = \langle start, \ldots, n\rangle$ *in* CFG_c *there is a node* n' *on p such that* $c[n']$ *acquires m for x, and if* n'' *follows* n' *on p then* $c[n'']$ *does not release m for x.*

Definition 19. *Let* $c \in Com$. *Then* $PDG^{\|}(CFG_c^{H,L}) = (N, E \cup E')$ *for* $(N, E) = PDG(CFG_c^{H,L})$ *and* $(n, n') \in E'$ *if and only if one of the following holds:*

1. $n = in$ *and there exist a variable* $x \in H \cap use_c(n')$, *a node* $n'' \in N$ *with* $x \notin modes_c(n'', asm\text{-}nowrite)$, *and a path* p *from* n'' *to* n' *in* $CFG_c^{H,L}$ *with* $x \notin def_c(n''')$ *for every node* n''' *on* p *with* $n''' \neq n''$ *and* $n''' \neq n'$,
2. $n' = out$ *and there exist a variable* $x \in L \cap def_c(n)$, *a node* $n'' \in N$ *with* $x \notin modes_c(n'', asm\text{-}noread)$, *and a path* p *from* n *to* n'' *in* $CFG_c^{H,L}$ *such that* $x \notin def_c(n''')$ *for every node* n''' *on* p *with* $n''' \neq n$ *and* $n''' \neq n''$, *or*
3. $n \in \{1, \dots, |c|\}$, $c[n] \in Exp$, *and* $n' = out$.

The edges defined in Items 1 and 2 are derived from the typing rules for assignments: The edge (in, n) in Item 1, where n uses a high variable whose value might have been written by another thread, captures that the high variable might contain secrets when being used at n. The edge (n, out) in Item 2, where n defines a low variable whose value might be eventually read by another thread, ensures that the command is rejected if the definition at n might depend on secret input (because then there is a path from in to n). The edges defined in Item 3 are derived from the typing rule for conditionals: If the guard represented by Node n depends on secrets (i.e., there is a path from in to n) then the command is rejected because together with the edge (n, out) there is a path from in to out.

Example 9. Consider the following command c where the security domains of variables are $dom(x) = l$ and $dom(y) = h$:

$$/\!\!/\mathsf{acq}(asm\text{-}noread, x)/\!\!/;\, x{:=}y;\, x{:=}0;\, /\!\!/\mathsf{rel}(asm\text{-}noread, x)/\!\!/$$

Then $PDG^{\|}(CFG_c^{H,L}) = PDG(CFG_c^{H,L})$, and c is accepted by the PDG-based analysis for multi-threaded programs. Let furthermore $c' = x{:=}y; x{:=}0$. Then $PDG^{\|}(CFG_{c'}^{H,L}) \neq PDG(CFG_c^{H,L})$, because the graph $PDG^{\|}(CFG_{c'}^{H,L})$ contains an edge from the node representing $x{:=}y$ to Node out (due to Item 2 in Definition 19). Hence, c' is not accepted, because $PDG^{\|}(CFG_{c'}^{H,L})$ contains a path from Node in to Node out via the node representing the assignment $x{:=}y$.

Not accepting c' is crucial for soundness because another thread executing $x'{:=}x$ could copy the intermediate secret value of x into a public variable x'.

Theorem 4. *If a multi-threaded program is accepted by the PDG-based analysis for multi-threaded programs then the program is accepted by the type-based analysis for multi-threaded programs.*

The proof is by contradiction; it exploits the connection between PDG-based and type-based analysis stated in Lemma 1. A detailed proof is available on the authors' website.[5]

Soundness of the PDG-based analysis follows directly from Theorem 4 and the soundness of the type-based analysis (see [15, Theorem 6]).

[5] The reverse direction of Theorem 4 does not hold, because the type-based analysis for multi-threaded programs classifies some programs with secret control conditions as secure that are not classified as secure by our PDG-based analysis.

6 Related Work

We focus on related work covering flow-sensitive type-based analysis, PDG-based analysis for concurrent programs, and connections between analysis techniques. For an overview on language-based information flow security we refer to [21].

Flow-sensitivity of type-based analyses. In contrast to PDG-based information flow analyses, many type-based information flow analyses are not flow-sensitive. The first flow-sensitive type-based information flow analysis is due to Hunt and Sands [11]. Based on the idea of flow-sensitive security types from [11], Mantel, Sands, and Sudbrock developed the first sound flow-sensitive security type system for concurrent programs [15].

PDG-based analyses for concurrent programs. Hammer [7] presents a PDG-based analysis for concurrent Java programs, where edges between the PDGs of the individual threads are added following the extension of PDGs to concurrent programs from [12]. However, there is no soundness result. In fact, since the construction of PDGs from [12] does not capture dependencies between nodes in the PDG that result from internal timing (cf. Example 8), the resulting PDG-based information flow analysis fails to detect some information leaks.

Giffhorn and Snelting [5] present a PDG-based information flow analysis for multi-threaded programs that does not accept programs with internal timing leaks. The analysis enforces an information flow property defined in the tradition of observational determinism [20,28], and, therefore, does not accept any programs that have nondeterministic public output. Hence, the analysis forbids useful nondeterminism, which occurs, for instance, when multiple threads append entries to the same log file. Our novel analysis (from Section 5) permits concurrent writes to public variables and, hence, accepts secure programs that are not accepted by the analysis from [5]. Moreover, in contrast to the analysis from [5] our novel analysis is compositional with respect to the parallel composition of threads.

Connections between different analysis techniques. Hunt and Sands show in [11] that the program logic from [1] is in fact equivalent to the type-based analysis from [11]. Rehof and Fähndrich [19] exploit concepts from PDGs (the computation of so-called summary edges for programs with procedures) in a type-based flow analysis. In this article, we go a step further by establishing and exploiting a formal connection between a type-based and a PDG-based analysis.

7 Conclusion

While security type systems are established as analysis technique for information flow security, information flow analyses based on program dependency graphs (PDGs) have only recently received increased attention. In this article, we investigated the relationship between these two alternative approaches.

As a main result, we showed that the precision of a prominent type-based information flow analysis is not only roughly similar to the precision of a prominent

PDG-based analysis, but that the precision is in fact exactly the same. Moreover, our result provides a bridge for transferring techniques and ideas from one tradition of information flow analysis to the other. This is an interesting possibility because there are other relevant attributes than the precision of an analysis (e.g., efficiency and the availability of tools). We showed at the example of rely-guarantee-style information flow analysis for multi-threaded programs that this bridge is suitable to facilitate learning by one sub-community from the other.

We hope that our results clarify the relationship between the two approaches. The established relationship could be used as a basis for communication between the sub-communities to learn from each other and to pursue joint efforts to make semantically justified information flow analysis more practical. For instance, our results give hope that results on controlling declassification with security type systems can be used to develop semantic foundations for PDG-based analyses that permit declassification. Though there are PDG-based analyses that permit declassification (e.g., [8]), all of them yet lack a soundness result, and, hence, it is unclear which noninterference-like property they certify.

Acknowledgment. We thank the anonymous reviewers for their valuable suggestions. This work was funded by the DFG under the project FM-SecEng in the Computer Science Action Program (MA 3326/1-3).

References

1. Amtoft, T., Banerjee, A.: Information Flow Analysis in Logical Form. In: Giacobazzi, R. (ed.) SAS 2004. LNCS, vol. 3148, pp. 100–115. Springer, Heidelberg (2004)
2. Boudol, G., Castellani, I.: Noninterference for Concurrent Programs and Thread Systems. Theoretical Computer Science 281(1-2), 109–130 (2002)
3. Cheng, J.: Slicing Concurrent Programs - A Graph-Theoretical Approach. In: Fritszon, P.A. (ed.) AADEBUG 1993. LNCS, vol. 749, pp. 223–240. Springer, Heidelberg (1993)
4. Ferrante, J., Ottenstein, K.J., Warren, J.D.: The Program Dependence Graph and its Use in Optimization. ACM Transactions on Programming Languages and Systems 9(3), 319–349 (1987)
5. Giffhorn, D., Snelting, G.: Probabilistic Noninterference Based on Program Dependence Graphs. Tech. Rep. 6, Karlsruher Institut für Technologie (KIT) (2012)
6. Goguen, J.A., Meseguer, J.: Security Policies and Security Models. In: IEEE Symposium on Security and Privacy, pp. 11–20 (1982)
7. Hammer, C.: Information Flow Control for Java. Ph.D. thesis, Universität Karlsruhe (TH) (2009)
8. Hammer, C., Snelting, G.: Flow-sensitive, Context-sensitive, and Object-sensitive Information Flow Control based on Program Dependence Graphs. International Journal of Information Security 8(6), 399–422 (2009)
9. Horwitz, S., Reps, T.W., Binkley, D.: Interprocedural Slicing Using Dependence Graphs. ACM Transactions on Programming Languages and Systems 12(1), 26–60 (1990)
10. Hsieh, C.S., Unger, E.A., Mata-Toledo, R.A.: Using Program Dependence Graphs for Information Flow Control. Journal of Systems and Software 17(3), 227–232 (1992)

11. Hunt, S., Sands, D.: On Flow-Sensitive Security Types. In: ACM Symposium on Principles of Programming Languages, pp. 79–90 (2006)
12. Krinke, J.: Advanced Slicing of Sequential and Concurrent Programs. Ph.D. thesis, Universität Passau (2003)
13. Mantel, H.: Information Flow and Noninterference. In: Encyclopedia of Cryptography and Security, 2nd edn., pp. 605–607. Springer (2011)
14. Mantel, H., Sands, D.: Controlled Declassification based on Intransitive Noninterference. In: Chin, W.-N. (ed.) APLAS 2004. LNCS, vol. 3302, pp. 129–145. Springer, Heidelberg (2004)
15. Mantel, H., Sands, D., Sudbrock, H.: Assumptions and Guarantees for Compositional Noninterference. In: IEEE Computer Security Foundations Symposium, pp. 218–232 (2011)
16. Myers, A.C.: JFlow: Practical Mostly-Static Information Flow Control. In: ACM Symposium on Principles of Programming Languages, pp. 228–241 (1999)
17. Naik, M., Palsberg, J.: A Type System Equivalent to a Model Checker. ACM Transactions on Programming Languages and Systems 30(5), 1–24 (2008)
18. Palsberg, J., O'Keefe, P.: A Type System Equivalent to Flow Analysis. In: ACM Symposium on Principles of Programming Languages, pp. 367–378 (1995)
19. Rehof, J., Fähndrich, M.: Type-Based Flow Analysis: From Polymorphic Subtyping to CFL-Reachability. In: ACM Symposium on Principles of Programming Languages, pp. 54–66 (2001)
20. Roscoe, A.W., Woodcock, J.C.P., Wulf, L.: Non-interference through Determinism. In: Gollmann, D. (ed.) ESORICS 1994. LNCS, vol. 875, pp. 33–53. Springer, Heidelberg (1994)
21. Sabelfeld, A., Myers, A.C.: Language-based Information-Flow Security. IEEE Journal on Selected Areas in Communication 21(1), 5–19 (2003)
22. Sabelfeld, A., Sands, D.: Probabilistic Noninterference for Multi-threaded Programs. In: IEEE Computer Security Foundations Workshop, pp. 200–215 (2000)
23. Smith, G., Volpano, D.: Secure Information Flow in a Multi-threaded Imperative Language. In: ACM Symposium on Principles of Programming Languages, pp. 355–364 (1998)
24. Volpano, D., Smith, G., Irvine, C.: A Sound Type System for Secure Flow Analysis. Journal of Computer Security 4(3), 1–21 (1996)
25. Volpano, D., Smith, G.: A Type-Based Approach to Program Security. In: Bidoit, M., Dauchet, M. (eds.) TAPSOFT 1997. LNCS, vol. 1214, pp. 607–621. Springer, Heidelberg (1997)
26. Wasserrab, D., Lochbihler, A.: Formalizing a Framework for Dynamic Slicing of Program Dependence Graphs in Isabelle/HOL. In: Mohamed, O.A., Muñoz, C., Tahar, S. (eds.) TPHOLs 2008. LNCS, vol. 5170, pp. 294–309. Springer, Heidelberg (2008)
27. Wasserrab, D., Lohner, D., Snelting, G.: On PDG-based Noninterference and its Modular Proof. In: ACM Workshop on Programming Languages and Analysis for Security, pp. 31–44 (2009)
28. Zdancewic, S., Myers, A.C.: Observational Determinism for Concurrent Program Security. In: IEEE Computer Security Foundations Workshop, pp. 29–43 (2003)

Galliwasp: A Goal-Directed Answer Set Solver

Kyle Marple and Gopal Gupta

University of Texas at Dallas
800 W. Campbell Road
Richardson, TX 75080, USA

Abstract. *Galliwasp* is a goal-directed implementation of answer set programming. Unlike other answer set solvers, *Galliwasp* computes partial answer sets which are provably extensible to full answer sets. *Galliwasp* can execute arbitrary answer set programs in a top-down manner similar to SLD resolution. *Galliwasp* generates candidate answer sets by executing *ordinary rules* in a top-down, goal-directed manner using *coinduction*. *Galliwasp* next checks if the candidate answer sets are consistent with restrictions imposed by *OLON rules*. Those that are consistent are reported as solutions. Execution efficiency is significantly improved by performing the consistency check incrementally, i.e., as soon as an element of the candidate answer set is generated. We discuss the design of the *Galliwasp* system and its implementation. *Galliwasp*'s performance figures, which are comparable to other popular answer set solvers, are also presented.

1 Introduction

As answer set programming (ASP) [8] has gained popularity, answer set solvers have been implemented using various techniques. These techniques range from simple guess-and-check based methods to those based on SAT solvers and complex heuristics. None of these techniques impart any operational semantics to the answer set program in the manner that SLD resolution does to Prolog; they can be thought of as being similar to bottom-up methods for evaluating logic programs. Earlier we described [10,17] how to find partial answer sets by means of a top-down, goal-directed method based on coinduction. The *Galliwasp* system [16], described in this paper, is the first efficient implementation of this goal-directed method. *Galliwasp* is consistent with the conventional semantics of ASP: no restrictions are placed on the queries and programs that can be executed. That is, any A-Prolog program can be executed on *Galliwasp*.

Galliwasp is an implementation of A-Prolog [7] that uses grounded normal logic programs as input. The underlying algorithm of *Galliwasp* leverages coinduction [23,10] to find partial answer sets containing a given query. Programs are executed in a top-down manner in the style of Prolog, computing partial answer sets through SLD-style call resolution and backtracking. Each partial answer set is provably extensible to a complete answer set [10,17].

The algorithm of *Galliwasp* uses call graphs to classify rules according to two attributes: (i) if a rule can be called recursively with an odd number of negations

E. Albert (Ed.): LOPSTR 2012, LNCS 7844, pp. 122–136, 2013.
© Springer-Verlag Berlin Heidelberg 2013

between the initial call and its recursive invocation, it is said to contain an *odd loop over negation* (OLON) and referred to as an OLON rule for brevity, and (ii) if a rule has at least one path in the call graph that will not result in such a call, it is called an ordinary rule. Our goal-directed method uses ordinary rules to generate candidate answer sets (via co-induction extended with negation) and OLON rules to reject invalid candidate sets. The procedure can be thought of as following the *generate and test* paradigm with ordinary rules being used for generation of candidate answer sets and OLON rules for testing that a candidate answer set is, indeed, a valid answer set.

The main contribution of this paper is an execution mechanism in which consistency checks are incrementally executed as soon as an element of the candidate answer set is generated. By interleaving the generation and testing of candidate answer sets in this manner, *Galliwasp* speeds up execution significantly. As soon as an element p is added to a candidate answer set, OLON rules that might reject the answer set due to presence of p in it are invoked.

In this paper we describe the design and implementation of *Galliwasp*. This includes an overview of goal-directed ASP, the design and implementation of *Galliwasp*'s components, and techniques developed to improve performance. Finally, the performance figures are presented and compared to other ASP solvers.

There are many advantages of a goal-directed execution strategy: (i) ASP can be extended to general predicates [18], i.e., to answer set programs that do not admit a finite grounding; (ii) extensions of ASP to constraints (in the style of CLP(R)) [12], probabilities (in the style of ProbLog [4]), etc., can be elegantly realized; (iii) Or-parallel implementations of ASP that leverage techniques from parallel Prolog implementations [11] can be developed; (iv) abductive reasoning [14] can be incorporated with greater ease. Work is in progress to extend *Galliwasp* in these directions.

2 Goal-Directed Answer Set Programming

The core of *Galliwasp* is a goal-directed execution method for computing answer sets which is sound and complete with respect to the Gelfond-Lifschitz method [10,17]. For convenience, two key aspects of our goal-directed method are summarized here: its handling of OLON rules and its use of coinduction.

2.1 OLON Rules and Ordinary Rules

Our goal-directed method requires that we identify and properly handle both ordinary and OLON rules, as defined in the introduction. Rules are classified by constructing and examining a call graph. Unlike in Prolog, literals can be either positive or negative, and the attributes of a rule are determined by the number of negations between any recursive calls in the call graph.

To build the call graph, each positive literal is treated as a node in the graph. For each rule, arcs are drawn from the node corresponding to the head of the rule to every node corresponding to the positive form of a literal in the body of the rule. If the literal in the body is positive, the arc is blue; if the literal is negative, the arc is red. At most one of each color arc is created between any two nodes, and each arc is labelled to identify all of the rules associated with it.

When the graph has been created, each path is examined in a bottom-up fashion, stopping when any literal in the path is repeated. If there is a path from a node `N` to itself such that the number of red arcs on the path is odd, then we call all of the rules for N's literal associated with the path OLON rules. If there is no path from a node `N` to itself, or if there is a path from `N` to itself such that the number of red arcs on the path is even, we call all of the rules for N's literal associated with the path ordinary rules.

Every rule will have at least one of the two attributes. Additionally, a rule can be both an OLON rule and an ordinary rule, as the example below illustrates:

```
p :- not q. ... (i)
q :- not p. ... (ii)
q :- not r. ... (iii)
r :- not p. ... (iv)
```

Rule (i) is both an OLON rule and an ordinary rule: ordinary because of rule (ii), OLON because of rules (iii) and (iv).

2.2 Coinductive Execution

The goal-directed method [10,17] generates candidates answer sets by executing ordinary rules via coinduction extended with negation. Given a query, an extended query is constructed to enforce the OLON rules. This extended query is executed in the style of SLD resolution with coinduction (co-SLD resolution) [23,10].

Under the operational semantics of coinduction, a call `p` succeeds if it unifies with one of its ancestor calls. Each call is remembered, and this set of ancestor calls forms the *coinductive hypothesis set* (CHS). Under our algorithm, the CHS also constitutes a candidate answer set. As such, it would be inconsistent for `p` and `not p` to be in the CHS at the same time: when this is about to happen, the system must backtrack. As a result, only ordinary rules can generate candidate answer sets. Any OLON rule that is not also an ordinary rule will fail during normal execution, as it will attempt to add the negation of its head to the CHS.

Candidate sets produced by ordinary rules must still be checked for consistency with the OLON rules in a program. To accomplish this, the query is extended to perform this check.

Before looking at this extension, let us consider an OLON rule of the form

```
p :- B, not p.
```

where `B` is a conjunction of goals. One of two cases must hold for an answer set to exist: (i) `p` is in the answer set through another rule in the program, (ii) at least one goal in `B` must fail. This is equivalent to saying that the negation of the rule, `not B` $\vee$ `p` must succeed. Note that checks of this form can be created for any OLON rule [10,17]. This includes both OLON rules of the form

```
p :- B.
```

where the call to `not p` is indirect, and headless rules of the form

```
:- B.
```

where the negation `not B` must succeed.

Our method handles OLON rules by creating a special rule, called the NMR check, which contains a sub-check for each OLON rule in a program. The body of the NMR check is appended to each query prior to evaluation, ensuring that any answer set returned is consistent with all of the OLON rules in the program. Prior to creating the sub-checks, a copy of each OLON rule is written in the form

```
p :- B, not p.
```

as shown above, with the negation of the rule head added to the body in cases where the call is indirect. Each sub-check is then created by negating the body of one of the copied rules. For instance, the sub-check for a rule

```
p :- q, not p.
```

will be

```
chk_p :- not q.
chk_p :- p.
```

The body of the NMR check will then contain the goal `chk_p`, ensuring that the OLON rule is properly applied.

Some modification to co-SLD resolution is necessary to remain faithful to ASP's stable model semantics. This is due to the fact that coinduction computes the greatest fixed point of a program, while the GL method computes a fixed point that is between the least fixed point (*lfp*) and greatest fixed point (*gfp*) of a program, namely, the *lfp* of the residual program.To adapt coinduction for use with ASP, our goal-directed method restricts coinductive success to cases in which there are an even, non-zero number of calls to `not` between a call and an ancestor to which it unifies. This prevents rules such as

```
p :- p.
```

from succeeding, as this would be the case under normal co-SLD resolution. To compute an answer set for a given program and query, our method performs co-SLD resolution using this modified criterion for coinductive success. When both a query and the NMR check have succeeded, the CHS will be a valid answer set [10,17]. Consider the program below:

```
p :- q.
q :- p.
```

Under normal co-SLD resolution, both `p` and `q` would be allowed to succeed, resulting in the incorrect answer set {`p, q`}. However, using our modified criterion, both rules will fail, yielding the correct answer set {`not p, not q`}.

3 The Galliwasp System

The *Galliwasp* system is an implementation of the above SLD resolution-style, goal-directed method of executing answer set programs. A number of issues arise that are discussed next. Improvements to the execution algorithm that make *Galliwasp* more efficient are also discussed.

3.1 Order of Rules and Goals

As with normal SLD resolution, rules for a given literal are executed in the order given in the program, with the body of each rule executed left to right. With the exception of OLON rules, once the body of a rule whose head matches a given literal has succeeded, the remaining rules do not need to be accessed unless failure and backtracking force them to be selected.

As in top-down implementations of Prolog, this means that the ordering of clauses and goals can directly impact the runtime performance of a program. In the best case scenario, this can allow *Galliwasp* to compute answers much faster than other ASP solvers, but in the worst case scenario *Galliwasp* may end up backtracking significantly more, and as a result take much longer.

One example of this can be found in an instance of the Schur Numbers benchmark used in Sect. 5. Consider the following clauses from the grounded instance for 3 partitions and 13 numbers:

```
_false :- not haspart(1).
_false :- not haspart(2).
_false :- not haspart(3).
_false :- not haspart(4).
_false :- not haspart(5).
_false :- not haspart(6).
_false :- not haspart(7).
_false :- not haspart(8).
_false :- not haspart(9).
_false :- not haspart(10).
_false :- not haspart(11).
_false :- not haspart(12).
_false :- not haspart(13).
```

During benchmarking, *Galliwasp* was able to find an answer set for the program containing the above clauses in 0.2 seconds. However, the same program with the order of only the above clauses reversed could be left running for several minutes without terminating.

Given that *Galliwasp* is currently limited to programs that have a finite grounding, its performance can be impacted by the grounder program that is used. In the example above, all 13 ground clauses are generated by the grounding of a single rule in the original program. Work is in progress to extend *Galliwasp* to allow direct execution of datalog-like ASP programs (i.e., those with only constants and variables) without grounding them first. In such a case, the user would have much more control over the order in which rules are tried.

3.2 Improving Execution Efficiency

The goal-directed method described in Sect. 2 can be viewed as following the *generate and test* paradigm. Given a query `Q, N` where `Q` represents the original user query, and `N` the NMR check, the goal directed procedure *generates* candidate answer sets through the execution of `Q` using ordinary rules. These candidate answer sets are *tested* by the NMR check `N` which rejects a candidate if the restrictions encoded in the OLON rules are violated.

Naive generate and test can be very inefficient, as the query `Q` may generate a large number of candidate answer sets, most of which may be rejected by the NMR check. This can lead to a significant amount of backtracking, slowing down the execution. Execution can be made significantly more efficient by generating and testing answer sets incrementally. This is done by interleaving the execution of `Q` and `N`. As soon as a literal, say `p`, is added to a candidate answer set, goals in NMR check that correspond to OLON rules that `p` may violate are invoked. We illustrate the incremental generate and test algorithm implemented in *Galliwasp*.

Consider the rules below:

```
p :- r.
p :- s.
q :- big_goal.
r :- not s.
s :- not r.
:- p, r.
```

Let us assume that this fragment is part of a program where `big_goal` will always succeed, but is computationally expensive and contains numerous choice points. For simplicity, let us also assume that the complete program contains no additional OLON rules, and no additional rules for the literals `p`, `q`, `r` or `s`. As the program contains only the OLON rule from the fragment above, the NMR check will have only one sub-check, which will consist of two clauses:

```
nmr_check :- chk_1.
chk_1 :- not p.
chk_1 :- not r.
```

Next, assume that the program is compiled and run with the following query:

```
?- p, q.
```

As with any query, the NMR check will be appended, resulting in the final query:

```
?- p, q, nmr_check.
```

The NMR check ensures that `p` and `r` are never present in the same answer set, so any valid answer set will contain `p`, `q`, `not r`, `s`, `big_goal`, and the literals upon which `big_goal` depends.

If the program is run without NMR check reordering, the first clause for `p` will initially succeed, adding `r` to the CHS. Failure will not occur until the NMR check is reached, at which point backtracking will ensue. Eventually, the initial call to `p` will be reached, the second clause will be selected, and the query will eventually succeed. However, the choicepoints in `big_goal` could result in a massive amount of backtracking, significantly delaying termination. Additionally, `big_goal` will have to succeed twice before a valid answer set is found.

Now let us consider the same program and query using incremental generate and test. The first clause for `p` will still initially succeed. However, as `r` succeeds, the clause of `chk_1` calling `not r` will be removed. Since only one clause will remain for `chk_1`, `chk_1` will be reordered, placing it immediately after the call to `p` and before the call to `q`. The call to `p` will then succeed as before, but failure and backtracking will occur almost immediately, as the call to `chk_1` will immediately call `not p`. As before, the call to `not p` will result in the second clause for `p` being selected and the query will eventually succeed. However, as failure occurs before `big_goal` is reached, the resulting backtracking is almost eliminated. Additionally, `big_goal` does not need to be called twice before a solution is found. As a result, execution with NMR check reordering will terminate much sooner than execution without reordering.

Adding incremental test and generation to *Galliwasp* leads to a considerable improvement in efficiency, resulting in the performance of the *Galliwasp* system becoming comparable to state-of-the-art solvers.

4 System Architecture of Galliwasp

The *Galliwasp* system consists of two components: a compiler and an interpreter. The compiler reads in a grounded instance of an ASP program and produces a compiled program, which is then executed by the interpreter to compute answer sets. An overview of the system architecture is shown in Fig. 4.

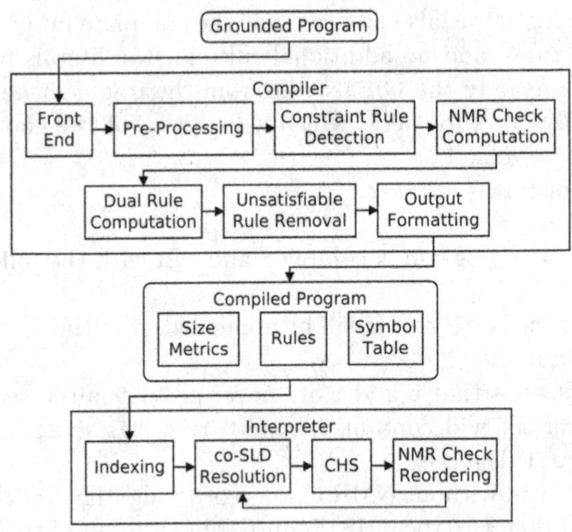

Fig. 1. *Galliwasp*'s system architecture

4.1 Compiler

The primary goal of *Galliwasp*'s design is to maximize the runtime performance of the interpreter. Because *Galliwasp* computes answer sets for a program based on user-supplied queries, answer sets cannot be computed until a program is actually run by the interpreter. However, many of the steps in our goal-directed algorithm, as well as parsing, program transformations and static analysis, are independent of the query. The purpose of *Galliwasp*'s compiler is to perform as much of the query-independent work as possible, leaving only the actual computation of answer sets to the interpreter. Because a single program can be run multiple times with different queries, compilation also eliminates the need to repeat the query-independent steps every time a program is executed. The most important aspects of the compiler are the input language accepted, the parsing and pre-processing performed, the construction of the NMR check, the program transformations applied, and the formatting of the compiled output.

Input Language. The *Galliwasp* compiler's input language is grounded A-Prolog [7], extended to allow partial compatibility with text-mode output of the *lparse* grounder. This allows grounded program instances to be obtained by invoking *lparse* with the `-t` switch, which formats the output as text.

Only a subset of *lparse*'s text output is supported by *Galliwasp*. The support is made possible by special handling of rules with the literal `_false` as their head, a convention used by *lparse* to add a head to otherwise headless rules, and support for *lparse*'s `compute` statement. Other features of *lparse*, such as constraint and weight literals, choice rules, and optimize statements, are not currently supported in *Galliwasp*. Work is in progress to support them.

When *lparse* encounters a headless rule, it produces a grounded rule with `_false` as the head and a compute statement containing the literal `not _false`. Because the literal `_false` is not present in the body of any rule, special handling is required to properly detect such rules as OLON rules.

The compute statements used by *lparse* are of the form

```
compute N { Q }.
```

where `N` specifies the number of answer sets to compute and `Q` is a set of literals that must be present in a valid answer set. Our system handles these statements by treating them as optional, hard-coded queries. If a compute statement is present, the interpreter may be run without user interaction, computing up to `N` answer sets using `Q` as the query. When the interpreter is run interactively, it ignores compute statements and executes queries entered by the user.

Parsing and Pre-processing. The compiler's front end and pre-processing stages prepare the input program for easy access and manipulation during the rest of the compilation process. The front end encompasses the initial lexical analysis and parsing of the input program, while the pre-processing stage handles additional formatting and simplification, and substitutes integers for literals.

After lexical analysis of the input program is performed, it is parsed into a list of rules and statements by a definite clause grammar (DCG). During parsing, a counter is used to number the rules as they are read, so that the relative order of rules for a given literal can be maintained.

The list of statements produced by the DCG is next converted into a list of rules and a single compute statement. Each rule is checked to remove duplicate goals. Any rule containing its head as a goal with no intervening negation will also be removed at this stage. The cases covered in this step should not normally occur, but as they are allowed by the language, they are addressed before moving on.

The next stage of pre-processing is integer substitution. Every propositional symbol `p` is mapped to a unique integer $N_p > 0$. A positive literal `p` is represented by N_p, while a negative literal `not p` is represented by $-N_p$. Since the interpreter must use the original names of propositions when it prints the answer, a table that maps each N_p to `p` is included in the compiled output.

Finally, the list of rules is sorted by head, maintaining the relative order of rules with the same head. This eliminates the need for repeated searching in subsequent stages of compilation. After sorting, compilation moves on to the next stage, the detection of OLON rules and construction of the NMR check.

Construction of the NMR Check. Construction of the NMR check begins with the detection of the OLON rules in the ASP program. This detection is performed by building and traversing a call graph similar to the one described in Sect. 2.1. These rules are then used to construct the individual checks that form the NMR check, as described in Sect. 2.2.

Clauses for the sub-checks are treated as any other rule in the program, and subject to program transformation in the next stage of compilation. However, the NMR check itself is not modified by the program transformation stage. Instead, if the modified sub-checks allow for immediate failure, this will be detected at runtime by the interpreter.

Program Transformation. The program transformation stage consists of computing dual rules, explained below, and removing rules when it can be trivially determined at compile time that they will never succeed. This stage of compilation improves performance without affecting the correctness of our algorithm.

Dual rules, i.e., rules for the negation of each literal, are computed as follows. For proposition `p` defined by the rules,

```
p :- B₁.
...
p :- Bₙ.
```

where each B_i is a conjunction of positive and negative literals, its dual

```
not p :- not B₁, ..., not Bₙ.
```

is added to the program. For any proposition `q` for which there are no rules whose head contains `q`, a fact is added for `not q`.

The addition of dual rules simplifies the design of the interpreter by removing the need to track the scope of negations: all rules can be executed in a uniform fashion, regardless of the number of negations encountered. When a negated calls are encountered, they are expanded using dual rules. While adding these rules may significantly increase the size of the program, this is not a problem: the interpreter performs indexing that allows it to access rules in constant time.

After the dual rules have been computed, the list is checked once more to remove simple cases of rules that can be trivially determined to always fail. This step simply checks to see if a fact exists for the negation of some goal in the rule body and removes the rule if this is the case. If the last rule for a literal is removed, a fact for the negation is added and subsequent rules calling the literal will also be removed.

Output Formatting. As with the rest of the compiler, the output produced by the compiler is designed to reduce the amount of work performed by the interpreter. This is done by including space requirements and sorting the rules by their heads before writing them to the output.

As a result of the output formatting, the interpreter is able to read in the input and create the necessary indexes in linear time with respect to the size of the compiled program. After indexing, all that remains is to execute the program, with all other work having been performed during compilation.

4.2 Interpreter

While the compiler was designed to perform a variety of time consuming tasks, the interpreter has been designed to maximize run-time performance, finding an answer set or determining failure as quickly as possible. Two modes of operation, interactive and automatic, are supported. When a program is run interactively, the user can input queries and accept or reject answer sets as can be done with answers to a query in Prolog. In automatic mode it executes a compute statement that is included in the program. In either mode, the key operations can be broken up into three categories: program representation and indexing, co-SLD resolution, and dynamic incremental enforcement of the NMR check.

Program Representation and Indexing. One of the keys to *Galliwasp*'s performance is the indexing performed prior to execution of the program. As a result, look-up operations during execution can be performed in constant time, much as in any implementation of Prolog.

As mentioned in Sect. 4.1, the format of the compiler's output allows the indexes to be created in time that is linear with respect to the size of the program. The compiled program includes the number of positive literals, which allows a hash table to be constructed using an extremely simple perfect hash function. For n positive literals, $2n$ entries are needed, as 0 is never used. For a literal L which may be positive or negative,

$$index(L) = \begin{cases} L & L > 0, \\ 2n + 1 + L & L < 0 \end{cases}$$

Because the program is sorted by the compiler, the main index can be created in a single pass, using fixed length arrays rather than dynamically allocated memory.

To allow for the NMR check interleaving and simplification discussed in Sect. 4.2, the query and NMR check are stored separately from the rest of the program. Whereas the rules of the program, including the sub-checks of the NMR check, are stored in ordinary arrays, the query and NMR check are stored in a linked list. This allows their goals to be reordered in constant time. Additional indexing is also performed, linking each literal to its occurrences in the NMR check. Together, these steps allow modification of the NMR check with respect to a given literal to be performed in linear time with respect to the number times the literal and its dual occurs in the sub-checks.

Co-SLD Resolution. Once the program has been indexed, answer sets are found by executing the query using coinduction, as described in Sect. 2.2. Each call encountered is checked against the CHS. If the call is not present, it is added to the CHS and expanded according to ordinary SLD resolution. If the call is already in the CHS, immediate success or failure occurs, as explained below.

As mentioned in Sect. 2.2, our algorithm requires that a call that unifies with an ancestor cannot succeed coinductively unless it is separated from that ancestor by an even, non-zero number of intervening negations. This is facilitated by our use of dual rules, discussed in Sect. 4.1.

When the current call is of the form `not p` and the CHS contains `p` (or vice versa), the current call must fail, lest the CHS become inconsistent. If the current call is identical to one of the elements of the CHS, then the number of intervening negations must have been even. However, it is not clear that it was also non-zero. This additional information is provided by a counter that tracks of the number of negations encountered between the root of the proof tree and the tip of the current branch. When a literal is added to the CHS, it is stored with the current value of the counter. A recursive call to the literal can coinductively succeed only if the counter has increased since the literal was stored.

Dynamic Incremental Enforcement of the NMR Check. Because recursive calls are never expanded, eventual termination is guaranteed, just as for any other ASP solver. However, since we use backtracking, the size of the search space can cause the program to execute for an unreasonable amount of time.

To alleviate this problem, we introduced the technique of incrementally enforcing the NMR check (cf. Sect. 3.2 above). Our technique can be viewed as a form of coroutining: steps of the testing and generation phases are interleaved. The technique consists of simplifying NMR sub-checks and of reordering the calls to sub-checks within the query. These modifications are done whenever a literal is added to the CHS, and are "undone" upon backtracking.

When a literal is added to the CHS, all occurrences of that literal are removed from every sub-check, as they would immediately succeed when treated as calls. If such a removal causes the body of a sub-check clause to become empty, the entire sub-check is removed, as it is now known to be satisfied.

The next step is to remove every sub-check clause that contains the negation of the literal: such a clause cannot be satisfied now. If this clause causes the entire sub-check to disappear, the literal obviously cannot be added to the CHS, and backtracking occurs. If the sub-check does not disappear, but is reduced to a single clause (i.e., it becomes deterministic), then a call to the sub-check is moved to the front of the current query: this ensures early testing of whether the addition of the literal is consistent with the relevant OLON rules. If the resultant sub-check contains more than one clause, there is no attempt to execute it earlier than usual, as that might increase the size of the search space.

As mentioned in Sect. 4.2, indexing allows these modifications to be performed efficiently. If the original search space is small, they may result in a small increase in runtime. In most non-trivial cases, however, the effect is a dramatic decrease in the size of the search space. It is this technique that enables *Galliwasp* to be as efficient as illustrated in the next section: early versions of the system used a simple generate-and-test approach, but many of the programs that now terminate in a fraction of a second ran for inordinate amounts of time.

5 Performance Results

In this section, we compare the performance of *Galliwasp* to that of *clasp* [5], *cmodels* [9] and *smodels* [20] using instances of several problems. The range of problems is limited somewhat by the availability of compatible programs. While projects such as Asparagus [2] have a wide variety of ASP problem instances

Table 1. Performance results; times in seconds

Problem	*Galliwasp*	*clasp*	*cmodels*	*smodels*
queens-12	0.033	0.019	0.055	0.112
queens-13	0.034	0.022	0.071	0.132
queens-14	0.076	0.029	0.098	0.362
queens-15	0.071	0.034	0.119	0.592
queens-16	0.293	0.043	0.138	1.356
queens-17	0.198	0.049	0.176	4.293
queens-18	1.239	0.059	0.224	8.653
queens-19	0.148	0.070	0.272	3.288
queens-20	6.744	0.084	0.316	47.782
queens-21	0.420	0.104	0.398	95.710
queens-22	69.224	0.112	0.472	N/A
queens-23	1.282	0.132	0.582	N/A
queens-24	19.916	0.152	0.602	N/A
mapclr-4x20	0.018	0.006	0.011	0.013
mapclr-4x25	0.021	0.007	0.014	0.016
mapclr-4x29	0.023	0.008	0.016	0.018
mapclr-4x30	0.026	0.008	0.016	0.019
mapclr-3x20	0.022	0.005	0.009	0.008
mapclr-3x25	0.065	0.006	0.011	0.010
mapclr-3x29	0.394	0.006	0.012	0.011
mapclr-3x30	0.342	0.007	0.012	0.011
schur-3x13	0.209	0.006	0.010	0.009
schur-2x13	0.019	0.005	0.007	0.006
schur-4x44	N/A	0.230	1.394	0.349
schur-3x44	7.010	0.026	0.100	0.076
pigeon-10x10	0.020	0.009	0.020	0.025
pigeon-20x20	0.050	0.048	0.163	0.517
pigeon-30x30	0.132	0.178	0.691	4.985
pigeon-8x7	0.123	0.072	0.089	0.535
pigeon-9x8	0.888	0.528	0.569	4.713
pigeon-10x9	8.339	4.590	2.417	46.208
pigeon-11x10	90.082	40.182	102.694	N/A

available, the majority use *lparse* features unsupported by *Galliwasp*. Work is in progress to support these features.

The times for *Galliwasp* in Table 1 are for the interpreter reading compiled versions of each problem instance. No queries are given, so the NMR check alone is used to find solutions. The times for the remaining solvers are for the solver reading problem instances from files in the *smodels* input language. A timeout of 600 seconds was used, with the processes being automatically killed after that time and the result being recorded as N/A in the table.

Galliwasp is outperformed by *clasp* in all but one case and outperformed by *cmodels*, in most cases, but the results are usually comparable. *Galliwasp*'s performance can vary significantly, even between instances of the same problem, depending on the amount of backtracking required. In the case of programs that timed out or took longer than a few seconds, the size and structure of the programs simply resulted in a massive amount of backtracking.

We believe that manually reordering clauses or changing the grounding method will improve performance in most cases. In particular, experimentation with manual clause reordering has resulted in execution times on par with *clasp*'s. However, the technique makes performance dependent on the query used to determine the optimal ordering. As a result, the technique is not general enough for use in performance comparisons, and no reordering of clauses was used to obtain the results reported here.

6 Related and Future Work

Galliwasp's goal-directed execution method is based on a previously published technique, but has been significantly refined [10,19]. In particular, the original algorithm was limited to ASP programs which were call-consistent or order-consistent [19]. Additionally, the implementation of the previous algorithm was written as a Prolog meta-interpreter, and incapable of providing results comparable to those of existing ASP solvers. *Galliwasp*, written in C, is the first goal-directed implementation capable of direct comparison with other ASP solvers.

Various other attempts have been made to introduce goal-directed execution to ASP. However, many of these methods rely on modifying the semantics or restricting the programs and queries which can be used, while *Galliwasp*'s algorithm remains consistent with stable model semantics and works with any arbitrary program or query. For example, the revised Stable Model Semantics [21] allows goal-directed execution [22], but does so by modifying the stable model semantics underlying ASP. SLD resolution has also been extended to ASP through credulous resolution [3], but with restrictions on the type of programs and queries allowed. Similar work may also be found in the use of ASP with respect to abduction [1], argumentation [13] and tableau calculi [6].

Our plans for future work focus on improving the performance of *Galliwasp* and extending its functionality. In both cases, we are planning to investigate several possible routes.

With respect to performance, we are looking into exercising better control over the grounding of programs, or eliminating of grounding altogether. The development of a grounder optimized for goal-directed execution is being investigated ([15], forthcoming), as is the extension of answer set programming to predicates (preliminary work can be found in [18]). Both have the potential to significantly improve *Galliwasp*'s performance by reducing the amount of backtracking that results from the use of grounded instances produced using *lparse*.

In the area of functionality, we plan to expand *Galliwasp*'s input language to support features commonly allowed by other solvers, such as constraint literals, weight literals and choice rules. Extensions of *Galliwasp* to incorporate constraint logic programming, probabilistic reasoning, abduction and or-parallelism are also being investigated. We believe that these extensions can be incorporated more naturally into *Galliwasp* given its goal-directed execution strategy.

The complete source for *Galliwasp* is available from the authors at [16].

7 Conclusion

In this paper we introduced the goal-directed ASP solver *Galliwasp* and presented *Galliwasp*'s underlying algorithm, limitations, design, implementation and performance. *Galliwasp* is the first approach toward solving answer sets that uses goal-directed execution for general answer set programs. Its performance is comparable to other state-of-the-art ASP solvers. *Galliwasp* demonstrates the viability of the top-down technique. Goal-directed execution can offer significant potential benefits. In particular, *Galliwasp*'s underlying algorithm paves the way for ASP over predicates [18] as well as integration with other areas of logic programming. Unlike in other ASP solvers, performance depends on the amount of backtracking required by a program: at least to some extent this can be controlled by the programmer. Future work will focus on improving performance and expanding functionality.

Acknowledgments. The authors are grateful to Feliks Kluźniak for many discussions and ideas.

References

1. Alferes, J.J., Pereira, L.M., Swift, T.: Abduction in Well-Founded Semantics and Generalized Stable Models via Tabled Dual Programs. Theory and Practice of Logic Programming 4, 383–428 (2004)
2. Asparagus (2012), `http://asparagus.cs.uni-potsdam.de`
3. Bonatti, P.A., Pontelli, E., Son, T.C.: Credulous Resolution for Answer Set Programming. In: Proceedings of the 23rd National Conference on Artificial Intelligence, AAAI 2008, vol. 1, pp. 418–423. AAAI Press (2008)
4. De Raedt, L., Kimmig, A., Toivonen, H.: ProbLog: A Probabilistic Prolog and Its Application in Link Discovery. In: Proceedings of the 20th International Joint Conference on Artifical Intelligence, IJCAI 2007, pp. 2468–2473. Morgan Kaufmann (2007)

5. Gebser, M., Kaufmann, B., Neumann, A., Schaub, T.: *clasp*: A Conflict-Driven Answer Set Solver. In: Baral, C., Brewka, G., Schlipf, J. (eds.) LPNMR 2007. LNCS (LNAI), vol. 4483, pp. 260–265. Springer, Heidelberg (2007)
6. Gebser, M., Schaub, T.: Tableau Calculi for Answer Set Programming. In: Etalle, S., Truszczyński, M. (eds.) ICLP 2006. LNCS, vol. 4079, pp. 11–25. Springer, Heidelberg (2006)
7. Gelfond, M.: Representing Knowledge in A-Prolog. In: Kakas, A.C., Sadri, F. (eds.) Computational Logic: Logic Programming and Beyond. LNCS (LNAI), vol. 2408, pp. 413–451. Springer, Heidelberg (2002)
8. Gelfond, M., Lifschitz, V.: The Stable Model Semantics for Logic Programming. In: Proceedings of the Fifth International Conference on Logic Programming, pp. 1070–1080. MIT Press (1988)
9. Giunchiglia, E., Lierler, Y., Maratea, M.: SAT-Based Answer Set Programming. In: Proceedings of the 19th National Conference on Artifical Intelligence, AAAI 2004, pp. 61–66. AAAI Press (2004)
10. Gupta, G., Bansal, A., Min, R., Simon, L., Mallya, A.: Coinductive Logic Programming and Its Applications. In: Dahl, V., Niemelä, I. (eds.) ICLP 2007. LNCS, vol. 4670, pp. 27–44. Springer, Heidelberg (2007)
11. Gupta, G., Pontelli, E., Ali, K.A., Carlsson, M., Hermenegildo, M.V.: Parallel Execution of Prolog Programs: A Survey. ACM Transactions on Programming Languages and Systems 23(4), 472–602 (2001)
12. Jaffar, J., Lassez, J.L.: Constraint Logic Programming. In: Proceedings of the 14th ACM SIGACT-SIGPLAN Symposium on Principles of Programming Languages, POPL 1987, pp. 111–119. ACM (1987)
13. Kakas, A., Toni, F.: Computing Argumentation in Logic Programming. Journal of Logic and Computation 9(4), 515–562 (1999)
14. Kakas, A.C., Kowalski, R.A., Toni, F.: Abductive Logic Programming. Journal of Logic and Computation 2(6), 719–770 (1992)
15. Marple, K.: Design and Implementation of a Goal-directed Answer Set Programming System. Ph.D. thesis, University of Texas at Dallas
16. Marple, K.: galliwasp, `http://www.utdallas.edu/~kbm072000/galliwasp/`
17. Marple, K., Bansal, A., Min, R., Gupta, G.: Goal-Directed Execution of Answer Set Programs. Tech. rep., University of Texas at Dallas (2012), `http://www.utdallas.edu/~kbm072000/galliwasp/publications/goaldir.pdf`
18. Min, R.: Predicate Answer Set Programming with Coinduction. Ph.D. thesis, University of Texas at Dallas (2010)
19. Min, R., Bansal, A., Gupta, G.: Towards Predicate Answer Set Programming via Coinductive Logic Programming. In: Iliadis, L., Vlahavas, I., Bramer, M. (eds.) AIAI 2009. IFIP, vol. 296, pp. 499–508. Springer, Boston (2009)
20. Niemelä, I., Simons, P.: Smodels - An Implementation of the Stable Model and Well-Founded Semantics for Normal Logic Programs. In: Fuhrbach, U., Dix, J., Nerode, A. (eds.) LPNMR 1997. LNCS, vol. 1265, pp. 420–429. Springer, Heidelberg (1997)
21. Pereira, L.M., Pinto, A.M.: Revised Stable Models - A Semantics for Logic Programs. In: Bento, C., Cardoso, A., Dias, G. (eds.) EPIA 2005. LNCS (LNAI), vol. 3808, pp. 29–42. Springer, Heidelberg (2005)
22. Pereira, L.M., Pinto, A.M.: Layered Models Top-Down Querying of Normal Logic Programs. In: Gill, A., Swift, T. (eds.) PADL 2009. LNCS, vol. 5418, pp. 254–268. Springer, Heidelberg (2009)
23. Simon, L.: Extending Logic Programming with Coinduction. Ph.D. thesis, University of Texas at Dallas (2006)

Computing More Specific Versions of Conditional Rewriting Systems*

Naoki Nishida[1] and Germán Vidal[2]

[1] Graduate School of Information Science, Nagoya University
Furo-cho, Chikusa-ku, 4648603 Nagoya, Japan
nishida@is.nagoya-u.ac.jp
[2] MiST, DSIC, Universitat Politècnica de València
Camino de Vera, s/n, 46022 Valencia, Spain
gvidal@dsic.upv.es

Abstract. Rewrite systems obtained by some automated transformation often have a poor syntactic structure even if they have good properties from a semantic point of view. For instance, a rewrite system might have overlapping left-hand sides even if it can only produce at most one constructor normal form (i.e., value). In this paper, we propose a method for computing "more specific" versions of deterministic conditional rewrite systems (i.e., typical functional programs) by replacing a given rule (e.g., an overlapping rule) with a finite set of instances of this rule. In some cases, the technique is able to produce a non-overlapping system from an overlapping one. We have applied the transformation to improve the systems produced by a previous technique for function inversion with encouraging results (all the overlapping systems were successfully transformed to non-overlapping systems).

1 Introduction

Rewrite systems [4] form the basis of several rule-based programming languages. In this work, we focus on the so called *deterministic* conditional rewrite systems (DCTRSs), which are typical functional programs with local declarations [23]. When the rewrite systems are automatically generated (e.g., by program inversion [2,14,15,17,22,28,27,29] or partial evaluation [1,7,8,35]), they often have a poor syntactic structure that might hide some properties. For instance, the rewriting systems generated by program inversion sometimes have overlapping left-hand sides despite the fact that they actually have the *unique normal form* property w.r.t. constructor terms — i.e., they can only produce at most one constructor normal form for every expression — or are even confluent.

* This work has been partially supported by the Spanish *Ministerio de Economía y Competitividad (Secretaría de Estado de Investigación, Desarrollo e Innovación)* under grant TIN2008-06622-C03-02, by the *Generalitat Valenciana* under grant PROMETEO/2011/052, and by *MEXT KAKENHI* #21700011.

E. Albert (Ed.): LOPSTR 2012, LNCS 7844, pp. 137–154, 2013.
© Springer-Verlag Berlin Heidelberg 2013

Consider, e.g., the following TRS (left) from [29] (where we use $[\_|\_]$ and nil as list constructors) and its inversion (right):

$$\begin{array}{rlrl} \mathsf{inc}(\mathsf{nil}) & \to [0] & \mathsf{inc}^{-1}([0]) & \to \mathsf{nil} \\ \mathsf{inc}([0|xs]) & \to [1|xs] & \mathsf{inc}^{-1}([1|xs]) & \to [0|xs] \\ \mathsf{inc}([1|xs]) & \to [0|\mathsf{inc}(xs)] & \mathsf{inc}^{-1}([0|ys]) & \to [1|\mathsf{inc}^{-1}(ys)] \end{array}$$

Now, observe that every instance of $\mathsf{inc}^{-1}(x)$ using a constructor term has a unique constructor normal form. However, this system is not confluent — consider, e.g., the reductions starting from $\mathsf{inc}^{-1}([0])$ — and, moreover some of the left-hand sides overlap, thus preventing us from obtaining a typical (deterministic) functional program. In this case, one can observe that the recursive call to inc^{-1} in the third rule can only bind variable ys to a non-empty list, say $[z|zs]$. Therefore, the system above could be transformed as follows:

$$\begin{array}{rl} \mathsf{inc}^{-1}([0]) & \to \mathsf{nil} \\ \mathsf{inc}^{-1}([1|xs]) & \to [0|xs] \\ \mathsf{inc}^{-1}([0,z|zs]) & \to [1|\mathsf{inc}^{-1}([z|zs])] \end{array}$$

Now, the transformed system is non-overlapping (and confluent) and, under some conditions, it is equivalent to the original system (roughly speaking, the derivations to constructor terms are the same).

A "more specific" transformation was originally introduced by Marriott et al. [24] in the context of logic programming. In this context, a *more specific version* of a logic program is a version of this program where each clause is further instantiated or removed while preserving the successful derivations of the original program. According to [24], the transformation increases the number of finitely failed goals (i.e., some infinite derivations are transformed into finitely failed ones), detects failure more quickly, etc. In general, the information about the allowed variable bindings which is hidden in the original program may be made explicit in a more specific version, thus improving the static analysis of the program's properties.

In this paper, we adapt the notion of a more specific program to the context of deterministic conditional term rewriting systems. In principle, the transformation may achieve similar benefits as in logic programming by making some variable bindings explicit (as illustrated in the transformation of function inc^{-1} above). Adapting this notion to rewriting systems, however, is far from trivial:

- First, there is no clear notion of *successful* reduction, and aiming at preserving all possible reductions would give rise to a meaningless notion. Consider, e.g., the rewrite systems $\mathcal{R} = \{\mathsf{f}(x) \to \mathsf{g}(x),\ \mathsf{g}(\mathsf{a}) \to \mathsf{b}\}$ and $\mathcal{R}' = \{\mathsf{f}(\mathsf{a}) \to \mathsf{g}(\mathsf{a}),\ \mathsf{g}(\mathsf{a}) \to \mathsf{b}\}$. Here, one would expect $\mathcal{R}'$ to be a correct more specific version of $\mathcal{R}$ (in the sense of $\to_\mathcal{R} = \to_{\mathcal{R}'}$). However, the reduction $\mathsf{f}(\mathsf{b}) \to_\mathcal{R} \mathsf{g}(\mathsf{b})$ is not possible in $\mathcal{R}'$.
- Second, different reduction *strategies* require different conditions in order to guarantee the correctness of the transformation.

– Finally, in contrast to [24], we often require computing a set of instances of a rewrite rule (rather than a single, more general instance) in order to produce non-overlapping left-hand sides. Consider the rewrite system $\mathcal{R} = \{\mathsf{f}(\mathsf{a}) \to \mathsf{a},\ \mathsf{f}(x) \to \mathsf{g}(x),\ \mathsf{g}(\mathsf{b}) \to \mathsf{b},\ \mathsf{g}(\mathsf{c}) \to \mathsf{c}\}$. If we aim at computing a single more specific version for the second rule, only the same rule is obtained. However, if a set of instances is allowed rather a single common instance, we can obtain the system $\mathcal{R}' = \{\mathsf{f}(\mathsf{a}) \to \mathsf{a},\ \mathsf{f}(\mathsf{b}) \to \mathsf{g}(\mathsf{b}),\ \mathsf{f}(\mathsf{c}) \to \mathsf{g}(\mathsf{c}),\ \mathsf{g}(\mathsf{b}) \to \mathsf{b},\ \mathsf{g}(\mathsf{c}) \to \mathsf{c}\}$, which is non-overlapping.

Apart from introducing the notions of *successful* reduction and *more specific* version (MSV), we provide a constructive algorithm for computing more specific versions, which is based on constructing finite (possibly incomplete) *narrowing* trees for the right-hand sides of the original rewrite rules. Here, narrowing [36,19], an extension of term rewriting by replacing pattern matching with unification, is used to *guess* the allowed variable bindings for a given rewrite rule. We prove the correctness of the algorithm (i.e., that it actually outputs a more specific version of the input system). We have tested the usefulness of the MSV transformation to improve the inverse systems obtained by the program inversion method of [28,27,29], and our preliminary results are very encouraging.

This paper is organized as follows. In Section 2, we briefly review some notions and notations of term rewriting and narrowing. Section 3 introduces the notions of more specific version of a rewriting system. Then, we introduce an algorithm for computing more specific versions in Section 4 and prove its correctness. Finally, Section 5 concludes and points out some directions for future research. A preliminary version of this paper — with a more restricted notion of more specific version and where only unconditional rules are considered — appeared in [31].

2 Preliminaries

We assume familiarity with basic concepts of term rewriting and narrowing. We refer the reader to, e.g., [4], [32], and [16] for further details.

Terms and Substitutions. A *signature* $\mathcal{F}$ is a set of function symbols. Given a set of variables $\mathcal{V}$ with $\mathcal{F} \cap \mathcal{V} = \varnothing$, we denote the domain of *terms* by $\mathcal{T}(\mathcal{F}, \mathcal{V})$. We assume that $\mathcal{F}$ always contains at least one constant $\mathsf{f}/0$. We use $\mathsf{f}, \mathsf{g}, \ldots$ to denote functions and $x, y, \ldots$ to denote variables. Positions are used to address the nodes of a term viewed as a tree. A *position* p in a term t is represented by a finite sequence of natural numbers, where ϵ denotes the root position. We let $t|_p$ denote the *subterm* of t at position p and $t[s]_p$ the result of *replacing the subterm* $t|_p$ by the term s. $\mathcal{V}ar(t)$ denotes the set of variables appearing in t. A term t is *ground* if $\mathcal{V}ar(t) = \varnothing$.

A *substitution* $\sigma : \mathcal{V} \mapsto \mathcal{T}(\mathcal{F}, \mathcal{V})$ is a mapping from variables to terms such that $\mathcal{D}\mathsf{om}(\sigma) = \{x \in \mathcal{V} \mid x \neq \sigma(x)\}$ is its domain. Substitutions are extended to morphisms from $\mathcal{T}(\mathcal{F}, \mathcal{V})$ to $\mathcal{T}(\mathcal{F}, \mathcal{V})$ in the natural way. We denote the application of a substitution σ to a term t by $t\sigma$ rather than $\sigma(t)$. The identity substitution is denoted by *id*. A *variable renaming* is a substitution that is a bijection on $\mathcal{V}$. A substitution σ is *more general* than a substitution θ, denoted by $\sigma \leqslant \theta$, if there is a substitution δ such that $\delta \circ \sigma = \theta$, where "$\circ$" denotes the composition of substitutions (i.e., $\sigma \circ \theta(x) = (x\theta)\sigma = x\theta\sigma$). The *restriction* $\theta\restriction_V$ of a substitution θ to a set of variables V is defined as follows: $x\theta\restriction_V = x\theta$ if $x \in V$ and $x\theta\restriction_V = x$ otherwise. We say that $\theta = \sigma$ $[V]$ if $\theta\restriction_V = \sigma\restriction_V$.

A term t_2 is an *instance* of a term t_1 (or, equivalently, t_1 is *more general* than t_2), in symbols $t_1 \leqslant t_2$, if there is a substitution σ with $t_2 = t_1\sigma$. Two terms t_1 and t_2 are *variants* (or equal up to variable renaming) if $t_2 = t_1\rho$ for some variable renaming ρ. A *unifier* of two terms t_1 and t_2 is a substitution σ with $t_1\sigma = t_2\sigma$; furthermore, σ is a *most general unifier* of t_1 and t_2, denoted by $\mathsf{mgu}(t_1, t_2)$, if, for every other unifier θ of t_1 and t_2, we have that $\sigma \leqslant \theta$.

TRSs and Rewriting. A set of rewrite rules $l \to r$ such that l is a nonvariable term and r is a term whose variables appear in l is called a *term rewriting system* (TRS for short); terms l and r are called the left-hand side and the right-hand side of the rule, respectively. We restrict ourselves to finite signatures and TRSs. Given a TRS $\mathcal{R}$ over a signature $\mathcal{F}$, the *defined* symbols $\mathcal{D}_\mathcal{R}$ are the root symbols of the left-hand sides of the rules and the *constructors* are $\mathcal{C}_\mathcal{R} = \mathcal{F} \setminus \mathcal{D}_\mathcal{R}$. *Constructor terms* of $\mathcal{R}$ are terms over $\mathcal{C}_\mathcal{R}$ and $\mathcal{V}$, denoted by $\mathcal{T}(\mathcal{C}_\mathcal{R}, \mathcal{V})$. We sometimes omit $\mathcal{R}$ from $\mathcal{D}_\mathcal{R}$ and $\mathcal{C}_\mathcal{R}$ if it is clear from the context. A substitution σ is a *constructor substitution* (of $\mathcal{R}$) if $x\sigma \in \mathcal{T}(\mathcal{C}_\mathcal{R}, \mathcal{V})$ for all variables x.

For a TRS $\mathcal{R}$, we define the associated rewrite relation $\to_\mathcal{R}$ as the smallest binary relation satisfying the following: given terms $s, t \in \mathcal{T}(\mathcal{F}, \mathcal{V})$, we have $s \to_\mathcal{R} t$ iff there exist a position p in s, a rewrite rule $l \to r \in \mathcal{R}$ and a substitution σ with $s|_p = l\sigma$ and $t = s[r\sigma]_p$; the rewrite step is often denoted by $s \to_{p,l\to r} t$ to make explicit the position and rule used in this step. The instantiated left-hand side $l\sigma$ is called a *redex*.

A term t is called *irreducible* or in *normal form* w.r.t. a TRS $\mathcal{R}$ if there is no term s with $t \to_\mathcal{R} s$. A substitution is called *normalized* w.r.t. $\mathcal{R}$ if every variable in the domain is replaced by a normal form w.r.t. $\mathcal{R}$. We sometimes omit "w.r.t. $\mathcal{R}$" if it is clear from the context. We denote the set of normal forms by $NF_\mathcal{R}$. A *derivation* is a (possibly empty) sequence of rewrite steps. Given a binary relation $\to$, we denote by $\to^*$ its reflexive and transitive closure. Thus $t \to^*_\mathcal{R} s$ means that t can be reduced to s in $\mathcal{R}$ in zero or more steps; we also use $t \to^n_\mathcal{R} s$ to denote that t can be reduced to s in exactly n rewrite steps.

A conditional TRS (CTRS) is a set of rewrite rules $l \to r \Leftarrow C$, where C is a sequence of equations. In particular, we consider only *oriented* equations of the form $s_1 \twoheadrightarrow t_1, \ldots, s_n \twoheadrightarrow t_n$. For a CTRS $\mathcal{R}$, we define the associated rewrite relation $\to_\mathcal{R}$ as follows: given terms $s, t \in \mathcal{T}(\mathcal{F}, \mathcal{V})$, we have $s \to_\mathcal{R} t$ iff there exist a position p in s, a rewrite rule $l \to r \Leftarrow s_1 \twoheadrightarrow t_1, \ldots, s_n \twoheadrightarrow t_n \in \mathcal{R}$ and a substitution σ such that $s|_p = l\sigma$, $s_i\sigma \to^*_\mathcal{R} t_i\sigma$ for all $i = 1, \ldots, n$, and $t = s[r\sigma]_p$.

Narrowing. The *narrowing* principle [36] mainly extends term rewriting by replacing pattern matching with unification, so that terms containing logic (i.e., *free*) variables can also be reduced by non-deterministically instantiating these variables. Conceptually, this is not significantly different from ordinary rewriting when TRSs contain *extra variables* (i.e., variables that appear in the right-hand side of a rule but not in its left-hand side), as noted in [3]. Formally, given a TRS $\mathcal{R}$ and two terms $s, t \in \mathcal{T}(\mathcal{F}, \mathcal{V})$, we have that $s \leadsto_{\mathcal{R}} t$ is a *narrowing step* iff there exist[1]

- a nonvariable position p of s,
- a variant $l \to r$ of a rule in $\mathcal{R}$,
- a substitution $\sigma = \mathsf{mgu}(s|_p, l)$ which is a most general unifier of $s|_p$ and l,

and $t = (s[r]_p)\sigma$. We often write $s \leadsto_{p,l\to r,\theta} t$ (or simply $s \leadsto_{\theta} t$) to make explicit the position, rule, and substitution of the narrowing step, where $\theta = \sigma\!\upharpoonright_{\mathcal{V}\mathsf{ar}(s)}$ (i.e., we label the narrowing step only with the bindings for the narrowed term). A *narrowing derivation* $t_0 \leadsto^*_\sigma t_n$ denotes a sequence of narrowing steps $t_0 \leadsto_{\sigma_1} \cdots \leadsto_{\sigma_n} t_n$ with $\sigma = \sigma_n \circ \cdots \circ \sigma_1$ (if $n = 0$ then $\sigma = id$). Given a narrowing derivation $s \leadsto^*_\sigma t$ with t a constructor term, we say that σ is a *computed answer* for s.

Example 1. Consider the TRS

$$\mathcal{R} = \left\{ \begin{array}{ll} \mathsf{add}(0, y) \to y & (R_1) \\ \mathsf{add}(\mathsf{s}(x), y) \to \mathsf{s}(\mathsf{add}(x, y)) & (R_2) \end{array} \right\}$$

defining the addition $\mathsf{add}/2$ on natural numbers built from $0/0$ and $\mathsf{s}/1$. Given the term $\mathsf{add}(x, \mathsf{s}(0))$, we have infinitely many narrowing derivations starting from $\mathsf{add}(x, \mathsf{s}(0))$, e.g.,

$$\begin{array}{lll} \mathsf{add}(x, \mathsf{s}(0)) \leadsto_{\epsilon,R_1,\{x\mapsto 0\}} & \mathsf{s}(0) & \\ \mathsf{add}(x, \mathsf{s}(0)) \leadsto_{\epsilon,R_2,\{x\mapsto \mathsf{s}(y_1)\}} & \mathsf{s}(\mathsf{add}(y_1, \mathsf{s}(0))) \leadsto_{1,R_1,\{y_1\mapsto 0\}} & \mathsf{s}(\mathsf{s}(0)) \\ \ldots & & \end{array}$$

with computed answers $\{x \mapsto 0\}$, $\{x \mapsto \mathsf{s}(0)\}$, etc.

Narrowing is naturally extended to deal with equations and CTRSs (see Section 4).

3 More Specific Conditional Rewrite Systems

In this section, we introduce the notion of a *more specific* conditional rewrite system. Intuitively speaking, we produce a more specific CTRS $\mathcal{R}'$ from a CTRS $\mathcal{R}$ by replacing a conditional rewrite rule $l \to r \Leftarrow C \in \mathcal{R}$ with a *finite* number of *instances* of this rule, i.e., $\mathcal{R}' = (\mathcal{R} \setminus \{l \to r \Leftarrow C\}) \cup \{(l \to r \Leftarrow C)\sigma_1, \ldots, (l \to r \Leftarrow C)\sigma_n\}$, such that $\mathcal{R}'$ is *semantically* equivalent to $\mathcal{R}$ under some conditions.

[1] We consider the so called *most general* narrowing, i.e., the mgu of the selected subterm and the left-hand side of a rule—rather than an arbitrary unifier—is computed at each narrowing step.

The key idea is that more specific versions should still allow the same reductions of the original system. However, as mentioned in Section 1, if we aimed at preserving *all* possible rewrite reductions, the resulting notion would be useless since it would never happen in practice. Therefore, to have a practically applicable technique, we only aim at preserving what we call *successful* reductions. In the following, given a CTRS $\mathcal{R}$, we denote by $\stackrel{s}{\to}_{\mathcal{R}}$ a generic conditional rewrite relation based on some strategy s (e.g., innermost conditional reduction $\stackrel{i}{\to}_{\mathcal{R}}$).

Definition 2 (successful reduction w.r.t. $\stackrel{s}{\to}$). *Let $\mathcal{R}$ be a CTRS and let $\stackrel{s}{\to}_{\mathcal{R}}$ be a conditional rewrite relation. A rewrite reduction $t \stackrel{s}{\to}^*_{\mathcal{R}} u$ where t is a term and u is a constructor term, is called a* successful *reduction w.r.t. $\stackrel{s}{\to}_{\mathcal{R}}$.*

Let us now introduce our notion of a *more specific version* of a rewrite rule:

Definition 3 (more specific version of a rule). *Let $\mathcal{R}$ be a CTRS and $\stackrel{s}{\to}_{\mathcal{R}}$ be a conditional rewrite relation. Let $l \to r \Leftarrow C \in \mathcal{R}$ be a rewrite rule. We say that the finite set of rewrite rules $\mathcal{R}_{msv} = \{l_1 \to r_1 \Leftarrow C_1, \ldots, l_n \to r_n \Leftarrow C_n\}$ is a* more specific version *of $l \to r \Leftarrow C$ in $\mathcal{R}$ w.r.t. $\stackrel{s}{\to}_{\mathcal{R}}$ if*

- *there are substitutions $\sigma_1, \ldots, \sigma_n$ such that $(l \to r \Leftarrow C)\sigma_i = l_i \to r_i \Leftarrow C_i$ for $i = 1, \ldots, n$, and*
- *for all terms t, u, we have that $t \stackrel{s}{\to}^*_{\mathcal{R}} u$ is successful in $\mathcal{R}$ iff $t \stackrel{s}{\to}^*_{\mathcal{R}'} u$ is successful in $\mathcal{R}'$, with $\mathcal{R}' = (\mathcal{R} \backslash \{l \to r \Leftarrow C\}) \cup \mathcal{R}_{msv}$; moreover, we require $t \stackrel{s}{\to}^*_{\mathcal{R}} u$ and $t \stackrel{s}{\to}^*_{\mathcal{R}'} u$ to apply the same rules to the same positions and in the same order, except for the rule $l \to r \Leftarrow C$ in $\mathcal{R}$ that is replaced with some rule $l_i \to r_i \Leftarrow C_i$, $i \in \{1, \ldots, n\}$, in $\mathcal{R}'$.*[2]

Note that a rewrite rule is always a more specific version of itself; therefore the existence of a more specific version of a given rule is always ensured. In general, however, the more specific version is not unique and, thus, there can be several strictly more specific versions of a rewrite rule.

The notion of a more specific version of a rule is extended to CTRSs in a stepwise manner: given a CTRS $\mathcal{R}$, we first replace a rule of $\mathcal{R}$ by its more specific version thus producing $\mathcal{R}'$, then we replace another rule of $\mathcal{R}'$ by its more specific version, and so forth. We denote each of these steps by $\mathcal{R} \mapsto_{\mathsf{more}} \mathcal{R}'$. We say that $\mathcal{R}'$ is a *more specific version* of $\mathcal{R}$ if there is a sequence of (zero or more) $\mapsto_{\mathsf{more}}$ steps leading from $\mathcal{R}$ to $\mathcal{R}'$. Note that, given a CTRS $\mathcal{R}$ and one of its more specific versions $\mathcal{R}'$, we have that $\to_{\mathcal{R}'} \subseteq \to_{\mathcal{R}}$ (i.e., $\mathit{NF}_{\mathcal{R}} \subseteq \mathit{NF}_{\mathcal{R}'}$), $\mathcal{D}_{\mathcal{R}} = \mathcal{D}_{\mathcal{R}'}$ and $\mathcal{C}_{\mathcal{R}} = \mathcal{C}_{\mathcal{R}'}$.

---

[2] This is required to prevent situations like the following one. Consider $\mathcal{R} = \{\mathsf{f}(\mathsf{0}) \to \mathsf{g}(\mathsf{0}), \mathsf{f}(x) \to \mathsf{g}(x), \mathsf{g}(\mathsf{0}) \to \mathsf{0}\}$. Here, any instance of rule $\mathsf{f}(x) \to \mathsf{g}(x)$ would be a more specific version if the last condition were not required (since the rule $\mathsf{f}(\mathsf{0}) \to \mathsf{g}(\mathsf{0})$ already belongs to $\mathcal{R}$ and could be used instead).

Example 4. Consider the following CTRS $\mathcal{R}$ (a fragment of a system obtained automatically by the function inversion technique of [29]):

$$\begin{array}{rll} \mathsf{inv}([\mathsf{left}|x_2]) \to (t, \mathsf{n}(x,y)) \Leftarrow & \mathsf{inv}(x_2) \twoheadrightarrow (x_1, x),\ \mathsf{inv}'(x_1) \twoheadrightarrow (t,y) & (R_1) \\ \mathsf{inv}([\mathsf{str}(x)|t]) \to (t, \mathsf{sym}(x)) & & (R_2) \\ \mathsf{inv}([\mathsf{left}, \mathsf{right}|t]) \to (t, \mathsf{bottom}) & & (R_3) \end{array}$$

where the definition of function inv' is not relevant for this example and we omit the tuple symbol (e.g., tp_2) from the tuple of two terms — we write (t_1, t_2) instead of $\mathsf{tp}_2(t_1, t_2)$. Here, we replace the first rule of $\mathcal{R}$ by the following two instances $\mathcal{R}_{msv}$:

$$\begin{array}{r} \mathsf{inv}([\mathsf{left}, \mathsf{left}|x_3]) \to (t, \mathsf{n}(x,y)) \Leftarrow \mathsf{inv}([\mathsf{left}|x_2]) \twoheadrightarrow (x_1, x), \mathsf{inv}'(x_1) \twoheadrightarrow (t,y) \\ \mathsf{inv}([\mathsf{left}, \mathsf{str}(x_3)|x_4]) \to (t, \mathsf{n}(x,y)) \Leftarrow \mathsf{inv}([\mathsf{str}(x_3)|x_4]) \twoheadrightarrow (x_1, x), \mathsf{inv}'(x_1) \twoheadrightarrow (t,y) \end{array}$$

using the substitutions $\sigma_1 = \{x_2 \mapsto [\mathsf{left}|x_3]\}$ and $\sigma_2 = \{x_2 \mapsto [\mathsf{str}(x_3)|x_4]\}$.

Observe that function inv in $(\mathcal{R}\backslash\{R_1\}) \cup \mathcal{R}_{msv}$ is now non-overlapping. Note that producing only a single instance would not work since the only common generalization of σ_1 and σ_2 is $\{x_2 \mapsto [x_3|x_4]\}$ so that the more specific version would be

$$\mathsf{inv}([\mathsf{left}, x_3|x_4]) \to (t, \mathsf{n}(x,y)) \Leftarrow\ \mathsf{inv}([x_3|x_4]) \twoheadrightarrow (x_1, x),\ \mathsf{inv}'(x_1) \twoheadrightarrow (t,y)$$

and thus, function inv would still be overlapping.

Now, we show a basic property of more specific versions of a CTRS. In the following, we say that a CTRS $\mathcal{R}$ *has the unique constructor normal form property w.r.t.* $\stackrel{s}{\to}_{\mathcal{R}}$ if, for all successful reductions $t \stackrel{s}{\to}^* u$ and $t \stackrel{s}{\to}^* u'$, with $u, u' \in \mathcal{T}(\mathcal{C}, \mathcal{V})$, we have $u = u'$.

Theorem 5. *Let $\mathcal{R}$ be a CTRS and let $\mathcal{R}'$ be a more specific version of $\mathcal{R}$ w.r.t. $\stackrel{s}{\to}_{\mathcal{R}}$. Then, $\mathcal{R}$ has the unique constructor normal form property w.r.t. $\stackrel{s}{\to}_{\mathcal{R}}$ iff so does $\mathcal{R}'$.*

Proof. The claim follows straightforwardly since derivations producing a constructor normal form are successful derivations, and they are preserved by the MSV transformation by definition. □

4 Computing More Specific Versions

In this section, we tackle the definition of a constructive method for computing more specific versions of a CTRS. For this purpose, we consider the following assumptions:

- We restrict the class of CTRSs to *deterministic* CTRSs (DCTRSs) [5,13,23]. Furthermore, we require them to be constructor systems (see below).
- We only consider constructor-based reductions (a particular case of innermost conditional reduction), i.e., reductions where the computed matching substitutions are constructor.

- We use a form of innermost conditional narrowing to approximate the potential successful reductions and, thus, compute its more specific version.

DCTRSs are *3-CTRSs* [5,13,23] (see also [32]) (i.e., CTRSs where extra variables are allowed as long as $\mathcal{V}ar(r) \subseteq \mathcal{V}ar(l) \cup \mathcal{V}ar(C)$ for all rules $l \to r \Leftarrow C$), where the conditional rules have the form

$$l \to r \Leftarrow s_1 \twoheadrightarrow t_1, \ldots, s_n \twoheadrightarrow t_n$$

with $s_1 \twoheadrightarrow t_1, \ldots, s_n \twoheadrightarrow t_n$ oriented equations, and such that

- $\mathcal{V}\mathsf{ar}(s_i) \subseteq \mathcal{V}\mathsf{ar}(l) \cup \mathcal{V}\mathsf{ar}(t_1) \cup \cdots \cup \mathcal{V}\mathsf{ar}(t_{i-1})$, for all $i = 1, \ldots, n$;
- $\mathcal{V}\mathsf{ar}(r) \subseteq \mathcal{V}\mathsf{ar}(l) \cup \mathcal{V}\mathsf{ar}(t_1) \cup \ldots \cup \mathcal{V}\mathsf{ar}(t_n)$.

Moreover, we assume that the DCTRSs are *constructor systems* where $l = \mathsf{f}(l_1, \ldots, l_n)$ with $l_i \in \mathcal{T}(\mathcal{C}, \mathcal{V})$, $i = 1, \ldots, m$, and $t_1, \ldots, t_n \in \mathcal{T}(\mathcal{C}, \mathcal{V})$.

In DCTRSs, extra variables in the conditions are not problematic since no redex contains extra variables when it is selected. Actually, as noted by [23], these systems are basically equivalent to functional programs since every rule

$$l \to r \Leftarrow s_1 \twoheadrightarrow t_1, \ldots, s_n \twoheadrightarrow t_n$$

can be seen in a functional language as

$$\begin{array}{rcl} l & = & \textbf{let } t_1 = s_1 \textbf{ in} \\ & & \quad \textbf{let } t_2 = s_2 \textbf{ in} \\ & & \quad \ldots \\ & & \qquad \textbf{let } t_n = s_n \textbf{ in } r \end{array}$$

Here, DCTRSs allow us to represent functional *local definitions* using oriented conditions.

Under these conditions, innermost reduction extends quite naturally to the conditional case. In particular, we follow Bockmayr and Werner's *conditional rewriting without evaluation of the premise* [6], adapted to our setting as follows:

Definition 6 (constructor-based conditional reduction). *Let $\mathcal{R}$ be a DCTRS. Constructor-based conditional reduction is defined as the smallest relation satisfying the following transition rules:*

$$(\mathsf{reduction})\ \frac{p = inn(s_1) \wedge l \to r \Leftarrow C \in \mathcal{R} \wedge s_1|_p = l\sigma \wedge \sigma \text{ is constructor}}{(s_1 \twoheadrightarrow t_1, \ldots, s_n \twoheadrightarrow t_n) \overset{c}{\to} (C\sigma, s_1[r\sigma]_p \twoheadrightarrow t_1, \ldots, s_n \twoheadrightarrow t_n)}$$

$$(\mathsf{matching})\ \frac{n > 1 \wedge s_1 \in \mathcal{T}(\mathcal{C}, \mathcal{V}) \wedge s_1 = t_1\sigma}{(s_1 \twoheadrightarrow t_1, \ldots, s_n \twoheadrightarrow t_n) \overset{c}{\to} (s_2 \twoheadrightarrow t_2, \ldots, s_n \twoheadrightarrow t_n)\sigma}$$

where $inn(s)$ selects the position of an innermost subterm s matchable with the left-hand side of a rule (i.e., a term $l'\sigma'$ of the form $f(c_1, \ldots, c_n)$ with $f \in \mathcal{D}$ and $c_1, \ldots, c_n \in \mathcal{T}(\mathcal{C}, \mathcal{V})$ for some $l' \to r' \Leftarrow C' \in \mathcal{R}$ and some σ'), e.g., the leftmost one.

Intuitively speaking, in order to reduce a sequence of equations $s_1 \twoheadrightarrow t_1, \ldots, s_n \twoheadrightarrow t_n$, we consider two possibilities:

- If the first oriented equation has some innermost subterm that matches the left-hand side of a rewrite rule, we perform a reduction step. Note that, in contrast to ordinary conditional rewriting, the conditions are not verified but just added to the sequence of equations (as in Bockmayr and Werner's reduction without evaluation of the premise).
- If the left-hand side of the first oriented equation is a constructor term, then we match both sides (note that the right-hand side is a constructor term by definition) and remove this equation from the sequence. In the original definition of [6], this matching substitution is computed when applying a given rule in order to verify the conditions. Our definition simply makes computing the substitution more operational by postponing it to the point when its computation is required.
  Note that this rule requires having more than one equation, since the initial equation should not be removed.

In order to reduce a ground term s, we consider an initial oriented equation $s \twoheadrightarrow x$, where x is a fresh variable not occurring in s, and reduce it as much as possible using the reduction and matching rules. If we reach an equation of the form $t \twoheadrightarrow x$, where t is a constructor term, we say that s reduces to t; actually, [6, Theorem 2] proves the equivalence between ordinary conditional rewriting and conditional rewriting without evaluation of the premise.

Example 7. Consider again the system $\mathcal{R}$ from Example 4 and the initial term $\mathsf{inv}([\mathsf{left}, \mathsf{str}(\mathsf{a})])$. We have, for instance, the following (incomplete) constructor-based reduction:

$$
\begin{aligned}
&(\mathsf{inv}([\mathsf{left}, \mathsf{str}(\mathsf{a})]) \twoheadrightarrow w)\\
&\quad \overset{c}{\to} (\mathsf{inv}([\mathsf{str}(\mathsf{a})]) \twoheadrightarrow (x_1, x), \mathsf{inv}'(x_1) \twoheadrightarrow (t, y), (t, \mathsf{n}(x, y)) \twoheadrightarrow w)\\
&\quad \overset{c}{\to} ((\mathsf{nil}, \mathsf{sym}(\mathsf{a})) \twoheadrightarrow (x_1, x), \mathsf{inv}'(x_1) \twoheadrightarrow (t, y), (t, \mathsf{n}(x, y)) \twoheadrightarrow w)\\
&\quad \overset{c}{\to} (\mathsf{inv}'(\mathsf{nil}) \twoheadrightarrow (t, y), (t, \mathsf{n}(\mathsf{sym}(\mathsf{a}), y)) \twoheadrightarrow w)
\end{aligned}
$$

In the following, given a rule $l \to r \Leftarrow C$, we introduce the use of conditional narrowing to automatically compute an approximation of the successful constructor-based reductions.

Our definition of constructor-based conditional narrowing (a special case of innermost conditional narrowing [9,12,18]), denoted by $\overset{c}{\leadsto}$, is defined as follows:

Definition 8 (constructor-based conditional narrowing). *Let $\mathcal{R}$ be a DC-TRS. Constructor-based conditional narrowing is defined as the smallest relation satisfying the following transition rules:*

$$
\textit{(narrowing)}\quad \frac{p = \mathit{inn}(s_1) \ \wedge \ l \to r \Leftarrow C \in \mathcal{R} \ \wedge \ \sigma = \mathsf{mgu}(s_1|_p, l)}{(s_1 \twoheadrightarrow t_1, \ldots, s_n \twoheadrightarrow t_n) \ \overset{c}{\leadsto}_\sigma \ (C, s_1[r]_p \twoheadrightarrow t_1, \ldots, s_n \twoheadrightarrow t_n)\sigma}
$$

$$
\textit{(unification)}\quad \frac{n > 1 \ \wedge \ s_1 \in \mathcal{T}(\mathcal{C}, \mathcal{V}) \ \wedge \ \sigma = \mathsf{mgu}(s_1, t_1)}{(s_1 \twoheadrightarrow t_1, \ldots, s_n \twoheadrightarrow t_n) \ \overset{c}{\leadsto}_\sigma \ (s_2 \twoheadrightarrow t_2, \ldots, s_n \twoheadrightarrow t_n)\sigma}
$$

where inn(s) selects the position of an innermost subterm whose proper subterms are constructor terms, and which is unifiable with the left-hand side of a rule, e.g., the leftmost one.

As it can be seen, our definition of constructor-based conditional narrowing for DCTRSs mimics the definition of constructor-based reduction but replaces matching with unification in both transition rules. Note, however, that the first rule is often non-deterministic since a given innermost subterm can unify with the left-hand sides of several rewrite rules.

We adapt the notion of successful derivations to rewriting and narrowing derivations of equations.

Definition 9. *Let $\mathcal{R}$ be a DCTRS and let $\stackrel{s}{\to}_{\mathcal{R}}$ be a conditional rewrite relation. A rewrite reduction $(C, t \twoheadrightarrow x) \stackrel{s}{\to}^*_{\mathcal{R}} (u \twoheadrightarrow x)$ where t is a term, x is a fresh variable, C is a (possibly empty) sequence of (oriented) equations, and $u \in \mathcal{T}(\mathcal{C}, \mathcal{V})$ is a constructor term, is called a* successful *reduction w.r.t. $\stackrel{s}{\to}_{\mathcal{R}}$.*

Note that $t \stackrel{s}{\to}^*_{\mathcal{R}} u$ is successful w.r.t. $\stackrel{s}{\to}_{\mathcal{R}}$ iff so is $(t \twoheadrightarrow x) \stackrel{s}{\to}^*_{\mathcal{R}} (u \twoheadrightarrow x)$, where x is a fresh variable.

Definition 10. *Consider a sequence of equations C, a term s and a fresh variable x. We say that a constructor-based conditional narrowing derivation of the form $(C, s \twoheadrightarrow x) \stackrel{c}{\leadsto}^*_{\sigma} (t \twoheadrightarrow x)$ is* successful *if $t \in \mathcal{T}(\mathcal{C}, \mathcal{V})$ is a constructor term.*

We say that a derivation of the form $(C, s \twoheadrightarrow x) \stackrel{c}{\leadsto}^*_{\sigma} (C', s' \twoheadrightarrow x)$ is a failing *derivation if no more narrowing steps can be applied and at least one of the following conditions holds:*

- *C' is not the empty sequence, or*
- *s' is not a constructor term.*

A finite derivation is called incomplete *when it is neither successful nor failing.*

Because of the non-determinism of rule narrowing, the computation of all narrowing derivations starting from a given term is usually represented by means of a narrowing *tree*:

Definition 11 (constructor-based conditional narrowing tree). *Let $\mathcal{R}$ be a DCTRS and C be a sequence of equations. A (possibly incomplete)* constructor-based conditional narrowing tree *for C in $\mathcal{R}$ is a (possibly infinite) directed rooted node- and edge-labeled graph τ built as follows:*

- *the root node of τ is labeled with C;*
- *every node C_1 is either a leaf (a node with no output edge) or it is unfolded as follows: there is an output edge from node C_1 to node C_2 labeled with σ for all constructor-based conditional narrowing steps $C_1 \stackrel{c}{\leadsto}_{\sigma} C_2$ for the selected innermost narrowable term;*
- *the root node is not a leaf — at least the root node should be unfolded.*

By abuse of notation, we will denote a finite constructor-based conditional narrowing tree τ *(and its subtrees) for* C *as a finite set with the constructor-based conditional narrowing derivations starting from* C *in this tree, i.e.,* $C \overset{c}{\leadsto}{}^{*}_{\theta} C' \in \tau$ *if there is a root-to-leaf path from* C *to* C' *in* τ *labeled with substitutions* $\theta_1, \theta_2, \ldots, \theta_n$ *such that* $C \overset{c}{\leadsto}_{\theta_1} \cdots \overset{c}{\leadsto}_{\theta_n} C'$ *is a constructor-based conditional narrowing derivation and* $\theta = \theta_n \circ \cdots \circ \theta_1$.

Example 12. Consider again the system $\mathcal{R}$ from Example 4 and the initial sequence of equations $C = (\mathsf{inv}(x_2) \twoheadrightarrow (x_1, x),\ \mathsf{inv}'(x_1) \twoheadrightarrow (t, y),\ (t, \mathsf{n}(x, y)) \twoheadrightarrow w)$, where w is a fresh variable (the reason to consider this initial sequence of equations will be clear in Definition 14 below). The depth-1 (i.e., derivations are stopped after one narrowing step) constructor-based conditional narrowing tree τ for C contains the following derivations:

$$
\begin{array}{ll}
C \overset{c}{\leadsto}_{\{x_2 \mapsto [\mathsf{left}|x_2']\}} & (\mathsf{inv}(x_2') \twoheadrightarrow (x_1', x'),\ \mathsf{inv}'(x_1') \twoheadrightarrow (t', y'), \\
& (t', \mathsf{n}(x', y')) \twoheadrightarrow (x_1, x),\ \mathsf{inv}'(x_1) \twoheadrightarrow (t, y),\ (t, \mathsf{n}(x, y)) \twoheadrightarrow w) \\
C \overset{c}{\leadsto}_{\{x_2 \mapsto [\mathsf{str}(x')|t']\}} & ((t', \mathsf{sym}(x')) \twoheadrightarrow (x_1, x),\ \mathsf{inv}'(x_1) \twoheadrightarrow (t, y),\ (t, \mathsf{n}(x, y)) \twoheadrightarrow w) \\
C \overset{c}{\leadsto}_{\{x_2 \mapsto [\mathsf{left}, \mathsf{right}|t']\}} & ((t', \mathsf{bottom}) \twoheadrightarrow (x_1, x),\ \mathsf{inv}'(x_1) \twoheadrightarrow (t, y),\ (t, \mathsf{n}(x, y)) \twoheadrightarrow w)
\end{array}
$$

Note that a constructor-based conditional narrowing tree can be *incomplete* in the sense that we do not require all unfoldable nodes to be unfolded (i.e., some finite derivations in the tree may be incomplete). Nevertheless, if a node is selected to be unfolded, it should be unfolded in all possible ways using constructor-based conditional narrowing (i.e., one cannot *partially* unfold a node by ignoring some unifying rules, which would give rise to incorrect results). In order to keep the tree finite, one can introduce a heuristics that determines when the construction of the tree should terminate. We consider the definition of a particular strategy for ensuring termination out of the scope of this paper; nevertheless, one could use a simple depth-k strategy (as in the previous example) or some more elaborated strategies based on well-founded or well-quasi orderings [10] (as in narrowing-driven partial evaluation [1]).

Let us now recall the notion of *least general generalization* [34] (also called *anti-unification* [33]), which will be required to compute a common generalization of all instances of a term by a set of narrowing derivations.

Definition 13 (least general generalization [34], *lgg*). *Given two terms* s *and* t*, we say that* w *is a generalization of* s *and* t *if* $w \leqslant s$ *and* $w \leqslant t$*; moreover, it is called the least general generalization of* s *and* t*, denoted by* $lgg(s,t)$*, if* $w' \leqslant w$ *for all other generalizations* w' *of* s *and* t*. This notion is extended to sets of terms in the natural way:* $lgg(\{t_1, \ldots, t_n\}) = lgg(t_1, lgg(t_2, \ldots lgg(t_{n-1}, t_n) \ldots))$ *(with* $lgg(\{t_1\}) = t_1$ *when* $n = 1$*).*

An algorithm for computing the least general generalization can be found, e.g., in [11]. Let us recall this algorithm for self-containedness. In order to compute

$\mathit{lgg}(s,t)$, this algorithm starts with a tuple $\langle \{s \sqcap_x t\},\ x\rangle$, where x is a fresh variable, and applies the following rules until no rule is applicable:

$$\langle \{f(s_1,\ldots,s_n) \sqcap_x f(t_1,\ldots,t_n)\} \cup P,\ w\rangle \Rightarrow \langle \{s_1 \sqcap_{x_1} t_1,\ldots,s_n \sqcap_{x_n} t_n\} \cup P,\ w\sigma\rangle$$
$$\text{where } \sigma \text{ is } \{x \mapsto f(x_1,\ldots,x_n)\} \text{ and } x_1,\ldots,x_n \text{ are fresh variables}$$
$$\langle \{s \sqcap_x t, s \sqcap_y t\} \cup P,\ w\rangle \Rightarrow \langle \{s \sqcap_y t\} \cup P,\ w\sigma\rangle$$
$$\text{where } \sigma \text{ is } \{x \mapsto y\}$$

Then, the second element of the final tuple is the computed least general generalization.

For instance, the computation of $\mathit{lgg}(\mathsf{f}(\mathsf{a},\mathsf{g}(\mathsf{a})),\mathsf{f}(\mathsf{b},\mathsf{g}(\mathsf{b})))$ proceeds as follows:

$$\begin{aligned}\langle \{\mathsf{f}(\mathsf{a},\mathsf{g}(\mathsf{a})) \sqcap_x \mathsf{f}(\mathsf{b},\mathsf{g}(\mathsf{b}))\},\ x\rangle &\Rightarrow \langle \{\mathsf{a} \sqcap_{x_1} \mathsf{b}, \mathsf{g}(\mathsf{a}) \sqcap_{x_2} \mathsf{g}(\mathsf{b})\},\ \mathsf{f}(x_1,x_2)\rangle\\ &\Rightarrow \langle \{\mathsf{a} \sqcap_{x_1} \mathsf{b}, \mathsf{a} \sqcap_{x_3} \mathsf{b}\},\ \mathsf{f}(x_1,\mathsf{g}(x_3))\rangle\\ &\Rightarrow \langle \{\mathsf{a} \sqcap_{x_3} \mathsf{b}\},\ \mathsf{f}(x_3,\mathsf{g}(x_3))\rangle\end{aligned}$$

Therefore, $\mathit{lgg}(\mathsf{f}(\mathsf{a},\mathsf{g}(\mathsf{a})),\mathsf{f}(\mathsf{b},\mathsf{g}(\mathsf{b}))) = \mathsf{f}(x_3,\mathsf{g}(x_3))$.

We now introduce a constructive algorithm to produce a more specific version of a rule:

Definition 14 (MSV algorithm for DCTRSs). *Let $\mathcal{R}$ be a DCTRS and let $l \to r \Leftarrow C \in \mathcal{R}$ be a conditional rewrite rule such that not all terms in r and C are constructor terms. Let τ be a finite (possibly incomplete) constructor-based conditional narrowing tree for $(C, r \twoheadrightarrow x)$ in $\mathcal{R}$, where x is a fresh variable, and $\tau' \subseteq \tau$ be the tree obtained from τ by excluding the failing derivations (if any). We compute a more specific version of $l \to r \Leftarrow C$ in $\mathcal{R}$, denoted by* $\mathrm{MSV}(\mathcal{R}, l \to r \Leftarrow C, \tau)$, *as follows:*

- *If $\tau' = \varnothing$, then $\mathrm{MSV}(\mathcal{R}, l \to r \Leftarrow C, \tau) = \bot$, where $\bot$ is used to denote that the rule is useless (i.e., no successful constructor-based reduction can use it) and can be removed from $\mathcal{R}$.*
- *If $\tau' \neq \varnothing$, then we let $\tau' = \tau_1 \uplus \ldots \uplus \tau_n$ be a partition of the set τ' such that $\tau_i \neq \varnothing$ for all $i = 1,\ldots,n$. Then,*[3]

$$\mathrm{MSV}(\mathcal{R}, l \to r \Leftarrow C, \tau) = \{(l \to r \Leftarrow C)\sigma_1,\ldots,(l \to r \Leftarrow C)\sigma_n\}$$

where

$$(l \to r \Leftarrow C)\sigma_i = \mathit{lgg}(\{(l \to r \Leftarrow C)\theta \mid (C, r \twoheadrightarrow x) \overset{c}{\leadsto}{}^{*}_{\theta} (C', r' \twoheadrightarrow x) \in \tau_i\})$$

with $\mathcal{D}\mathsf{om}(\sigma_i) \subseteq \mathcal{V}\mathsf{ar}(C) \cup \mathcal{V}\mathsf{ar}(r)$, $i = 1,\ldots,n$.[4]

[3] The lgg operator is trivially extended to equations by considering them as terms, e.g., the sequence $s_1 \twoheadrightarrow t_1, s_2 \twoheadrightarrow t_2$ is considered as the term $\wedge(\twoheadrightarrow(s_1,t_1), \twoheadrightarrow(s_2,t_2))$, where $\twoheadrightarrow$ and $\wedge$ are binary function symbols.

[4] By definition of constructor-based conditional narrowing, it is clear that $\sigma_1,\ldots,\sigma_n$ are constructor substitutions.

Regarding the partitioning of the derivations in the tree τ', i.e., computing $\tau_1, \ldots, \tau_n$ such that $\tau' = \tau_1 \uplus \ldots \uplus \tau_n$, we first apply a simple pre-processing to avoid trivial overlaps between the generated rules: we remove from τ' those derivations $(C, r \twoheadrightarrow x) \overset{c}{\leadsto}{}^*_{\theta} (C', r' \twoheadrightarrow x)$ such that there exists another derivation $(C, r \twoheadrightarrow x) \overset{c}{\leadsto}{}^*_{\theta'} (C'', r'' \twoheadrightarrow x)$ with $\theta' \leqslant \theta$. In this way, we avoid the risk of having such derivations in different subtrees, τ_i and τ_j, thus producing overlapping rules. Once these derivations have been removed, we could proceed as follows:

- No partition ($n = 1$). This is what is done in the original transformation for logic programs [24] and gives good results for most examples (i.e., produces a non-overlapping system).
- Consider a maximal partitioning, i.e., each τ_i just contains a single derivation. This strategy might produce poor results when the computed substitutions overlap, since overlapping instances would then be produced (even if the considered function was originally non-overlapping).
- Use a heuristics that tries to produce a non-overlapping system whenever possible. Basically, it would proceed as follows. Assume we want to apply the MSV transformation to a function f. Let k be a natural number greater than the maximum depth of the left-hand sides of the rules defining f. Then, we partition the tree τ' as $\tau_1, \ldots, \tau_n$ so that it satisfies the following condition (while keeping n as small as possible): for each $(C, r \twoheadrightarrow x) \overset{c}{\leadsto}{}^*_{\theta} C_1$ and $(C, r \twoheadrightarrow x) \overset{c}{\leadsto}{}^*_{\theta'} C_2$ in τ_i, we have that
  - $l\theta$ and $l\theta'$ are unifiable,[5] or
  - $top_k(l\theta)$ and $top_k(l\theta')$ are equivalent up to variable renaming,

  where $l \to r \Leftarrow C$ is the considered rule. Here, given a fresh constant $\top$, top_k is defined as follows: $top_0(t) = \top$, $top_k(x) = x$ for $k > 0$, $top_k(f(t_1, \ldots, t_m)) = f(top_{k-1}(t_1), \ldots, top_{k-1}(t_m))$ for $k > 0$, i.e., $top_k(t)$ returns the topmost symbols of t up to depth k and replaces the remaining subterms by the fresh constant $\top$.

Example 15. Let us apply the MSV algorithm to the first rule of the system $\mathcal{R}$ introduced in Example 4. Here, we consider the depth-1 constructor-based conditional narrowing tree τ with the derivations shown in Example 12. We first remove derivations labeled with less general substitutions, so that from the first and third derivations of Example 12, only the first one remains. Therefore, we only consider the first and second derivations of Example 12:

$$C \overset{c}{\leadsto}_{\{x_2 \mapsto [\mathsf{left}|x_2']\}} \; (\mathsf{inv}(x_2') \twoheadrightarrow (x_1', x'),\; \mathsf{inv}'(x_1') \twoheadrightarrow (t', y'),$$
$$(t', \mathsf{n}(x', y')) \twoheadrightarrow (x_1, x),\; \mathsf{inv}'(x_1) \twoheadrightarrow (t, y),\; (t, \mathsf{n}(x, y)) \twoheadrightarrow w)$$
$$C \overset{c}{\leadsto}_{\{x_2 \mapsto [\mathsf{str}(x')|t']\}} \; (t', \mathsf{sym}(x')) \twoheadrightarrow (x_1, x),\; \mathsf{inv}'(x_1) \twoheadrightarrow (t, y),\; (t, \mathsf{n}(x, y)) \twoheadrightarrow w$$

[5] Suppose that θ and θ' belong to different partitions, e.g., τ_i and τ_j, resp., and let σ_i and σ_j be the substitutions obtained from τ_i and τ_j, resp., in Definition 14. Then, $\sigma_i \leq \theta$ and $\sigma_j \leq \theta'$ and, thus, $l\sigma_i$ and $l\sigma_j$ are unifiable. Therefore, MSV generates an overlapping system.

Now, either by considering a maximal partitioning or the one based on function top_k, we compute the following partitions:

$$\begin{aligned}\tau_1 =\\ \{C \overset{c}{\leadsto}_{\{x_2 \mapsto [\mathsf{left}|x'_2]\}} & (\mathsf{inv}(x'_2) \twoheadrightarrow (x'_1, x'),\ \mathsf{inv}'(x'_1) \twoheadrightarrow (t', y'),\\ & (t', \mathsf{n}(x', y')) \twoheadrightarrow (x_1, x),\ \mathsf{inv}'(x_1) \twoheadrightarrow (t, y),\ (t, \mathsf{n}(x, y)) \twoheadrightarrow w)\}\end{aligned}$$

$$\begin{aligned}\tau_2 =\\ \{C \overset{c}{\leadsto}_{\{x_2 \mapsto [\mathsf{str}(x')|t']\}} ((t', \mathsf{sym}(x')) \twoheadrightarrow (x_1, x),\ \mathsf{inv}'(x_1) \twoheadrightarrow (t, y),\ (t, \mathsf{n}(x, y)) \twoheadrightarrow w)\}\end{aligned}$$

so that the computed more specific version, $\mathcal{R}_{msv}$, contains the two rules already shown in Example 4.

Now, we consider the correctness of the MSV transformation.

Theorem 16. *Let $\mathcal{R}$ be a DCTRS, $l \to r \Leftarrow C \in \mathcal{R}$ be a rewrite rule such that not all terms in r and C are constructor terms. Let τ be a finite (possibly incomplete) constructor-based conditional narrowing tree for $(C, r \twoheadrightarrow x)$ in $\mathcal{R}$. Then,*

- *If $\mathrm{MSV}(\mathcal{R}, l \to r \Leftarrow C, \tau) = \mathcal{R}_{msv}$, then $\mathcal{R}_{msv}$ is a more specific version of $l \to r \Leftarrow C$ in $\mathcal{R}$ w.r.t. $\overset{c}{\to}_{\mathcal{R}}$.*
- *If $\mathrm{MSV}(\mathcal{R}, l \to r \Leftarrow C, \tau) = \bot$, then $l \to r \Leftarrow C$ is not used in any successful reduction in $\mathcal{R}$ w.r.t. $\overset{c}{\to}_{\mathcal{R}}$.*

This theorem can be proved by using a *lifting lemma* (cf. [25, Lemmas 3.4, 6.11, and 9.4]).

Lemma 17 (lifting lemma). *Let $\mathcal{R}$ be a DCTRS, S, T sequences of equations, θ a constructor substitution, and V a set of variables such that $\mathcal{V}ar(S) \cup \mathcal{D}om(\theta) \subseteq V$ and $T = S\theta$. If $T \overset{c}{\to}{}^* T'$ then there exists a sequence of equations S' and substitutions θ', σ such that*

- *$S \overset{c}{\leadsto}{}^*_\sigma S'$,*
- *$S'\theta' = T'$,*
- *$\theta' \circ \sigma = \theta\ [V]$, and*
- *θ' is a constructor substitution.*

Furthermore, one may assume that the narrowing derivation $S \overset{c}{\leadsto}{}^*_\sigma S'$ and the rewrite sequence $T \overset{c}{\to}{}^*_{\mathcal{R}} T'$ employ the same rewrite rules at the same positions in the corresponding equations.

Proof. We prove this lemma by induction on the length k of $T \overset{c}{\to}{}^* T'$. The case with $k = 0$ is trivial, so let $k > 0$. Let $S = (s \twoheadrightarrow t, S_1)$ and $T = (s\theta \twoheadrightarrow t\theta, S_1\theta) \overset{c}{\to} T_1 \overset{c}{\to}{}^* T'$. We make a case analysis depending on which transition rule is applied at the first step.

- If the reduction rule is applied, there exist a conditional rewrite rule $l \to r \Leftarrow C \in \mathcal{R}$ and a constructor substitution σ such that $\mathcal{V}ar(l, r, C) \cap V = \varnothing$, $s\theta = s\theta[l\sigma]_p$, $T_1 = (C\sigma, s\theta[r\sigma]_p \twoheadrightarrow t\theta, S_1\theta) = (C, s[r] \twoheadrightarrow t, S_1)(\sigma \cup \theta)$. Since

θ is a constructor substitution, the root symbol of $s|_p$ is a defined symbol, and thus, $s|_p$ is not a variable. It follows from $s|_p\theta = l\sigma$ that $s|_p$ and l are unifiable. Let δ be an mgu of $s|_p$ and l. Then, we have that $\delta \leqslant (\sigma \cup \theta)$, and hence δ is a constructor substitution. Now, by applying narrowing to $S = (s \twoheadrightarrow t, S_1)$, we have that $S \overset{c}{\rightsquigarrow}_\delta (C, s[r] \twoheadrightarrow t, S_1)\delta$. Let δ' be a constructor substitution such that $\delta' \circ \delta = (\sigma \cup \theta)$. By the induction hypothesis, we have that $(C, s[r] \twoheadrightarrow t, S_1)\delta \overset{c}{\rightsquigarrow}^*_{\delta''} S'$ with $S'\theta' = T'$, $\theta' \circ \delta'' = \delta'$ for some constructor substitution θ'. Now we have that $S \overset{c}{\rightsquigarrow}^*_{\delta''\circ\delta} S'$, $S'\theta' = T'$, and $\theta' \circ (\delta'' \circ \delta) = \theta\ [V]$.
- If rule matching is applied, there exists a constructor substitution σ such that $s\theta = t\theta\sigma$, and hence s and t are unifiable. Let δ be an mgu of s and t. Then, we have that $\delta \leqslant (\sigma \circ \theta)$, and hence δ is a constructor substitution. By applying unification to S, we have that $S \overset{c}{\rightsquigarrow}_\delta S_1\delta$. Let δ' be a constructor substitution such that $\delta' \circ \delta = (\sigma \circ \theta)$. By the induction hypothesis, we have that $S_1\delta \overset{c}{\rightsquigarrow}^*_{\delta''} S'$ with $S'\theta' = T'$, $\theta' \circ \delta'' = \delta'$ for some constructor substitution θ'. Now we have that $S \overset{c}{\rightsquigarrow}^*_{\delta''\circ\delta} S'$, $S'\theta' = T'$, and $\theta' \circ (\delta'' \circ \delta) = \theta\ [V]$. □

The correctness of the MSV transformation can now be proved by using the lifting lemma:

Proof (Theorem 16). Let $\mathcal{R}_{msv} = \{(l \to r \Leftarrow C)\sigma_1, \ldots, (l \to r \Leftarrow C)\sigma_n\}$ and $\mathcal{R}' = (\mathcal{R} \setminus \{l \to r \Leftarrow C\}) \cup \mathcal{R}_{msv}$. It suffices to show that for a constructor terms t and a sequence S of equations with a fresh variable x, $S \overset{c}{\to}_\mathcal{R} (t \twoheadrightarrow x)$ is successful in $\mathcal{R}$ iff $S \overset{c}{\to}_{\mathcal{R}'} (t \twoheadrightarrow x)$ is successful in $\mathcal{R}'$.

First, we note that if $S \overset{c}{\to}^*_{\mathcal{R}'} (t \twoheadrightarrow x)$ is successful in $\mathcal{R}'$ then $S \overset{c}{\to}^*_\mathcal{R} (t \twoheadrightarrow x)$ is successful in $\mathcal{R}$: it follows from the definition of $\overset{c}{\to}$ that the set of rules used in $S \overset{c}{\to}^*_{\mathcal{R}'} T$ is a DCTRS such that every rule either appears in $\mathcal{R}$ or it is a constructor instance of some rule in $\mathcal{R}$. Hence $\overset{c}{\to}_{\mathcal{R}'} \subseteq \overset{c}{\to}_\mathcal{R}$. Thus, we only show that if $S \overset{c}{\to}^*_\mathcal{R} (t \twoheadrightarrow x)$ is successful in $\mathcal{R}$ then $S \overset{c}{\to}^*_{\mathcal{R}'} (t \twoheadrightarrow x)$ is successful in $\mathcal{R}'$. We prove this claim by induction on the length k of $S \overset{c}{\to}^*_\mathcal{R} t \twoheadrightarrow x$. The case that $k = 0$ is trivial, so let $k > 0$. Let $S \overset{c}{\to}^*_\mathcal{R} (t \twoheadrightarrow x)$ be

$$
\begin{aligned}
S = (s[l'\theta] \twoheadrightarrow s', S') &\overset{c}{\to}_{l'\to r'\Leftarrow C'} (C'\theta, s[r'\theta] \twoheadrightarrow s', S') \\
&\overset{c}{\to}^*_\mathcal{R} (s[t_1] \twoheadrightarrow s', S') \overset{c}{\to}^*_\mathcal{R} (t \twoheadrightarrow x)
\end{aligned}
$$

for a term s, a constructor term s', a constructor substitution θ and a constructor term $t_1 \in \mathcal{T}(\mathcal{C}, \mathcal{V})$. We make a case analysis whether $l' \to r' \Leftarrow C'$ is $l \to r \Leftarrow C$ or not.

- The case that $l' \to r' \Leftarrow C' \neq l \to r \Leftarrow C$. Since $l' \to r' \Leftarrow C' \in \mathcal{R}'$, we have that $S = (s[l'\theta] \twoheadrightarrow s', S') \overset{c}{\to}_{\mathcal{R}'} (C'\theta, s[r'\theta] \twoheadrightarrow s', S')$.
- Otherwise, by definition of $\overset{c}{\to}$, we have that $(l\theta \twoheadrightarrow y) \overset{c}{\to}_\mathcal{R} (C\theta, r\theta \twoheadrightarrow y) \overset{c}{\to}^*_\mathcal{R} (t_1 \twoheadrightarrow y)$ for some fresh variable y. Thus, it follows from Lemma 17 that $(C, r \twoheadrightarrow y) \overset{c}{\rightsquigarrow}^*_\delta (t_1 \twoheadrightarrow y)$ and $\delta \leqslant \theta$ for some constructor substitution δ. By construction of τ, there exist a sequence T' of equations and constructor substitutions σ', σ'' such that $(C, r \twoheadrightarrow y) \overset{c}{\rightsquigarrow}^*_{\sigma'} T' \overset{c}{\rightsquigarrow}^*_{\sigma''} (t_1 \twoheadrightarrow y)$ and $(C, r \twoheadrightarrow$

$y) \overset{c}{\leadsto}^*_{\sigma'} T' \in \tau$, i.e., $\delta = \sigma' \circ \sigma''$ and $\tau \neq \varnothing$ — this means that if $l \to r \Leftarrow C$ is used in a successful derivation, then $\mathrm{MSV}(R, l \to r \Leftarrow C, \tau) \neq \bot$. By the construction of $\sigma_1, \ldots, \sigma_n$, we have that $\sigma_i \leqslant \sigma'$ for some i, and hence $\sigma_i \leqslant \theta$. Thus, we have that $S = (s[l'\theta] \twoheadrightarrow s', S') \overset{c}{\to}_{(l \to r \Leftarrow C)\sigma_i} (C'\theta, s[r'\theta] \twoheadrightarrow s', S')$.

By the induction hypothesis, we have that $(C'\theta, s[r'\theta] \twoheadrightarrow s', S') \overset{c}{\to}^*_{\mathcal{R}'} (t \twoheadrightarrow x)$, and hence $S \overset{c}{\to}^*_{\mathcal{R}'} (t \twoheadrightarrow x)$. □

5 Conclusion and Future Work

We have introduced the notion of a *more specific* version of a rewrite rule in the context of conditional rewrite systems with some restrictions (i.e., typical functional programs). The transformation is useful to produce non-overlapping systems from overlapping ones while preserving the so called successful reductions. We have introduced an automatic algorithm for computing more specific versions and have proved its correctness.

We have undertaken the extension of the implemented program inverter of [29] with a post-process based on the MSV transformation. We have tested the resulting transformation with the 15 program inversion benchmarks of [20].[6] In nine of these benchmarks (out of fifteen) an overlapping system was obtained by inversion while the remaining six are non-overlapping. By applying the MSV transformation to all overlapping rules — except for a predefined operator du — using a depth-3 constructor-based conditional narrowing tree in all examples, we succeeded in improving the nine overlapping systems. These promising results point out the usefulness of the approach to improve the quality of inverted systems.

As for future work, we plan to explore the use of the MSV transformation to improve the accuracy of non-termination analyses.

Acknowledgements. We thank the anonymous reviewers for their useful comments and suggestions to improve this paper.

References

1. Albert, E., Vidal, G.: The narrowing-driven approach to functional logic program specialization. New Generation Computing 20(1), 3–26 (2002)
2. Almendros-Jiménez, J.M., Vidal, G.: Automatic partial inversion of inductively sequential functions. In: Horváth, Z., Zsók, V., Butterfield, A. (eds.) IFL 2006. LNCS, vol. 4449, pp. 253–270. Springer, Heidelberg (2007)

[6] Unfortunately, the site shown in [20] is not accessible now. Some of the benchmarks can be found in [14,15,21]. All the benchmarks are reviewed in [26] and also available from the following URL: `http://www.trs.cm.is.nagoya-u.ac.jp/repius/`. The two benchmarks pack and packbin define non-tail-recursive functions which include some tail-recursive rules; we transform them into pure tail recursive ones using the method introduced in [30] (to avoid producing non-operationally terminating programs), that are also inverted using the approach of [29].

3. Antoy, S., Hanus, M.: Overlapping rules and logic variables in functional logic programs. In: Etalle, S., Truszczyński, M. (eds.) ICLP 2006. LNCS, vol. 4079, pp. 87–101. Springer, Heidelberg (2006)
4. Baader, F., Nipkow, T.: Term Rewriting and All That. Cambridge University Press (1998)
5. Bertling, H., Ganzinger, H.: Completion-time optimization of rewrite-time goal solving. In: Dershowitz, N. (ed.) RTA 1989. LNCS, vol. 355, pp. 45–58. Springer, Heidelberg (1989)
6. Bockmayr, A., Werner, A.: LSE narrowing for decreasing conditional term rewrite systems. In: Dershowitz, N., Lindenstrauss, N. (eds.) CTRS 1994. LNCS, vol. 968, pp. 51–70. Springer, Heidelberg (1995)
7. Bondorf, A.: Towards a self-applicable partial evaluator for term rewriting systems. In: Bjørner, D., Ershov, A., Jones, N. (eds.) Proceedings of the International Workshop on Partial Evaluation and Mixed Computation, pp. 27–50. North-Holland, Amsterdam (1988)
8. Bondorf, A.: A self-applicable partial evaluator for term rewriting systems. In: Díaz, J., Orejas, F. (eds.) TAPSOFT 1989. LNCS, vol. 352, pp. 81–95. Springer, Heidelberg (1989)
9. Bosco, P., Giovannetti, E., Moiso, C.: Narrowing vs. SLD-resolution. Theoretical Computer Science 59, 3–23 (1988)
10. Dershowitz, N.: Termination of rewriting. Journal of Symbolic Computation 3(1&2), 69–115 (1987)
11. Dershowitz, N., Jouannaud, J.-P.: Rewrite systems. In: van Leeuwen, J. (ed.) Handbook of Theoretical Computer Science. Formal Models and Semantics, vol. B, pp. 243–320. Elsevier, Amsterdam (1990)
12. Fribourg, L.: SLOG: a logic programming language interpreter based on clausal superposition and rewriting. In: Proceedings of the Symposium on Logic Programming, pp. 172–185. IEEE Press (1985)
13. Ganzinger, H.: Order-sorted completion: The many-sorted way. Theoretical Computer Science 89(1), 3–32 (1991)
14. Glück, R., Kawabe, M.: A program inverter for a functional language with equality and constructors. In: Ohori, A. (ed.) APLAS 2003. LNCS, vol. 2895, pp. 246–264. Springer, Heidelberg (2003)
15. Glück, R., Kawabe, M.: A method for automatic program inversion based on LR(0) parsing. Fundamenta Informaticae 66(4), 367–395 (2005)
16. Hanus, M.: The integration of functions into logic programming: From theory to practice. Journal of Logic Programming 19&20, 583–628 (1994)
17. Harrison, P.G.: Function inversion. In: Proceedings of the International Workshop on Partial Evaluation and Mixed Computation, pp. 153–166. North-Holland, Amsterdam (1988)
18. Hölldobler, S.: Foundations of Equational Logic Programming. LNCS (LNAI), vol. 353. Springer, Heidelberg (1989)
19. Hullot, J.-M.: Canonical forms and unification. In: Bibel, W., Kowalski, R.A. (eds.) Automated Deduction. LNCS, vol. 87, pp. 318–334. Springer, Heidelberg (1980)
20. Kawabe, M., Futamura, Y.: Case studies with an automatic program inversion system. In: Proceedings of the 21st Conference of Japan Society for Software Science and Technology, 6C-3, 5 pages (2004)
21. Kawabe, M., Glück, R.: The program inverter LRinv and its structure. In: Hermenegildo, M.V., Cabeza, D. (eds.) PADL 2004. LNCS, vol. 3350, pp. 219–234. Springer, Heidelberg (2005)

22. Khoshnevisan, H., Sephton, K.M.: InvX: An automatic function inverter. In: Dershowitz, N. (ed.) RTA 1989. LNCS, vol. 355, pp. 564–568. Springer, Heidelberg (1989)
23. Marchiori, M.: On deterministic conditional rewriting. Technical Report MIT-LCS-TM-405, MIT Laboratory for Computer Science (1997), `http://www.w3.org/People/Massimo/papers/MIT-LCS-TM-405.pdf`
24. Marriott, K., Naish, L., Lassez, J.-L.: Most specific logic programs. Annals of Mathematics and Artificial Intelligence 1, 303–338 (1990), `http://www.springerlink.com/content/k3p11kl316m73764/`
25. Middeldorp, A., Hamoen, E.: Completeness results for basic narrowing. Applicable Algebra in Engineering, Communication and Computing
26. Nishida, N., Sakai, M.: Completion after program inversion of injective functions. Electronic Notes in Theoretical Computer Science 237, 39–56 (2009)
27. Nishida, N., Sakai, M., Sakabe, T.: Generation of inverse computation programs of constructor term rewriting systems. IEICE Transactions on Information and Systems J88-D-I(8), 1171–1183 (2005) (in Japanese)
28. Nishida, N., Sakai, M., Sakabe, T.: Partial inversion of constructor term rewriting systems. In: Giesl, J. (ed.) RTA 2005. LNCS, vol. 3467, pp. 264–278. Springer, Heidelberg (2005)
29. Nishida, N., Vidal, G.: Program inversion for tail recursive functions. In: Schmidt-Schauß, M. (ed.) Proceedings of the 22nd International Conference on Rewriting Techniques and Applications. Leibniz International Proceedings in Informatics, vol. 10, pp. 283–298. Schloss Dagstuhl - Leibniz-Zentrum für Informatik (2011)
30. Nishida, N., Vidal, G.: Conversion to first-order tail recursion for improving program inversion (2012) (submitted for publication)
31. Nishida, N., Vidal, G.: More specific term rewriting systems. In: The 21st International Workshop on Functional and (Constraint) Logic Programming, Nagoya, Japan (2012), Informal proceedings `http://www.dsic.upv.es/~gvidal/german/wflp12/paper.pdf`
32. Ohlebusch, E.: Advanced topics in term rewriting. Springer, UK (2002)
33. Pfenning, F.: Unification and anti-unification in the calculus of constructions. In: Proceedings of the Sixth Annual Symposium on Logic in Computer Science, pp. 74–85. IEEE Computer Society (1991)
34. Plotkin, G.: Building-in equational theories. Machine Intelligence 7, 73–90 (1972)
35. Ramos, J.G., Silva, J., Vidal, G.: Fast Narrowing-Driven Partial Evaluation for Inductively Sequential Systems. In: Danvy, O., Pierce, B.C. (eds.) Proceedings of the 10th ACM SIGPLAN International Conference on Functional Programming, pp. 228–239. ACM Press (2005)
36. Slagle, J.R.: Automated theorem-proving for theories with simplifiers, commutativity and associativity. Journal of the ACM 21(4), 622–642 (1974)

Improving Determinization of Grammar Programs for Program Inversion*

Minami Niwa, Naoki Nishida, and Masahiko Sakai

Graduate School of Information Science, Nagoya University
Furo-cho, Chikusa-ku, 4648603 Nagoya, Japan
{mniwa@sakabe.i., nishida@, sakai@}is.nagoya-u.ac.jp

Abstract. The inversion method proposed by Glück and Kawabe uses grammar programs as intermediate results that comprise sequences of operations (data generation, matching, etc.). The determinization method used in the inversion method fails for a grammar program of which the collection of item sets causes a conflict even if there exists a deterministic program equivalent to the grammar program. In this paper, by ignoring shift/shift conflicts, we improve the determinization method so as to cover grammar programs causing shift/shift conflicts. Moreover, we propose a method to eliminate infeasible definitions from unfolded grammar programs and show that the method succeeds in determinizing some grammar programs for which the original method fails. By using the method as a post-process of the original inversion method, we make the original method strictly more powerful.

1 Introduction

Inverse computation for an n-ary function f is, given an output t of f, the calculation of (all) possible inputs $t_1, \ldots, t_n$ of f such that $f(t_n, \ldots, t_n) = t$. Methods of inverse computation are distinguished into two approaches [30, 31]: *inverse interpretation* [5, 2, 1, 3, 10][1] and *program inversion*[21, 11, 18, 12, 30, 17, 31, 13, 34, 7, 25, 24, 8, 9, 16, 27, 20]. Methods of program inversion take the definition of f as input and compute a definition of the inverse function for f. Surprisingly, the essential ideas of the existing inversion methods for functional programs are almost the same,[2] except for [27, 20]. In general, function definitions for inverses that are generated by inversion methods are non-deterministic w.r.t. the application of function definitions even if the target functions are injective. For this reason, determinization of such inverted programs is one of the most interesting and challenging topics in developing inversion methods and has been investigated in several ways [7, 8, 16, 4, 23].

* This work has been partially supported by *MEXT KAKENHI* #21700011.

[1] *Narrowing* [14] which is a method for equational unification is also classified into inverse interpretation for term rewriting systems.

[2] Roughly speaking, $f(\overrightarrow{p_0}) =$ `let` $t_1 = f_1(\overrightarrow{p_1})$ `in` ... `let` $t_n = f_n(\overrightarrow{p_n})$ `in` t_0 is transformed into $f^{-1}(t_0) =$ `let` $(\overrightarrow{p_n}) = f_n^{-1}(t_n)$ `in` ... `let` $(\overrightarrow{p_1}) = f_1^{-1}(t_1)$ `in` $(\overrightarrow{p_0})$.

E. Albert (Ed.): LOPSTR 2012, LNCS 7844, pp. 155–175, 2013.
© Springer-Verlag Berlin Heidelberg 2013

The inversion method LRinv proposed by Glück and Kawabe [7, 8, 16] adopts an interesting approach to determinization. A functional program is first translated into a context-free grammar, called a *grammar program*, whose language is the sequences of *atomic operations* (data generation, matching, etc.), each of which corresponds to a possible execution of the original program. The semantics of operations in grammar programs is defined as a simple stack machine where data terms are stored in the stack (input arguments, intermediate results, and outputs), and the meanings of atomic operations (terminal symbols) are defined as *pop* and *push* operations on the stack. The grammar program is inverted by reversing the bodies of production rules and by replacing each operation by its opposite operation. Then, the inverted grammar program is determinized by a method based on *LR(0) parsing* so that the resulting grammar is easily convertible into a functional program without overlaps between function definitions.

LRinv deals with functional programs that cannot express non-deterministic definitions of functions. For this reason, for an input whose inverted grammar program causes conflicts in its item collection, LRinv fails and does not produce any program.

The mechanism of the determinization method in LRinv is complicated but very interesting. For example, some grammar programs corresponding to non-terminating and overlapping *rewriting systems*[3] are transformed into terminating and non-overlapping rewriting systems which correspond to terminating functional programs. Moreover, the determinization method is independent of the inversion principle, and thus, the method is useful as a post-process of any other inversion methods.

Unfortunately, the determinization method is not successful for all grammar programs obtained by the inversion phase — there exist some examples for which this method fails but other inversion methods (e.g., [27]) succeed. The failure of the determinization method is mainly caused by *shift/shift conflicts* that are branches of shift relations in the collection of *item sets* with two or more different atomic operations, not all of which are matching operations. In the determinization method, a grammar program causing such conflicts does not proceed to the code generation step even if there exists a deterministic program that is equivalent to the grammar program.

In this paper, by ignoring shift/shift conflicts, we improve the determinization method, proposing it as a *semi-determinization* method so as to cover grammar programs causing shift/shift conflicts. To make the scope of LRinv larger, as target programs, we deal with *conditional term rewriting systems* — one of well studied computation models of functional programs — instead of functional programs in [7, 8, 16]. Inevitably, we adapt LRinv to the setting of term rewriting.

For any grammar program that causes no conflict, the semi-determinization method works as well as the original method since we do not modify the transformation itself. We show that the semi-determinization method succeeds in determinizing some grammar programs causing shift/shift conflicts. This means that

---

[3] Unlike functional programs used in [8], rewriting systems can express non-deterministic computation.

the semi-determinization is strictly more powerful than the original determinization method. We also show that the semi-determinization method converts some grammar programs whose corresponding rewriting systems are not terminating, into ones whose corresponding rewriting systems are terminating.

The language generated by a grammar program may contain *infeasible* sequences, in which a term $c(t_1, \ldots, t_n)$ is generated and pushed onto the stack and later its root symbol is examined by an unsuccessful matching operation that examines whether the root symbol is a constructor other than c. Such sequences never happen in completed executions of functions, and thus, we can exclude function definitions containing infeasible sequences while preserving semantic equivalence. Unfortunately, such infeasible sequences do not explicitly appear in function definitions obtained by the semi-determinization method. However, they sometimes appear explicitly after unfolding functions. In this paper, we adapt the unfolding method of functional programs to grammar programs, and propose a method to eliminate infeasible definitions from unfolded grammar programs. We show that the method succeeds in determinizing some grammar programs which both the original and semi- determinization methods fail to determinize.

The main contributions of this paper are to adapt LRinv to the setting of term rewriting, and to make LRinv strictly more powerful.

This paper is organized as follows. In Section 2, we recall rewriting systems and grammar programs. In Section 3, we adapt translations between functional and grammar programs to rewriting systems. In Section 4, we introduce the overview of LRinv by means of an example. In Section 5, we show that the semi-determinization method is meaningful for some examples. In Section 6, we propose a method to eliminate infeasible definitions from unfolded grammar programs. In Section 7, we summarize and describe future work of this research.

2 Preliminaries

In this section, we recall rewriting systems and grammar programs.

2.1 Rewriting Systems

Throughout this paper, we use $\mathcal{V}$ as a set of variables, and $\mathcal{C}$ and $\mathcal{F}$ for finite sets of *constructor* and *function* symbols that appear in programs, where each symbol f has a fixed arity n and coarity m, represented by f/n or $f/n/m$ — if $m > 0$, then f returns a tuple of m values. For terms $t_1, \ldots, t_n$, we denote the set of variables in $t_1, \ldots, t_n$ by $\mathcal{V}ar(t_1, \ldots, t_n)$. We assume that $\mathcal{C}$ contains constructor symbols to represent tuples $(t_1, \ldots, t_m)$, e.g., $\mathsf{tp}_m/m/m$, and we omit such symbols from $\mathsf{tp}_m(t_1, \ldots, t_m)$. The set of *constructor terms* over $\mathcal{C}$ is denoted by $T(\mathcal{C}, \mathcal{V})$ and the set of ground constructor terms is denoted by $T(\mathcal{C})$. For the sake of readability and expressiveness, functional programs are written as *deterministic conditional term rewriting systems* [4] (DCTRSs) [29], which are sets of conditional rewrite rules of

[4] The terminology "deterministic" in the field of conditional rewriting means that the conditional part can be evaluated from left to right. Thus, the terminology is different from the one used for "deterministic computation" of functional programs.

the form $l \to r \Leftarrow s_1 \twoheadrightarrow t_1, \ldots, s_k \twoheadrightarrow t_k$ such that $\mathcal{V}ar(r) \subseteq \mathcal{V}ar(l, t_1, \ldots, t_k)$ and $\mathcal{V}ar(s_i) \subseteq \mathcal{V}ar(l, t_1, \ldots, t_{i-1})$ for all $i > 0$. A DCTRS is called a *term rewriting system* (TRS) if every rule in the DCTRS is unconditional.

A DCTRS R is called a *pure constructor systems* [22] (a *normalized* CTRS in [4]) if every conditional rule is of the form

$$f_0(u_{0,1}, \ldots, u_{0,n_0}) \to r_0 \Leftarrow f_1(u_{1,1}, \ldots, u_{1,n_1}) \twoheadrightarrow r_1, \ldots, f_k(u_{k,1}, \ldots, u_{k,n_k}) \twoheadrightarrow r_k$$

where $f_i \in \mathcal{F}$ and $u_{i,1}, \ldots, u_{i,n_i}, r_i \in T(\mathcal{C}, \mathcal{V})$ for all $i \geq 0$. R is called *ultra-non-erasing* [26] if $\mathcal{V}ar(l, t_1, \ldots, t_k) \subseteq \mathcal{V}ar(s_1, \ldots, s_k, r)$ for every conditional rule $l \to r \Leftarrow s_1 \twoheadrightarrow t_1, \ldots, s_k \twoheadrightarrow t_k$. R is called *ultra-linear* [26] if both the tuples $(l, t_1, \ldots, t_k)$ and $(s_1, \ldots, s_k, r)$ are linear for every conditional rule $l \to r \Leftarrow s_1 \twoheadrightarrow t_1, \ldots, s_k \twoheadrightarrow t_k$.

This paper assumes that DCTRSs are ultra-non-erasing and ultra-linear pure constructor systems — conditional rules of such DCTRSs can be seen in a functional language as

$$\begin{aligned} f_0(u_{0,1}, \ldots, u_{0,n_0}) \;=\; & \texttt{let } r_1 = f_1(u_{1,1}, \ldots, u_{1,n_1}) \texttt{ in} \\ & \cdots \\ & \texttt{let } r_k = f_k(u_{k,1}, \ldots, u_{k,n_k}) \texttt{ in } r_0 \end{aligned}$$

This paper also assumes that the main function defined by a given system is injective. For a DCTRS R, the reduction relation we consider is the leftmost *constructor-based* reduction [33, 28], exactly the usual *call-by-value* evaluation of functional programs — only constructor instances of the left-hand sides of rules are reduced.

Two different rewrite rules $l_1 \to r_1 \Leftarrow C_1$ and $l_2 \to r_2 \Leftarrow C_2$ are called *overlapping* if l_1 and a renamed variant of l_2 are unifiable; otherwise, they are *non-overlapping.* Note that there is no inner overlaps since we restrict DCTRSs to be pure constructor systems. A DCTRS is called *overlapping* if it has overlapping rewrite rules; otherwise, the DCTRS is *non-overlapping.*

2.2 Grammar Programs

In the following, to represent sequences of objects, we may use the usual list representations, e.g., the sequence 1 2 3 is denoted by [1, 2, 3], [1]@[2, 3], etc.

Let $\mathcal{O}p = \{c! \mid c \in \mathcal{C}\} \cup \{c? \mid c \in \mathcal{C}\} \cup \{(i_1 \ldots i_j) \mid j > 0,\ \{i_1, \ldots, i_j\} = \{1, \ldots, j\}\}$, the set of *alphabets.* A *grammar program* is a *context-free grammar* $(\mathcal{F}, \mathcal{O}p, S, G)$ [8], where we consider $\mathcal{F}$ the set of *non-terminal symbols,* $\mathcal{O}p$ the set of *terminal symbols,* S the initial non-terminal symbol (not used in this paper), and G the set of *production rules* (*function definition*) of the form $f \to \alpha$ with $f \in \mathcal{F}$ and $\alpha \in (\mathcal{F} \cup \mathcal{O}p)^*$.[5] In the following, instead of the

[5] For an operation a, $\mathbf{pop}(a)$ and $\mathbf{push}(a)$ denote the numbers of values popped and pushed by a, resp.: $\mathbf{pop}(f) = n$ and $\mathbf{push}(f) = m$ for $f/n/m \in \mathcal{F}$; $\mathbf{pop}(c?) = \mathbf{push}(c!) = 1$ and $\mathbf{push}(c?) = \mathbf{pop}(c!) = n$ for $c/n \in \mathcal{C}$; $\mathbf{pop}((i_1 \ldots i_j)) = \mathbf{push}((i_1 \ldots i_j)) = j$. Note that $f \to a_1 \ldots a_m$ with $a_1, \ldots, a_m \in \mathcal{F} \cup \mathcal{O}p$ should be consistent, i.e., $\mathbf{pop}(f) + \sum_{i=1}^{m}(\mathbf{push}(a_i) - \mathbf{pop}(a_i)) = \mathbf{push}(f)$ and $\mathbf{pop}(f) + \sum_{i=n}^{k-1}(\mathbf{push}(a_i) - \mathbf{pop}(a_i)) \geq \mathbf{pop}(a_k)$ for all $1 \leq k \leq m-1$. These (in)equalities are necessary to guess the arity and coarity of generated auxiliary functions (non-terminals).

quadruple $(\mathcal{F}, \mathcal{O}p, S, G)$ for a context-free grammar, we only write G as the grammar program. We consider only grammar programs that can be translated back into functional programs (DCTRSs in this paper). For the sake of readability, this paper does not deal with the *duplication/equality* operator $\lfloor \_ \rfloor$ shown in [8], but the techniques in this paper also extend to the setting with the duplication/equality operator.

Non-terminal and terminal symbols are called *operations*, and terminal symbols are called *atomic operations* (i.e., $\mathcal{O}p$ is the set of atomic operations). For a constructor $c \in \mathcal{C}$, the operation $c!$ is called a *constructor application* of c, and the operation $c?$ is called a *pattern matching* of c. Operations $(i_1 \ldots i_j)$ in $\mathcal{O}p$ are called *permutations*. For a grammar program G with a function $f \in \mathcal{F}$, the notation $L(G, f)$ represents the set of atomic-operation sequences generated from the non-terminal f.

The semantics $\to_G$ of a grammar program G is defined over pairs of operation- and term-sequences — a binary relation over $\mathcal{O}p^* \times T(\mathcal{C})^*$ — as follows:

$$
\begin{array}{rll}
([f] @ \alpha, \vec{t}) \to_G (\beta @ \alpha, \vec{t}) & & \text{where } f \to \beta \in G \\
([c!] @ \alpha, [u_1, \ldots, u_n] @ \vec{t}) \to_G (\alpha, [c(u_1, \ldots, u_n)] @ \vec{t}) & & \text{where } c/n \in \mathcal{C} \\
([c?] @ \alpha, [c(u_1, \ldots, u_n)] @ \vec{t}) \to_G (\alpha, [u_1, \ldots, u_n] @ \vec{t}) & & \text{where } c/n \in \mathcal{C} \\
([(i_1 \ldots i_j)] @ \alpha, [u_1, \ldots, u_j] @ \vec{t}) \to_G (\alpha, [u_{i_1}, \ldots, u_{i_j}] @ \vec{t}) & &
\end{array}
$$

where $\vec{t}$ denotes a sequence of terms. When computing $f(t_1, \ldots, t_n)$ with $f/n/m \in \mathcal{F}$, we start with $([f], [t_1, \ldots, t_n])$ and obtain the result $(u_1, \ldots, u_m)$ if $([f], [t_1, \ldots, t_n]) \to_G^* ([\,], [u_1, \ldots, u_m])$. The second component of the pairs plays a role of a *stack* that stores input, intermediate, and resulting terms. For $c/n \in \mathcal{C}$, the constructor application $c!$ pops the n topmost values $t_1, \ldots, t_n$ from the stack and pushes the value $c(t_1, \ldots, t_n)$ onto the stack; the pattern matching $c?$ pops the topmost value $c(t_1, \ldots, t_n)$ from the stack and then pushes values $t_1, \ldots, t_n$ onto the stack. A permutation $(i_1 \ldots i_j)$ reorders the j topmost values on the stack by moving in parallel the k-th element to the i_k-th position. We merge sequences of permutations in G into single optimized ones such as (2 1) (3 1 2) into (3 2 1), and drop identity permutations such as (1 2 3).

3 Translation between DCTRSs and Grammar Programs

Translations between functional and grammar programs can be found in [8]. In this section, we adapt the translations to DCTRSs. In the translations, a conditional rewrite rule is transformed into a function definition, and vice versa.

First, we show a translation of DCTRSs into grammar programs. Recall that DCTRSs in this paper are ultra-non-erasing and ultra-linear. The translation below relies on these properties.

Definition 1. *Let R be a DCTRS. We assume that any conditional rewrite rule in R is normalized to the following form:*

$$
\begin{array}{rl}
f_0(u_{0,1}, \ldots, u_{0,n_0}) \to (r_{0,1}, \ldots, r_{0,m_0}) \Leftarrow & f_1(u_{1,1}, \ldots, u_{1,n_1}) \twoheadrightarrow (r_{1,1}, \ldots, r_{1,m_1}), \\
& \vdots \\
& f_k(u_{k,1}, \ldots, u_{k,n_k}) \twoheadrightarrow (r_{k,1}, \ldots, r_{k,m_k})
\end{array}
$$

where $f_i/n_i/m_i \in \mathcal{F}$ *and* $u_{i,1}, \ldots, u_{i,n_i}, r_{i,1}, \ldots, r_{i,m_i} \in T(\mathcal{C}, \mathcal{V})$ *for all* $i \geq 0$. *The translation GRM of DCTRSs into grammar programs is defined as follows:*

$$
\begin{aligned}
GRM(R) = \{ f \to \alpha \mid\ & f(u_{0,1}, \ldots, u_{0,n_0}) \to (r_{0,1}, \ldots, r_{0,m_0}) \Leftarrow C \in R, \\
& (\alpha_1, s_1) = Grm_?([u_{0,1}, \ldots, u_{0,n_0}], [\,], [\overbrace{\bot, \ldots, \bot}^{n_0}]), \\
& (\alpha_2, s_2) = Grm_{\mathsf{c}}(C, \alpha_1, s_1), \\
& (\alpha, [r_{0,1}, \ldots, r_{0,m_0}]) = Grm_!([r_{0,m_0}, \ldots, r_{0,1}], \alpha_2, s_2) \,\}
\end{aligned}
$$

where $\bot$ *is a fresh constant used as dummy elements in a stack, and* Grm_{c}, $Grm_!$, *and* $Grm_?$ *are functions with type* $\mathcal{C}ond^* \times \mathcal{O}p^* \times T(\mathcal{C}, \mathcal{V})^* \to \mathcal{O}p^* \times T(\mathcal{C}, \mathcal{V})^*$ *and* $T(\mathcal{C}, \mathcal{V})^* \times \mathcal{O}p^* \times T(\mathcal{C}, \mathcal{V})^* \to \mathcal{O}p^* \times T(\mathcal{C}, \mathcal{V})^*$,[6] *which are defined as follows:*

- $Grm_{\mathsf{c}}([\,], \alpha, s) = (\alpha, s)$,
- $Grm_{\mathsf{c}}([f(l_1, \ldots, l_n) \twoheadrightarrow (r_1, \ldots, r_m)] @ C, \alpha, s) = Grm_{\mathsf{c}}(C, \alpha_2, s_2)$, *where*
  - $(\alpha_1, [l_1, \ldots, l_n] @ s_1) = Grm_!([l_n, \ldots, l_1], \alpha, s)$, *and*
  - $(\alpha_2, s_2) = Grm_?([r_1, \ldots, r_m], \alpha_1 @ [f], [\overbrace{\bot, \ldots, \bot}^{n}] @ s_1)$,
- $Grm_?([\,], \alpha, s) = (\alpha, s)$,
- $Grm_?([x] @ \vec{t}, \alpha, [u_1, \ldots, u_{i-1}, \bot] @ s) = Grm_?(\vec{t}, \alpha, [u_1, \ldots, u_{i-1}, x] @ s)$,
- $Grm_?([c(l_1, \ldots, l_n)] @ \vec{t}, \alpha, [u_1, \ldots, u_{i-1}, \bot] @ s) = Grm_?([l_1, \ldots, l_n] @ \vec{t}, \alpha @ [(i\ 1 \ldots i-1), c?], [\overbrace{\bot, \ldots, \bot}^{n}] @ s)$,
- $Grm_!([\,], \alpha, s) = (\alpha, s)$,
- $Grm_!([x] @ \vec{t}, \alpha, [u_1, \ldots, u_{i-1}, x] @ s) = Grm_!(\vec{t}, \alpha @ [(i\ 1 \ldots i-1)], [x, u_1, \ldots, u_{i-1}] @ s)$, *and*
- $Grm_!([c(l_1, \ldots, l_n)] @ \vec{t}, \alpha, s) = Grm_!(\vec{t}, \alpha_1 @ [c!], [c(l_1, \ldots, l_n)] @ s_1)$, *where* $(\alpha_1, [l_1, \ldots, l_n] @ s_1) = Grm_!([l_n, \ldots, l_1], \alpha, s)$

where x *is a variable and* $u_j \neq \bot$ *for all* $j < i$. *Sequences of permutations are optimized during (or after) the translation.*

Example 2. Let us consider the function snoc defined by the following TRS:

$$
R_{\mathsf{snoc}} = \left\{ \begin{array}{c} \mathsf{snoc}(\mathsf{nil}, y) \to \mathsf{cons}(y, \mathsf{nil}) \\ \mathsf{snoc}(\mathsf{cons}(x_1, x_2), y) \to \mathsf{cons}(x_1, \mathsf{snoc}(x_2, y)) \end{array} \right\}
$$

where $\mathcal{C} = \{\mathsf{nil}/0, \mathsf{cons}/2, \ldots\}$ and $\mathcal{F} = \{\mathsf{snoc}/2/1\}$. The function snoc appends the second argument to the end of the first argument, e.g., $\mathsf{snoc}(\mathsf{cons}(1, \mathsf{cons}(2, \mathsf{nil})), 3) = \mathsf{cons}(1, \mathsf{cons}(2, \mathsf{cons}(3, \mathsf{nil})))$. R_{snoc} is normalized as follows:

$$
\left\{ \begin{array}{lr} (\mathrm{A}) & \mathsf{snoc}(\mathsf{nil}, y) \to (\mathsf{cons}(y, \mathsf{nil})) \\ (\mathrm{B}) & \mathsf{snoc}(\mathsf{cons}(x_1, x_2), y) \to (\mathsf{cons}(x_1, y)) \Leftarrow \mathsf{snoc}(x_2, y) \twoheadrightarrow (y) \end{array} \right\}
$$

6 $\mathcal{C}ond$ is the set of conditions. The first argument stores terms or conditions to be transformed, the second stores generated operations, and the third stores the state of the stack that is used in executing operations.

We have that $GRM(\{(A)\}) = \{$ snoc $\rightarrow$ nil? nil! (2 1) cons! $\}$ since

- $Grm_?([\mathsf{nil}, y], [\,], [\bot, \bot]) = Grm_?([y], \mathsf{nil?}, [\bot]) = Grm_?([\,], \mathsf{nil?}, [y]) = (\mathsf{nil?}, [y])$,
- $Grm_\mathsf{c}([\,], \mathsf{nil?}, [y]) = (\mathsf{nil?}, [y])$,
- $Grm_!([\mathsf{cons}(y, \mathsf{nil})], \mathsf{nil?}, [y]) = Grm_!([\,], \mathsf{nil?\ nil!\ (2\ 1)\ cons!}, [\mathsf{cons}(y, \mathsf{nil})]) = (\mathsf{nil?\ nil!\ (2\ 1)\ cons!}, [\mathsf{cons}(y, \mathsf{nil})])$, where $Grm_!([\mathsf{nil}, y], \mathsf{nil?}, [y]) = Grm_!([y], \mathsf{nil?\ nil!}, [\mathsf{nil}, y]) = Grm_!([\,], \mathsf{nil?\ nil!\ (2\ 1)}, [y, \mathsf{nil}]) = (\mathsf{nil?\ nil!\ (2\ 1)}, [y, \mathsf{nil}])$.

We also have that $GRM(\{(B)\}) = \{$ snoc $\rightarrow$ cons? (2 3 1) snoc (2 1) cons! $\}$. Therefore, R_{snoc} is translated into the following grammar program:

$$G_{\mathsf{snoc}} = \left\{ \begin{array}{l} \mathsf{snoc} \rightarrow \mathsf{nil?\ nil!\ (2\ 1)\ cons!} \\ \mathsf{snoc} \rightarrow \mathsf{cons?\ (2\ 3\ 1)\ snoc\ (2\ 1)\ cons!} \end{array} \right\}$$

Next, we show a translation of grammar programs into DCTRSs. We assume that the arity and coarity of the main function defined by a given grammar program are given. Before the translation, we need to determine the arities and coarities of all auxiliary functions (non-terminal symbols) in grammar programs. To this end, we solve equalities and inequalities to keep consistency of stacks (see Footnote 5). Solutions of the (in)equalities are not unique but we take one of them, in which the arity and coarity of auxiliary functions are as small as possible.

Definition 3. *Let G be a grammar program. The translation FCT of grammar programs into DCTRSs is defined as follows:*

$$FCT(G) = \{\ \theta(f(x_1, \ldots, x_n) \rightarrow (r_1, \ldots r_m) \Leftarrow C) \mid f \rightarrow \alpha \in G,\ f/n/m \in \mathcal{F},\ ([r_1, \ldots, r_m], C, \theta) = Fct(\alpha, [x_1, \ldots, x_n], [\,], \emptyset)\ \}$$

where $x_1, \ldots, x_n$ are fresh variables, and Fct is a function with type $Op^* \times \mathcal{T}(\mathcal{C}, \mathcal{V})^* \times \mathcal{C}ond^* \times \mathcal{S}ubst \rightarrow \mathcal{T}(\mathcal{C}, \mathcal{V})^* \times \mathcal{C}ond^* \times \mathcal{S}ubst$,[7] which is defined as follows:

- $Fct([\,], s, C, \theta) = (s, C, \theta)$,
- $Fct([c?] @ \alpha, [x] @ s, C, \theta) = (\alpha, [x_1, \ldots, x_n] @ s, C, \theta \circ \{x \mapsto c(x_1, \ldots, x_n)\})$,
- $Fct([c!] @ \alpha, [u_1, \ldots, u_n] @ s, C, \theta) = (\alpha, [c(u_1, \ldots, u_n)] @ s, C, \theta)$,
- $Fct([f] @ \alpha, [u_1, \ldots, u_n] @ s, C, \theta) = (\alpha, [y_1, \ldots, y_m] @ s, C\, @ [f(u_1, \ldots, u_n) \rightarrow (y_1, \ldots, y_m)], \theta)$, *where $f/n/m \in \mathcal{F}$, and*
- $Fct([(i_1 \ \ldots\ i_j)] @ \alpha, [u_1, \ldots, u_j] @ s, C, \theta) = (\alpha, [u_{i_1}, \ldots, u_{i_j}] @ s, C, \theta)$,

where x is a variable, $c/n \in \mathcal{C}$, and $x_1, \ldots, x_n, y_1, \ldots, y_m$ are fresh variables.

We will show an example of the application of FCT to a grammar program later (see Example 5).

[7] $\mathcal{S}ubst$ denotes the set of substitutions. The first argument of Fct stores operation sequences we translate; the second stores the state of the stack that is used in executing operations; the third stores generated conditions; the fourth stores substitutions generated by matching operators.

As mentioned above, given a grammar program G with a function f, if the arity and coarity of the main function are fixed and we determine the arities and coarities of auxiliary functions (non-terminals) to be as small as possible, then each function definition in G is translated into a unique conditional rewrite rule and G is translated back into a unique DCTRS, up to variable renaming. Viewed in this light, for any grammar program, the corresponding DCTRSs are unique. We call a grammar program *deterministic* if the corresponding DCTRS is *non-overlapping.* Two different function definitions are called *non-overlapping* if the corresponding rewrite rules are non-overlapping.

4 Overview of LRinv

In this section, we briefly recall LRinv [7, 8, 16] by using an example. We also recall the concrete definitions of item sets and conflicts related to item sets, called *shift/shift conflicts,* which influence the success or failure of the determinization method (i.e., the power of LRinv).

LRinv assumes that functions to be inverted are injective over inductive data structures. To make the scope of LRinv larger, we deal with DCTRSs instead of the original source language shown in [7, 8, 16].

As the first step, we translate a given DCTRS into a grammar program G. To invert the grammar program G, each function definition $f \to a_1 \dots a_n \in G$ is inverted to the one $f^{-1} \to (a_n)^{-1} \dots (a_1)^{-1}$, where $(g)^{-1} = g^{-1}$ for $g \in \mathcal{F}$, $(c?)^{-1} = c!$ and $(c!)^{-1} = c?$ for $c \in \mathcal{C}$, and $((i_1 \dots i_m))^{-1} = (j_1 \dots j_m)$ with $k = i_{j_k}$ for a permutation $(i_1 \dots i_m)$ [8]. Note that the inverted grammar can be translated back into a DCTRS, and thus, the inverted grammar is a grammar program, where the arity of f^{-1} is the coarity of f, and the coarity of f^{-1} is the arity of f.

Example 4. The grammar program G_{snoc} in Example 2 is inverted as follows:

$$G_{\mathsf{snoc}^{-1}} = \left\{ \begin{array}{l} \mathsf{snoc}^{-1} \to \mathsf{cons?}\ (2\ 1)\ \mathsf{nil?}\ \mathsf{nil!} \\ \mathsf{snoc}^{-1} \to \mathsf{cons?}\ (2\ 1)\ \mathsf{snoc}^{-1}\ (3\ 1\ 2)\ \mathsf{cons!} \end{array} \right\}$$

$G_{\mathsf{snoc}^{-1}}$ is translated back into the following overlapping DCTRS:

$$R_{\mathsf{snoc}^{-1}} = \left\{ \begin{array}{l} \mathsf{snoc}^{-1}(\mathsf{cons}(x, \mathsf{nil})) \to (\mathsf{nil}, x) \\ \mathsf{snoc}^{-1}(\mathsf{cons}(x, y)) \to (\mathsf{cons}(x, z), w) \Leftarrow \mathsf{snoc}^{-1}(y) \twoheadrightarrow (z, w) \end{array} \right\}$$

Note that this DCTRS is also obtained by other inversion methods, e.g., [27].

When the inverted grammar programs are deterministic (i.e., the corresponding DCTRSs are non-overlapping), program inversion is completed and returns the corresponding non-overlapping DCTRSs. In the case of Example 4, however, $G_{\mathsf{snoc}^{-1}}$ is not deterministic (i.e., $R_{\mathsf{snoc}^{-1}}$ is overlapping). Thus, we proceed to the determinization method based on LR(0) parsing as follows:

Step 1 (collection of item sets). Given a grammar program, the collection of LR(0) item sets is computed by a closure operation (see below). If the collection has no conflict (shift/reduce, reduce/reduce, or shift/shift conflicts), then we proceed to Step 2; otherwise, the determinization method fails and returns the overlapping DCTRS corresponding to the given grammar program.

Step 2 (code generation). Given the conflict-free collection of LR(0) item sets, the grammar program is converted into a deterministic one. For lack of space, we do not describe the details of this step.

In the following, we recall how to construct from a grammar program the *canonical LR(0) collection.* An *LR(0) parse item* (*item*, for short) is a function definition with a dot $\cdot$ at some position on the right-hand side. An item indicates how much of a sequence of operations has been performed at a certain point during the evaluation of a grammar program. For example, the item $\mathsf{snoc}^{-1} \to \mathsf{cons?}\ (2\ 1) \cdot \mathsf{nil?}\ \mathsf{nil!}$ indicates that we have successfully performed a pattern matching cons? and a permutation (2 1), and that we hope to find a value nil on the top of the stack. We group items together into *LR(0) item sets* (*item sets*, for short) which represent the set of all possible operations that a computation can take at a certain point during evaluation. For a grammar program G with the collection $\mathcal{I}$ of its item sets, to calculate the collection $\mathcal{I}_G$ of all reachable item sets, we introduce two relations, *shift* and *reduce*, which correspond to determining the parse action in *LR(0) parsing*:

shift. $I_1 \rightsquigarrow^a I_2$ if $I_2 = \{f \to \alpha_1\ a \cdot \alpha_2 \mid f \to \alpha_1 \cdot a\ \alpha_2 \in \mathit{closure}(I_1)\}$
reduce. $I \hookrightarrow f$ if $f \to \alpha\ \cdot \in \mathit{closure}(I)$

where $a \in \mathcal{F} \cup \mathcal{O}p$ and the function $\mathit{closure} : \mathcal{I} \to \mathcal{I}$ and its auxiliary function $\mathit{cls} : \mathcal{I} \times \mathcal{F} \to \mathcal{I}$ are defined as follows:

- $\mathit{closure}(I) = \mathit{cls}(I, \emptyset)$,
- $\mathit{cls}(\emptyset, F) = \emptyset$, and
- $\mathit{cls}(I, F) = I \cup \mathit{cls}(\{f \to \cdot\ \alpha \mid f \to \alpha \in G, f \in F'\}, F \cup F')$ where $F' = \{f \mid f' \to \alpha_1 \cdot f\ \alpha_2 \in I, f \notin F\}$.

The shift relation $\rightsquigarrow^a$ with an operation a transforms item set I_1 to item set I_2 under a, the reduce relation $\hookrightarrow$ is to return from item set I after n operations of function f were shifted, and $\mathit{closure}(I)$ calculates a new item set I of grammar program G. Intuitively, item $f \to \alpha_1 \cdot a\ \alpha_2 \in \mathit{closure}(I)$ indicates that, at some point during evaluation of program G, we may perform operation a. The closure calculation terminates since there are only a finite number of different items for every grammar program. The set of all reachable item sets of G, the *canonical collection* $\mathcal{I}_G$, is defined as follows: $\mathcal{I}_G = \{I \mid I_0 \rightsquigarrow^* I\}$ where $\mathit{main}(G)$ denotes the main (or considered) function of G, start is a fresh non-terminal symbol, $I_0 = \{\mathsf{start} \to \cdot\ \mathit{main}(G)\}$, and $\rightsquigarrow^*$ is the reflexive and transitive closure of $\rightsquigarrow$. The set $\mathcal{I}_G$ is finite since there are only a finite number of different item sets.

As mentioned before, at the end of Step 1, we examine whether there is a conflict in the collection or not — if there is no conflict, we proceed to Step 2. In this paper, we focus on *shift/shift conflicts* only since such conflicts are much more important for the determinization method than other conflicts (*shift/reduce* and *reduce/reduce* conflicts). We say that a grammar program G causes a *shift/shift*

$$
\begin{aligned}
I_0 &= \{\ \mathsf{start} \to \cdot\, \mathsf{snoc}^{-1}\ \} \\
I_1 &= \left\{ \begin{array}{l} \mathsf{snoc}^{-1} \to \mathsf{cons?} \cdot (2\ 1)\ \mathsf{nil?}\ \mathsf{nil!} \\ \mathsf{snoc}^{-1} \to \mathsf{cons?} \cdot (2\ 1)\ \mathsf{snoc}^{-1}\ (3\ 1\ 2)\ \mathsf{cons!} \end{array} \right\} \\
I_2 &= \left\{ \begin{array}{l} \mathsf{snoc}^{-1} \to \mathsf{cons?}\ (2\ 1) \cdot \mathsf{nil?}\ \mathsf{nil!} \\ \mathsf{snoc}^{-1} \to \mathsf{cons?}\ (2\ 1) \cdot \mathsf{snoc}^{-1}\ (3\ 1\ 2)\ \mathsf{cons!} \end{array} \right\} \\
I_3 &= \{\ \mathsf{snoc}^{-1} \to \mathsf{cons?}\ (2\ 1)\ \mathsf{nil?} \cdot \mathsf{nil!}\ \} \\
I_4 &= \{\ \mathsf{snoc}^{-1} \to \mathsf{cons?}\ (2\ 1)\ \mathsf{nil?}\ \mathsf{nil!}\ \cdot\ \} \\
&\ \vdots \\
I_8 &= \{\ \mathsf{start} \to \mathsf{snoc}^{-1} \cdot\ \}
\end{aligned}
$$

$$
\begin{array}{llllll}
I_0 \rightsquigarrow^{\mathsf{snoc}^{-1}} I_8 & I_1 \rightsquigarrow^{(2\ 1)} I_2 & I_2 \rightsquigarrow^{\mathsf{nil?}} I_3 & I_3 \rightsquigarrow^{\mathsf{nil!}} I_4 & \ldots & I_8 \hookrightarrow^{1} \mathsf{start} \\
I_0 \rightsquigarrow^{\mathsf{cons?}} I_1 & I_2 \rightsquigarrow^{\mathsf{snoc}^{-1}} I_5 & I_2 \rightsquigarrow^{\mathsf{cons?}} I_1 & & &
\end{array}
$$

Fig. 1. The item sets of $G_{\mathsf{snoc}^{-1}}$ and their shift and reduce relations

conflict if its collection causes two shift relations $I \rightsquigarrow^{a_1} I_1$ and $I \rightsquigarrow^{a_2} I_2$ such that $a_1, a_2 \in \mathcal{O}p$, $a_1 \neq a_2$, and neither a_1 nor a_2 is a matching operator — if an item set has two or more shift actions labeled with atomic operations, all of them must be matching operations. For instance, $I \rightsquigarrow^{\mathsf{nil!}} I'$ and $I \rightsquigarrow^{\mathsf{cons?}} I''$ are in shift/shift conflict, while $I \rightsquigarrow^{\mathsf{nil?}} I'$ and $I \rightsquigarrow^{\mathsf{cons?}} I''$ are not.

Example 5. Consider the grammar program $G_{\mathsf{snoc}^{-1}}$ in Example 4 again. The item sets and their shift and reduce relations are illustrated in Fig. 1. Since this collection is conflict-free, we proceed to Step 2 of the determinization method, transforming the grammar program $G_{\mathsf{snoc}^{-1}}$ into the following grammar program:

$$
G'_{\mathsf{snoc}^{-1}} = \left\{ \begin{array}{lrl} \text{(C)} & \mathsf{snoc}^{-1} & \to \mathsf{cons?}\ (2\ 1)\ \mathsf{f} \\ \text{(D)} & \mathsf{f} & \to \mathsf{nil?}\ \mathsf{nil!} \\ \text{(E)} & \mathsf{f} & \to \mathsf{cons?}\ (2\ 1)\ \mathsf{f}\ (3\ 1\ 2)\ \mathsf{cons!} \end{array} \right\}
$$

where f is a non-terminal symbol introduced by means of the code generation. The arity and coarity of snoc^{-1} are 1 and 2, resp., since the arity and coarity of snoc are 2 and 1, resp. Solving the (in)equalities in Footnote 5, we determine that both the arity and coarity of f are 2. We have that $FCT(\{(\text{C})\}) = \{\ \mathsf{snoc}^{-1}(\mathsf{cons}(x, v)) \to (z, y) \Leftarrow \mathsf{f}(v, x) \twoheadrightarrow (z, y)\ \}$ since

$$
\begin{aligned}
Fct(\mathsf{cons?}\ (2\ 1)\ \mathsf{f}, [x_1], [\,], \emptyset) &= Fct((2\ 1)\ \mathsf{f}, [x, v], [\,], \{x_1 \mapsto \mathsf{cons}(x, v)\}) \\
&= Fct(\mathsf{f}, [v, x], [\,], \{x_1 \mapsto \mathsf{cons}(x, v)\}) \\
&= Fct([\,], [z, y], [\mathsf{f}(v, x) \twoheadrightarrow (z, y)], \{x_1 \mapsto \mathsf{cons}(x, v)\}) \\
&= ([z, y], [\mathsf{f}(v, x) \twoheadrightarrow (z, y)], \{x_1 \mapsto \mathsf{cons}(x, v)\})
\end{aligned}
$$

The rewrite rule $\mathsf{snoc}^{-1}(\mathsf{cons}(x, v)) \to (z, y) \Leftarrow \mathsf{f}(v, x) \twoheadrightarrow (z, y)$ is simplified to $\mathsf{snoc}^{-1}(\mathsf{cons}(x, v)) \to \mathsf{f}(v, x)$. We also have that

- $FCT(\{(\text{D})\}) = \{\ \mathsf{f}(\mathsf{nil}, y) \to (\mathsf{nil}, y)\ \}$, and
- $FCT(\{(\text{E})\}) = \{\ \mathsf{f}(\mathsf{cons}(x, v), y) \to (\mathsf{cons}(y, z), w) \Leftarrow \mathsf{f}(v, x) \twoheadrightarrow (z, w)\ \}$.

Therefore, the resulting grammar program $G'_{\mathsf{snoc}^{-1}}$ is translated back into the following non-overlapping DCTRS:

$$R'_{\mathsf{snoc}^{-1}} = FCT(G'_{\mathsf{snoc}^{-1}}) = \left\{ \begin{array}{r} \mathsf{snoc}^{-1}(\mathsf{cons}(x,v)) \to \mathsf{f}(v,x) \\ \mathsf{f}(\mathsf{nil},y) \to (\mathsf{nil},y) \\ \mathsf{f}(\mathsf{cons}(x,v),y) \to (\mathsf{cons}(y,z),w) \Leftarrow \mathsf{f}(v,x) \twoheadrightarrow (z,w) \end{array} \right\}$$

One may think that transformations of context-free grammars into Greibach normal forms (GNF) are sufficient for determinization. For example, $G'_{\mathsf{snoc}^{-1}}$ in Example 5 is easily converted into GNF. Roughly speaking, LRinv converts grammar programs into a grammar in GNF (and then minimizes the grammar), but not all the transformations of context-free grammars into GNF are useful for determinizing grammar programs.

As seen in Example 5, we always succeed in determinizing conflict-free grammar programs. On the other hand, the determinization method fails for all grammar programs with conflicts. Viewed in this light, a necessary condition to succeed in determinizing by this method is that grammar programs do not cause any conflict. However, the non-existence of shift/shift conflicts requires the resulting grammar programs to be of a restricted form:

$$f \to \alpha\ c_1?\ \alpha_1 \qquad \dots \qquad f \to \alpha\ c_n?\ \alpha_n$$

where $\alpha \in \mathcal{O}p^*$ and $c_1, \dots, c_n$ are pairwise different. The grammar program above is translated back into the following rewrite rules:

$$f(p_1,\dots,p_i,c_1(\dots),\dots) \to r_1 \qquad \dots \qquad f(p_1,\dots,p_i,c_n(\dots),\dots) \to r_n$$

It is clear that not all deterministic grammar programs and non-overlapping DCTRSs are in these forms. For example, the following grammar program causes a shift/shift conflict, and thus, the program is not of the form above:

$$\left\{ \begin{array}{l} \mathsf{f} \to (2\ 1)\ \mathsf{cons?}\ \mathsf{suc?}\ (2\ 1)\ \mathsf{cons?}\ \mathsf{cons!}\ (2\ 1)\ \mathsf{cons!}\ (2\ 1)\ \mathsf{cons!} \\ \mathsf{f} \to \mathsf{suc?}\ (2\ 1)\ \mathsf{cons?}\ (2\ 1)\ \mathsf{nil?}\ \mathsf{nil!}\ (2\ 1)\ \mathsf{cons!}\ (2\ 1)\ \mathsf{cons!} \end{array} \right\}$$

This grammar program is deterministic since the corresponding TRS is non-overlapping, where f is assumed to be binary:

$$\left\{ \begin{array}{r} \mathsf{f}(x,\mathsf{cons}(\mathsf{suc}(y),\mathsf{cons}(z_1,z_2))) \to \mathsf{cons}(x,\mathsf{cons}(y,\mathsf{cons}(z_1,z_2))) \\ \mathsf{f}(\mathsf{suc}(x),\mathsf{cons}(y,\mathsf{nil})) \to \mathsf{cons}(x,\mathsf{cons}(y,\mathsf{nil})) \end{array} \right\}$$

Note that exchanging the arguments of f is not a solution to transform the TRS to the form mentioned above.

Roughly speaking, shift/shift conflicts are not crucial for the determinization method while the non-existence of shift/shift conflicts ensures in advance that the determinization method succeeds. In the next section, we show a grammar program which can be determinized by the code generation (Step 2) although the grammar program causes shift/shift conflicts.

5 Semi-determinization of Grammar Programs with Shift/Shift Conflicts

In this section, by means of two examples, we observe what happens in applying the code generation (Step 2) to grammar programs causing shift/shift conflicts.

Example 6. Consider the following TRS, a variant of unbin shown in [15]:[8]

$$R_{\mathsf{unbin2}} = \left\{ \begin{array}{c} \mathsf{unbin2}(x) \to \mathsf{ub}(x, \mathsf{nil}) \\ \mathsf{ub}(\mathsf{suc}(\mathsf{zero}), y) \to y \\ \mathsf{ub}(\mathsf{suc}(\mathsf{suc}(x)), y) \to \mathsf{ub}(\mathsf{suc}(x), \mathsf{inc}(y)) \\ \mathsf{inc}(\mathsf{nil}) \to \mathsf{cons}(0, \mathsf{nil}) \\ \mathsf{inc}(\mathsf{cons}(0, y)) \to \mathsf{cons}(1, y) \\ \mathsf{inc}(\mathsf{cons}(1, y)) \to \mathsf{cons}(0, \mathsf{inc}(y)) \end{array} \right\}$$

where $\mathcal{C} = \{\mathsf{nil}/0, \mathsf{cons}/2, 0/0, 1/0, \mathsf{zero}/0, \mathsf{suc}/1, \ldots\}$ and $\mathcal{F} = \{\mathsf{unbin2}/1/1, \mathsf{ub}/2/1, \mathsf{inc}/1/1\}$. The function unbin2 converts positive natural numbers represented as suc(zero), suc(suc(zero)), suc(suc(suc(zero))), ... into binary-numeral expressions nil, cons(0, nil), cons(1, nil), ...[9] This TRS is translated and inverted to the following grammar program:

$$G_{\mathsf{unbin2}^{-1}} = \left\{ \begin{array}{l} \mathsf{unbin2}^{-1} \to \mathsf{ub}^{-1}\ (2\ 1)\ \mathsf{nil?} \\ \quad\ \mathsf{ub}^{-1} \to \mathsf{zero!}\ \mathsf{suc!} \\ \quad\ \mathsf{ub}^{-1} \to \mathsf{ub}^{-1}\ \mathsf{suc?}\ (2\ 1)\ \mathsf{inc}^{-1}\ (2\ 1)\ \mathsf{suc!}\ \mathsf{suc!} \\ \quad \mathsf{inc}^{-1} \to \mathsf{cons?}\ 0?\ \mathsf{nil?}\ \mathsf{nil!} \\ \quad \mathsf{inc}^{-1} \to \mathsf{cons?}\ 1?\ 0!\ \mathsf{cons!} \\ \quad \mathsf{inc}^{-1} \to \mathsf{cons?}\ 0?\ \mathsf{inc}^{-1}\ 1!\ \mathsf{cons!} \end{array} \right\}$$

This grammar program corresponds to the following DCTRS:

$$R_{\mathsf{unbin2}^{-1}} = \left\{ \begin{array}{l} \mathsf{unbin2}^{-1}(x) \to y \Leftarrow \mathsf{ub}^{-1}(x) \twoheadrightarrow (y, \mathsf{nil}) \\ \qquad \mathsf{ub}^{-1}(x) \to (\mathsf{suc}(\mathsf{zero}), x) \\ \qquad \mathsf{ub}^{-1}(x) \to (\mathsf{suc}(\mathsf{suc}(y)), \mathsf{inc}^{-1}(z)) \Leftarrow \mathsf{ub}^{-1}(x) \twoheadrightarrow (\mathsf{suc}(y), z) \\ \qquad\qquad\qquad \vdots \end{array} \right\}$$

This conditional TRS is not *operationally terminating* [19] since $\mathsf{ub}^{-1}(x)$ is called in the conditional part of the third rule which defines $\mathsf{ub}^{-1}(x)$. The item sets and their shift relations for $G_{\mathsf{unbin2}^{-1}}$ are illustrated in Fig. 2. For item set I_1, we have two shift relations $I_1 \leadsto^{(2\ 1)} I_3$ and $I_1 \leadsto^{\mathsf{suc?}} I_4$. These relations cause a shift/shift conflict. Therefore, we cannot proceed to the code generation (Step 2). However,

[8] All the examples shown in [15] are mentioned in [27] and they can be found in `http://www.trs.cm.is.nagoya-u.ac.jp/repius/`.

[9] To make unbin2 simpler, following the original unbin in [15], we use this coding instead of the usual style cons(0, nil) (as 0), cons(1, nil) (as 1), cons(1, cons(0, nil)) (as 2), ...

$$
\begin{aligned}
I_0 &= \{\ \mathsf{start} \to \cdot\ \mathsf{unbin2}^{-1}\ \}\\
I_1 &= \left\{ \begin{array}{r} \mathsf{unbin2}^{-1} \to \mathsf{ub}^{-1} \cdot (2\ 1)\ \mathsf{nil?} \\ \mathsf{ub}^{-1} \to \mathsf{ub}^{-1} \cdot \mathsf{suc?}\ (2\ 1)\ \mathsf{inc}^{-1}\ (2\ 1)\ \mathsf{suc!}\ \mathsf{suc!} \end{array} \right\}\\
I_2 &= \{\ \mathsf{ub}^{-1} \to \mathsf{zero!} \cdot \mathsf{suc!}\ \}\\
I_3 &= \{\ \mathsf{unbin2}^{-1} \to \mathsf{ub}^{-1}\ (2\ 1) \cdot \mathsf{nil?}\ \}\\
I_4 &= \{\ \mathsf{ub}^{-1} \to \mathsf{ub}^{-1}\ \mathsf{suc?} \cdot (2\ 1)\ \mathsf{inc}^{-1}\ (2\ 1)\ \mathsf{suc!}\ \mathsf{suc!}\ \}\\
&\ \vdots
\end{aligned}
$$

$$
I_0 \leadsto^{\mathsf{unbin2}^{-1}} I_{22} \quad I_0 \leadsto^{\mathsf{zero!}} I_2 \quad I_0 \leadsto^{\mathsf{ub}^{-1}} I_1 \quad I_1 \leadsto^{(2\ 1)} I_3 \quad I_1 \leadsto^{\mathsf{suc?}} I_4 \quad \cdots
$$

Fig. 2. The item sets of $G_{\mathsf{unbin2}^{-1}}$ and their shift relations

by ignoring the shift/shift conflict and by proceeding to the code generation, we obtain the following grammar program:

$$
\left\{
\begin{array}{r@{\ }l}
\mathsf{unbin2}^{-1} & \to \mathsf{zero!}\ \mathsf{suc!}\ \mathsf{f} \\
\mathsf{f} & \to (2\ 1)\ \mathsf{nil?} \\
\mathsf{f} & \to \mathsf{suc?}\ (2\ 1)\ \mathsf{cons?}\ \mathsf{g}\ (2\ 1)\ \mathsf{suc!}\ \mathsf{suc!}\ \mathsf{f} \\
\mathsf{g} & \to \mathsf{0?}\ \mathsf{h} \\
\mathsf{g} & \to \mathsf{1?}\ \mathsf{0!}\ \mathsf{cons!} \\
\mathsf{h} & \to \mathsf{nil?}\ \mathsf{nil!} \\
\mathsf{h} & \to \mathsf{cons?}\ \mathsf{g}\ \mathsf{1!}\ \mathsf{cons!}
\end{array}
\right\}
$$

This grammar program is translated back into the following non-overlapping and terminating TRS:

$$
R'_{\mathsf{unbin2}^{-1}} = \left\{
\begin{array}{r@{\ }l}
\mathsf{unbin2}^{-1}(x) & \to \mathsf{f}(\mathsf{suc}(\mathsf{zero}), x) \\
\mathsf{f}(x, \mathsf{nil}) & \to x \\
\mathsf{f}(\mathsf{suc}(x), \mathsf{cons}(y, z)) & \to \mathsf{f}(\mathsf{suc}(\mathsf{suc}(x)), \mathsf{g}(y, z)) \\
\mathsf{g}(\mathsf{0}, y) & \to \mathsf{h}(y) \\
\mathsf{g}(\mathsf{1}, y) & \to \mathsf{cons}(\mathsf{0}, y) \\
\mathsf{h}(\mathsf{nil}) & \to \mathsf{nil} \\
\mathsf{h}(\mathsf{cons}(y, z)) & \to \mathsf{cons}(\mathsf{1}, \mathsf{g}(y, z))
\end{array}
\right\}
$$

This DCTRS is desirable as an inverse of unbin2. Note that (operational) termination of the examples in this paper was proved by the termination tools AProVE 1.2 [6] and VMTL [32].

In Example 6, by ignoring shift/shift conflicts, we succeeded in determinizing $G_{\mathsf{unbin2}^{-1}}$. However, not for all grammar programs causing shift/shift conflicts, such a determinization method succeeds in converting the programs into deterministic ones.

Example 7. Consider the following TRS, a variant of unbin2 in Example 6:

$$
R_{\mathsf{unbin3}} = \left\{
\begin{array}{r@{\ }l}
\mathsf{unbin3}(x) & \to \mathsf{ub}(x, \mathsf{cons}(\mathsf{0}, \mathsf{nil})) \\
\mathsf{ub}(\mathsf{suc}(\mathsf{zero}), y) & \to y \\
\mathsf{ub}(\mathsf{suc}(\mathsf{suc}(x)), y) & \to \mathsf{ub}(\mathsf{suc}(x), \mathsf{inc}(y)) \\
& \vdots
\end{array}
\right\}
$$

This TRS is inverted and translated back into the following overlapping DCTRS:

$$R_{\mathsf{unbin3}^{-1}} = \left\{ \begin{array}{l} \mathsf{unbin3}^{-1}(x) \to z \Leftarrow \mathsf{ub}^{-1}(x) \twoheadrightarrow (z, \mathsf{cons}(0, \mathsf{nil})) \\ \quad \mathsf{ub}^{-1}(x) \to (\mathsf{suc}(\mathsf{zero}), x) \\ \quad \mathsf{ub}^{-1}(x) \to (\mathsf{suc}(\mathsf{suc}(y), \mathsf{inc}^{-1}(z)) \Leftarrow \mathsf{ub}^{-1}(x) \twoheadrightarrow (\mathsf{suc}(y), z) \\ \qquad \vdots \end{array} \right\}$$

This DCTRS is not operationally terminating. The corresponding grammar program causes a shift/shift conflict, but we ignore the shift/shift conflict, proceeding to the code generation. Then, we obtain the following grammar program:

$$G'_{\mathsf{unbin3}^{-1}} = \left\{ \begin{array}{l} \mathsf{unbin3}^{-1} \to \mathsf{zero!}\ \mathsf{suc!}\ \mathsf{f} \\ \qquad \mathsf{f} \to (2\ 1)\ \mathsf{cons?}\ \mathsf{0?}\ \mathsf{nil?} \\ \qquad \mathsf{f} \to \mathsf{suc?}\ (2\ 1)\ \mathsf{cons?}\ \mathsf{g}\ (2\ 1)\ \mathsf{suc!}\ \mathsf{suc!}\ \mathsf{f} \\ \qquad \mathsf{g} \to \mathsf{0?}\ \mathsf{h} \\ \qquad \mathsf{g} \to \mathsf{1?}\ \mathsf{0!}\ \mathsf{cons!} \\ \qquad \mathsf{h} \to \mathsf{nil?}\ \mathsf{nil!} \\ \qquad \mathsf{h} \to \mathsf{cons?}\ \mathsf{g}\ \mathsf{1!}\ \mathsf{cons!} \end{array} \right\}$$

This grammar program corresponds to the following TRS:

$$R'_{\mathsf{unbin3}^{-1}} = \left\{ \begin{array}{c} \mathsf{unbin3}^{-1}(x) \to \mathsf{f}(\mathsf{suc}(\mathsf{zero}), x) \\ \mathsf{f}(x, \mathsf{cons}(0, \mathsf{nil})) \to x \\ \mathsf{f}(\mathsf{suc}(x), \mathsf{cons}(y, z))) \to \mathsf{f}(\mathsf{suc}(\mathsf{suc}(x)), \mathsf{g}(y, z)) \\ \mathsf{g}(0, y) \to \mathsf{h}(y) \\ \mathsf{g}(1, y) \to \mathsf{cons}(0, y) \\ \mathsf{h}(\mathsf{nil}) \to \mathsf{nil} \\ \mathsf{h}(\mathsf{cons}(y, z)) \to \mathsf{cons}(1, \mathsf{g}(y, z)) \end{array} \right\}$$

This TRS is still overlapping but terminating. The determinization method fails to produce a non-overlapping DCTRS but transforms a non-operationally-terminating DCTRS to an operationally terminating DCTRS. Therefore, proceeding to the code generation regardless shift/shift conflicts is meaningful for this example even though it fails to produce a non-overlapping system.

It can be easily decided whether a DCTRS is non-overlapping, i.e., whether a grammar program is deterministic. For this reason, it is not so important to check the non-existence of shift/shift conflicts in advance. Therefore, we do not have to take care of shift/shift conflicts.

The determinization method with ignoring shift/shift conflicts is no longer a *determinization* (see Example 7). We call such a method *semi-determinization.* Our semi-determinization method is summarized as follows: given a grammar program G, (1) we compute the collection of item sets of G, and (2) if the collection has neither shift/reduce nor reduce/reduce conflicts, then we apply the code generation to G.

6 Elimination of Infeasible Sequences

For a grammar program G defining f, the language $L(G, f)$ may contain *infeasible* sequence (e.g., nil! cons?). In such cases, by removing function definitions that produce infeasible sequences only, we sometimes succeed in determinizing the grammar program. In this section, we first formalize infeasibility. Then, as a post-process of the semi-determinization method, we propose a method to eliminate infeasible definitions from unfolded grammar programs.

We first formalize the notion of infeasibility for atomic-operation sequences and function definitions.

Definition 8. *Let G be a grammar program. An atomic-operation sequence $\alpha \in \mathcal{O}p^*$ is called* feasible *if there exist sequences $\vec{t}, \vec{u} \in T(\mathcal{C})^*$ such that $(\alpha, \vec{t}) \to_G^* ([\,], \vec{u})$; otherwise, α is* infeasible. *A function definition $f \to \alpha' \in G$ is called* infeasible *if α' contains an infeasible sequence.*

Note that it is decidable whether an atomic-operation sequence is infeasible or not: we adapt *narrowing* [14] to computation of atomic-operation sequences as follows:

- $([c!] @ \alpha, [u_1, \ldots, u_n] @ s) \rightsquigarrow (\alpha, [c(u_1, \ldots, u_n)] @ s)$,
- $([c!] @ \alpha, [u_1, \ldots, u_i] @ X) \rightsquigarrow (\alpha, [c(u_1, \ldots, u_i, x_{i+1}, \ldots, x_n)] @ Y)$,
- $([c?] @ \alpha, [c(u_1, \ldots, u_n)] @ s) \rightsquigarrow (\alpha, [u_1, \ldots, u_n] @ s)$,
- $([c?] @ \alpha, [x] @ s) \rightsquigarrow (\alpha, [x_1, \ldots, x_n] @ s)$,
- $([c?] @ \alpha, X) \rightsquigarrow (\alpha, [x_1, \ldots, x_n] @ Y)$,
- $([(i_1 \ldots i_n)] @ \alpha, [u_1, \ldots, u_n] @ s) \rightsquigarrow (\alpha, [u_{i_1}, \ldots, u_{i_n}] @ s)$,
- $([(i_1 \ldots i_n)] @ \alpha, [u_1, \ldots, u_i] @ X) \rightsquigarrow (\alpha, [u'_{i_1}, \ldots, u'_{i_n}] @ Y)$, where $u'_j = u_j$ if $j \leq i$, and otherwise, $u'_j = x_j$,

where $c/n \in \mathcal{C}$, $i < n$, x is a variable, $x_1, \ldots, x_n, y$ are fresh variables, X is a variable to represent a sequence, and Y is a fresh variable to represent a sequence. Note that, given a variable X, a $\rightsquigarrow$-derivation starting from (α, X) is a naive analysis for the capability of the execution of α. Note also that $\rightsquigarrow$ is a deterministic and terminating reduction. An atomic-operation sequence α is feasible iff $(\alpha, X) \rightsquigarrow ([\,], \vec{u})$ for some sequence $\vec{u} \in T(\mathcal{C}, \mathcal{V})$.

Example 9. Let us consider the grammar program $G'_{\mathsf{unbin3}^{-1}}$ in Example 7 again. The language $L(G'_{\mathsf{unbin3}^{-1}}, \mathsf{f})$ contains the following sequence:

$$\mathsf{suc?}\ (2\ 1)\ \mathsf{cons?}\ \mathsf{0?}\ \mathsf{nil?}\ \mathsf{nil!}\ (2\ 1)\ \mathsf{suc!}\ \mathsf{suc!}\ (2\ 1)\ \mathsf{cons?}\ \ldots$$

In the execution of this sequence, the term nil is generated by nil! and examined by the second occurrence of cons?. This can be seen in the following derivation:

$$\begin{array}{l}(\mathsf{suc?}\ (2\ 1)\ \mathsf{cons?}\ \mathsf{0?}\ \mathsf{nil?}\ \mathsf{nil!}\ (2\ 1)\ \mathsf{suc!}\ \mathsf{suc!}\ (2\ 1)\ \mathsf{cons?}\ \ldots, X) \\ \quad \rightsquigarrow^+ (\mathsf{cons?}\ \ldots, [\mathsf{nil}, \mathsf{suc}(\mathsf{suc}(x_1))] @ Y)\end{array}$$

This examination fails, and thus, the above sequence is infeasible.

Infeasible function definitions are useless for grammar programs since they are never used in completed computations $([f], \vec{t}) \to_G^* ([], \vec{u})$. For this reason, we may eliminate infeasible function definitions from grammar programs.

Definition 10. *Let G and G' be grammar programs defining $f/n/m \in \mathcal{F}$. We say that G and G' are* semantically equivalent w.r.t. f *if for all terms $t_1, \ldots, t_n$, $u_1, \ldots, u_m$, $([f], [t_1, \ldots, t_n]) \to_G^* ([], [u_1, \ldots, u_m])$ iff $([f], [t_1, \ldots, t_n]) \to_{G'}^* ([], [u_1, \ldots, u_m])$.*

We denote the set of feasible atomic-operation sequences in $L(G, f)$ by $L_{fe}(G, f)$.

Proposition 11. *Let G and G' be grammar programs defining $f/n/m \in \mathcal{F}$. Then, G and G' are semantically equivalent w.r.t. f if $L_{fe}(G, f) = L_{fe}(G', f)$.*

The converse of Proposition 11 does not hold in general. For example, $G = \{f \to$ 0? cons?$\}$ and $G' = \{f \to$ (2 1) cons? (3 1 2) 0?$\}$ are semantically equivalent, but $L_{fe}(G, f) \neq L_{fe}(G', f)$. By definition, the following theorem holds.

Theorem 12. *Let G be a grammar program such that $G = G' \uplus \{g \to \alpha\}$ and $g \to \alpha$ is infeasible. Then, for any function f defined in G, $L_{fe}(G, f) = L_{fe}(G', f)$. That is, G and G' are semantically equivalent w.r.t. f.*

Consider $G'_{\mathsf{unbin3}^{-1}}$ in Example 7 again. Although $L(G'_{\mathsf{unbin3}^{-1}}, \mathsf{f})$ contains infeasible sequences, there is no infeasible function definition in $G'_{\mathsf{unbin3}^{-1}}$. Thus, unfortunately, we cannot drop any function definitions. Roughly speaking, any infeasible sequence is of the form α_1 $c_1!$ α_2 $c_2?$ α_3, where $c_1 \neq c_2$ and the term generated by $c_1!$ is examined by $c_2?$. From the viewpoint of DCTRSs obtained from grammar programs, constructor application operations correspond to generated constructors in the right-hand side of the rewrite rule, and constructor matching operations correspond to pattern matching in the left-hand side. Thus, in general, infeasible sequences do not appear explicitly in function definitions even if the language contains infeasible sequences. This means that we rarely find infeasible definitions in initial grammar programs. However, by unfolding function definitions, infeasible definitions sometimes appear explicitly.

Example 13. Let us consider the grammar program $G'_{\mathsf{unbin3}^{-1}}$ in Example 7 again. The first and second definitions of f are overlapping.

(1) f → (2 1) cons? 0? nil?
(2) f → suc? (2 1) cons? g (2 1) suc! suc! f

Now, we focus on the second one since the first one has no function call in the right-hand side. Let us unfold the leftmost function call g:

(2a) f → suc? (2 1) cons? 0? h (2 1) suc! suc! f
(2b) f → suc? (2 1) cons? 1? 0! cons! (2 1) suc! suc! f

The definition (2a) and the first definition (1) of f are still overlapping, while the definition (2b) is no longer overlapping with any other definitions (1) and (2a). Let us unfold the leftmost function call h in (2a):

(2a-i) f → suc? (2 1) cons? 0? nil? nil! (2 1) suc! suc! f
(2a-ii) f → suc? (2 1) cons? 0? cons? g 1! cons! (2 1) suc! suc! f

Again, the definition (2a-i) and the first definition (1) of f are still overlapping, while the definition (2a-ii) is no longer overlapping with any other definitions (1), (2b), and (2a-i). Let us unfold the leftmost function call f in (2a-i):

(2a-i-1) f → suc? (2 1) cons? 0? nil? nil! (2 1) suc! suc! (2 1) cons? 0? nil?
(2a-i-2) f → suc? (2 1) cons? 0? nil? nil! (2 1) suc! suc! suc? (2 1) cons? 0? nil? nil! (2 1) suc! suc! f
(2a-i-3) f → suc? (2 1) cons? 0? nil? nil! (2 1) suc! suc! suc? (2 1) cons? 0? cons? g 1! cons! (2 1) suc! suc! f

Finally, all of the definitions (2a-i-1), (2a-i-2), and (2a-i-3) are infeasible (see Example 9), and thus, we obtain the following grammar program:

$$G''_{\mathsf{unbin3}^{-1}} = \left\{ \begin{array}{l} \mathsf{unbin3}^{-1} \rightarrow \mathsf{zero!\ suc!\ f} \\ \quad\quad \mathsf{f} \rightarrow \mathsf{(2\ 1)\ cons?\ 0?\ nil?} \\ \quad\quad \mathsf{f} \rightarrow \mathsf{suc?\ (2\ 1)\ cons?\ 0?\ cons?\ g\ 1!\ cons!\ (2\ 1)\ suc!\ suc!\ f} \\ \quad\quad \mathsf{f} \rightarrow \mathsf{suc?\ (2\ 1)\ cons?\ 1?\ 0!\ cons!\ (2\ 1)\ suc!\ suc!\ f} \\ \quad\quad \vdots \end{array} \right\}$$

$G''_{\mathsf{unbin3}^{-1}}$ is translated into the following non-overlapping and terminating TRS:

$$R''_{\mathsf{unbin3}^{-1}} = \left\{ \begin{array}{r} \mathsf{unbin3}^{-1}(x) \rightarrow \mathsf{f}(\mathsf{suc}(\mathsf{zero}), x) \\ \mathsf{f}(x, \mathsf{cons}(0, \mathsf{nil})) \rightarrow x \\ \mathsf{f}(\mathsf{suc}(x), \mathsf{cons}(0, \mathsf{cons}(y, z))) \rightarrow \mathsf{f}(\mathsf{suc}(\mathsf{suc}(x)), \mathsf{cons}(1, \mathsf{g}(y, z))) \\ \mathsf{f}(\mathsf{suc}(x), \mathsf{cons}(1, y)) \rightarrow \mathsf{f}(\mathsf{suc}(\mathsf{suc}(x)), \mathsf{cons}(0, y)) \\ \vdots \end{array} \right\}$$

This TRS is desirable as an inverse of unbin3.

In the rest of this section, we formalize a way of unfolding function definitions and how to eliminate infeasible definitions from unfolded grammar programs.

Definition 14. *Let G be a grammar program and $f \rightarrow \alpha_1\ g\ \alpha_2$ be a function definition in G such that α_1 is an atomic-operation sequence. We define Unfold as follows: $\mathit{Unfold}_G(f \rightarrow \alpha_1\ g\ \alpha_2) = \{f \rightarrow \alpha_1\ \beta\ \alpha_2 \mid g \rightarrow \beta \in G\}$. Note that $\mathit{Unfold}_G(f \rightarrow \alpha) = \{f \rightarrow \alpha\}$ if α does not contain any function call. We extend Unfold_G to grammar programs in order to unfold overlapping definitions:*

$$\mathit{Unfold}(G) = (G \setminus G') \cup \textstyle\bigcup_{f \rightarrow \alpha \in G'} \mathit{Unfold}_G(f \rightarrow \alpha)$$

where G' is the maximum subset of G such that each definition in G' is overlapping with another definition in G. Note that no definition in $G \setminus G'$ is overlapping with any other definition in G. We define a function RmInfe that eliminates infeasible definitions from unfolded grammar programs:

$$\mathit{RmInfe}(G) = \{f \rightarrow \alpha \in \mathit{Unfold}(G) \mid f \rightarrow \alpha \text{ is not infeasible}\}$$

We define the k times application of RmInfe as RmInfe_k: $\mathit{RmInfe}_0(G) = G$ and $\mathit{RmInfe}_k(G) = \mathit{RmInfe}(\mathit{RmInfe}_{k-1}(G))$ for $k > 0$.

It is almost impossible to know what k is enough to obtain deterministic grammar programs or whether such k exists. Thus, we heuristically specify k in advance and try to obtain deterministic grammar programs. One of the heuristics is the number of function definitions. Of course, we may stop applying *RmInfe* before applying it k times if we obtain a deterministic grammar program.

Example 15. Consider $G'_{\mathsf{unbin3}^{-1}}$ and $G''_{\mathsf{unbin3}^{-1}}$ in Examples 7 and 9 again. We have that $\mathit{RmInfe}_3(G'_{\mathsf{unbin3}^{-1}}) = G''_{\mathsf{unbin3}^{-1}}$.

Finally, we show correctness of *RmInfe*.

Theorem 16. *Let G be a grammar program. Then, for any $f \in \mathcal{F}$, G and $\mathit{RmInfe}(G)$ are semantically equivalent w.r.t. f, and thus, for any $k \geq 0$, G and $\mathit{RmInfe}_k(G)$ are semantically equivalent w.r.t. f.*

Proof. This theorem follows from Theorem 12 and the following facts:

- for all grammar programs G_1 and G_2, if $L(G_1, f) = L(G_2, f)$, then $L_{fe}(G_1, f) = L_{fe}(G_2, f)$, and
- for any function definition $g \to \alpha \in G$ and for any symbol $f' \in \mathcal{F}$, $L(G, f') = L((G \setminus \{g \to \alpha\}) \cup \mathit{Unfold}_G(g \to \alpha), f')$. □

7 Conclusion and Future Work

In this paper, we have adapted translations between functional and grammar programs to rewriting systems, and have shown that the semi-determinization method succeeds for some grammar programs causing shift/shift conflicts. We have also shown that the semi-determinization method converts some grammar programs whose corresponding DCTRSs are not terminating, into ones whose corresponding DCTRSs are terminating. Moreover, we have proposed the method to eliminate infeasible sequences from unfolded grammar programs. We have shown that the method converts some grammar programs into deterministic ones. By using the method as a post-process of the semi-determinization method, we made the original inversion method LRinv strictly more powerful. We will implement a tool to deal with grammar programs, and then we will introduce both LRinv and the method in this paper to the tool. As future work, we will characterize the difference between grammar programs for which the semi-determinization succeeds/fails to generate deterministic programs, clarifying a sufficient condition for producing deterministic grammar programs.

The method of eliminating infeasible sequences from unfolded grammar programs looks similar to the computation of *more specific versions* shown in [28]. However, the method is a bit different from the MSV transformation in [28], e.g., the MSV transformation converts $R'_{\mathsf{unbin3}^{-1}}$ in Example 7 into the following TRS which is different from $R''_{\mathsf{unbin3}^{-1}}$ in Example 13:

$$\left\{ \begin{array}{r} \mathsf{unbin3}^{-1}(x) \to \mathsf{f}(\mathsf{suc}(\mathsf{zero}), x) \\ \mathsf{f}(x, \mathsf{cons}(0, \mathsf{nil})) \to x \\ \mathsf{f}(\mathsf{suc}(x), \mathsf{cons}(0, \mathsf{cons}(y, z))) \to \mathsf{f}(\mathsf{suc}(\mathsf{suc}(x)), \mathsf{g}(0, \mathsf{cons}(y, z))) \\ \mathsf{f}(\mathsf{suc}(x), \mathsf{cons}(1, y)) \to \mathsf{f}(\mathsf{suc}(\mathsf{suc}(x)), \mathsf{g}(1, y)) \\ \vdots \end{array} \right\}$$

For a given rewrite rule, the MSV transformation produces some instances of the given rewrite rule — the transformation does not unfold the rewrite rule. On the other hand, our method unfolds the rewrite rule — the method instantiates the rule and proceeds the execution of the right-hand side. Viewed in this light, the results of our method and the MSV transformation are different. However, they looks equivalent in the sense of (semi-)determinization. One of the advantages of our method over the MSV transformation is simpleness since our method is simpler than the MSV transformation and since grammar programs are more abstract than rewriting systems — grammar programs have no variables and the semantics is defined as a simple stack machine. On the other hand, our method is not applicable to rewriting systems with erasing rules while the MSV transformation is applicable to them. We will compare the method in this paper with the MSV transformation, from a theoretical point of view.

As comparison of program inversion, we should compare LRinv combined with both our semi-determinization method and the elimination method of infeasible definitions, with other inversion methods (e.g., [27]). In general, the resulting programs of the inversion methods are different. For this reason, it is very difficult to compare the inversion methods from a theoretical point of view. Thus, we will compare the inversion methods by means of the performance to the benchmarks shown in several papers on program inversion.

Acknowledgements. We thank the anonymous reviewers very much for their helpful comments and suggestions to improve this paper.

References

1. Abramov, S.M., Glück, R.: Principles of inverse computation in a functional language. In: APLAS 2000, pp. 141–152 (2000)
2. Abramov, S., Glück, R.: The universal resolving algorithm: Inverse computation in a functional language. In: Backhouse, R., Oliveira, J.N. (eds.) MPC 2000. LNCS, vol. 1837, pp. 187–212. Springer, Heidelberg (2000)
3. Abramov, S.M., Glück, R.: The universal resolving algorithm and its correctness: inverse computation in a functional language. Sci. Comput. Program. 43(2-3), 193–229 (2002)
4. Almendros-Jiménez, J.M., Vidal, G.: Automatic partial inversion of inductively sequential functions. In: Horváth, Z., Zsók, V., Butterfield, A. (eds.) IFL 2006. LNCS, vol. 4449, pp. 253–270. Springer, Heidelberg (2007)
5. Dershowitz, N., Mitra, S.: Jeopardy. In: Narendran, P., Rusinowitch, M. (eds.) RTA 1999. LNCS, vol. 1631, pp. 16–29. Springer, Heidelberg (1999)
6. Giesl, J., Schneider-Kamp, P., Thiemann, R.: AProVE 1.2: Automatic termination proofs in the dependency pair framework. In: Furbach, U., Shankar, N. (eds.) IJCAR 2006. LNCS (LNAI), vol. 4130, pp. 281–286. Springer, Heidelberg (2006)
7. Glück, R., Kawabe, M.: A program inverter for a functional language with equality and constructors. In: Ohori, A. (ed.) APLAS 2003. LNCS, vol. 2895, pp. 246–264. Springer, Heidelberg (2003)
8. Glück, R., Kawabe, M.: A method for automatic program inversion based on LR(0) parsing. Fundam. Inform. 66(4), 367–395 (2005)

9. Glück, R., Kawabe, M.: Revisiting an automatic program inverter for Lisp. SIGPLAN Notices 40(5), 8–17 (2005)
10. Glück, R., Kawada, Y., Hashimoto, T.: Transforming interpreters into inverse interpreters by partial evaluation. In: Proceedings of Partial Evaluation and Semantics-based Program Manipulation, pp. 10–19. ACM Press (2003)
11. Gries, D.: The Science of Programming. Springer, Heidelberg (1981)
12. Harrison, P.G.: Function inversion. In: Proceedings of the IFIP TC2 Workshop on Partial Evaluation and Mixed Computation, pp. 153–166. North-Holland (1988)
13. Harrison, P.G., Khoshnevisan, H.: On the synthesis of function inverses. Acta Inf. 29(3), 211–239 (1992)
14. Hullot, J.M.: Canonical forms and unification. In: Bibel, W., Kowalski, R. (eds.) CADE 1980. LNCS, vol. 87, pp. 318–334. Springer, Heidelberg (1980)
15. Kawabe, M., Futamura, Y.: Case studies with an automatic program inversion system. In: Proceedings of the 21st Conference of Japan Society for Software Science and Technology, 6C-3, pp. 1–5 (2004)
16. Kawabe, M., Glück, R.: The program inverter LRinv and its structure. In: Hermenegildo, M.V., Cabeza, D. (eds.) PADL 2005. LNCS, vol. 3350, pp. 219–234. Springer, Heidelberg (2005)
17. Khoshnevisan, H., Sephton, K.M.: InvX: An automatic function inverter. In: Dershowitz, N. (ed.) RTA 1989. LNCS, vol. 355, pp. 564–568. Springer, Heidelberg (1989)
18. Korf, R.E.: Inversion of applicative programs. In: Proceedings of the 7th International Joint Conference on Artificial Intelligence, pp. 1007–1009. William Kaufmann (1981)
19. Lucas, S., Marché, C., Meseguer, J.: Operational termination of conditional term rewriting systems. Inf. Process. Lett. 95(4), 446–453 (2005)
20. Matsuda, K., Mu, S.-C., Hu, Z., Takeichi, M.: A grammar-based approach to invertible programs. In: Gordon, A.D. (ed.) ESOP 2010. LNCS, vol. 6012, pp. 448–467. Springer, Heidelberg (2010)
21. McCarthy, J.: The inversion of functions defined by Turing machines. In: Automata Studies, pp. 177–181. Princeton University Press (1956)
22. Nagashima, M., Sakai, M., Sakabe, T.: Determinization of conditional term rewriting systems. Theor. Comput. Sci. 464, 72–89 (2012)
23. Nishida, N., Sakai, M.: Completion after program inversion of injective functions. In: Proceedings of the 8th International Workshop on Reduction Strategies in Rewriting and Programming. ENTCS, vol. 237, pp. 39–56 (2009)
24. Nishida, N., Sakai, M., Sakabe, T.: Generation of inverse computation programs of constructor term rewriting systems. The IEICE Trans. Inf. & Syst. J88-D-I(8), 1171–1183 (2005) (in Japanese)
25. Nishida, N., Sakai, M., Sakabe, T.: Partial inversion of constructor term rewriting systems. In: Giesl, J. (ed.) RTA 2005. LNCS, vol. 3467, pp. 264–278. Springer, Heidelberg (2005)
26. Nishida, N., Sakai, M., Sakabe, T.: Soundness of unravelings for conditional term rewriting systems via ultra-properties related to linearity. Logical Methods in Computer Science 8(3), 1–49 (2012)
27. Nishida, N., Vidal, G.: Program inversion for tail recursive functions. In: Schmidt-Schauß, M. (ed.) RTA 2011. LIPIcs, vol. 10, pp. 283–298. Schloß Dagstuhl–Leibniz-Zentrum für Informatik (2011)
28. Nishida, N., Vidal, G.: Computing more specific versions of conditional rewriting systems. In: Albert, E. (ed.) LOPSTR 2012. LNCS, vol. 7844, pp. 137–154. Springer, Heidelberg (2013)

29. Ohlebusch, E.: Advanced topics in term rewriting. Springer, Heidelberg (2002)
30. Romanenko, A.: The generation of inverse functions in Refal. In: Proceedings of the IFIP TC2 Workshop on Partial Evaluation and Mixed Computation, pp. 427–444. North-Holland (1988)
31. Romanenko, A.: Inversion and metacomputation. In: Proceedings of the Symposium on Partial Evaluation and Semantics-Based Program Manipulation, pp. 12–22. ACM Press (1991)
32. Schernhammer, F., Gramlich, B.: VMTL–A modular termination laboratory. In: Treinen, R. (ed.) RTA 2009. LNCS, vol. 5595, pp. 285–294. Springer, Heidelberg (2009)
33. Schneider-Kamp, P., Giesl, J., Serebrenik, A., Thiemann, R.: Automated termination proofs for logic programs by term rewriting. ACM Trans. on Comput. Log. 11(1), 1–52 (2009)
34. Secher, J.P., Sørensen, M.H.: From checking to inference via driving and dag grammars. In: Proceedings of Partial Evaluation and Semantics-Based Program Manipulation, pp. 41–51. ACM Press (2002)

A Framework for Guided Test Case Generation in Constraint Logic Programming

José Miguel Rojas[1] and Miguel Gómez-Zamalloa[2]

[1] Technical University of Madrid, Spain
[2] DSIC, Complutense University of Madrid, Spain

Abstract. Performing test case generation by symbolic execution on large programs becomes quickly impracticable due to the path explosion problem. A common limitation that this problem poses is the generation of unnecessarily large number of possibly irrelevant or redundant test cases even for medium-size programs. Tackling the path explosion problem and selecting high quality test cases are considered major challenges in the software testing community. In this paper we propose a constraint logic programming-based framework to guide symbolic execution and thus test case generation towards a more relevant and potentially smaller subset of paths in the program under test. The framework is realized as a tool and empirical results demonstrate its applicability and effectiveness. We show how the framework can help to obtain high quality test cases and to alleviate the scalability issues that limit most symbolic execution-based test generation approaches.

Keywords: Constraint Logic Programming, Guided Test Case Generation, Software Testing, Symbolic Execution, Trace-abstraction.

1 Introduction

Testing remains a mostly manual stage within the software development process [4]. Test Case Generation (TCG) is a research field devoted to the automation of a crucial part of the testing process, the generation of input data. Symbolic Execution is nowadays one of the predominant techniques to automate the generation of input data. It is the underlying technique of several popular testing tools, both in academia and software industry [4].

Symbolic execution [10] executes a program with the contents of variables being symbolic formulas over the input arguments rather than concrete values. The outcome is a set of equivalence classes of inputs, each of them consisting of the constraints that characterize a set of feasible concrete executions of a program that takes the same path. A *test suite* is the set of *test cases* obtained from such path constraints by symbolically executing a program using a particular coverage criterion. Concrete instantiations of the test cases can be generated to obtain actual test inputs for the program, amenable for further validation by testing tools.

In spite of its popularity, it is well known that symbolic execution of large programs can become quickly impracticable due to the large number and the size

E. Albert (Ed.): LOPSTR 2012, LNCS 7844, pp. 176–193, 2013.
© Springer-Verlag Berlin Heidelberg 2013

of paths that need to be explored. This issue is considered a major challenge in the fields of symbolic execution and TCG [12]. Furthermore, a common limitation of TCG by symbolic execution is that it tends to produce an unnecessarily large number of test cases even for medium-size programs.

In previous work [1,8], we have developed a glass-box Constraint Logic Programming (CLP)-based approach to TCG for imperative object-oriented programs, which consists of two phases: First, the imperative program is translated into an equivalent CLP program by means of partial evaluation [7]. Second, symbolic execution is performed on the CLP-translated program, controlled by a termination criterion (in this context also known as coverage criterion), relying on CLP's constraint solving facilities and its standard evaluation mechanism, with extensions for handling dynamic memory allocation.

In this paper we develop a framework to *guide* symbolic execution in CLP-based TCG, dubbed Guided TCG. Guided TCG can serve different purposes. It can be used to discover bugs in a program, to analyze reachability of certain parts of a program, to lead symbolic execution to stress more interesting parts of the program, etc. This paper targets selective and unit testing. Selective testing aims at testing only specific paths of a program. Unit testing is a widely used software engineering methodology, where units of code (e.g. methods) are tested in isolation to validate their correctness. Incorporating the notion of selection criteria in our TCG framework represents one step towards fully supporting both unit and integration testing, a different methodology, where all the pieces of a system must be tested as a single unit.

Our Guided TCG is a heuristics that aims at steering symbolic execution, and thus TCG, towards specific program paths to generate more relevant test cases and filter less interesting ones with respect to a given selection criterion. The goal is to improve on scalability and efficiency by achieving a high degree of control over the coverage criterion and hence avoiding the exploration of unfeasible paths. In particular, we develop two instances of the framework: one for covering all the local paths of a method, and the other to steer TCG towards a selection of program points in the program under test. Both instances have been implemented and we provide experimental results to substantiate their applicability and effectiveness.

The structure of the paper is as follows. Section 2 conveys the essentials of our CLP-based approach to TCG. Section 3 introduces the framework for guided TCG. Section 4 presents an instantiation of the framework based on trace-abstractions and targeting structural coverage criteria. Section 5 reports on the implementation and empirical evaluation of the approach. Section 6 discusses a complementary strategy to further optimize the framework. Finally, Section 7 situates our work in the existing research space, sketches ongoing and future work and concludes.

2 CLP-Based Test Case Generation

Our CLP-based approach to TCG for imperative object-oriented programs essentially consists of two phases: (1) the program is translated into an equivalent

CLP counterpart through partial evaluation; and (2) symbolic execution is performed on the CLP-translated program by relying on the CLP standard evaluation mechanisms. Details on the methodology can be found elsewhere [1,7,8].

2.1 CLP-Translated Programs

All features of the imperative object-oriented program under test are covered by its equivalent *executable* CLP-translated counterpart. Essentially, there exists a one-to-one correspondence between blocks in the control flow of the original program and rules in the CLP counterpart:

Definition 1 (CLP-translated Program). *A CLP-translated program consists of a set of predicates, each of them defined by a set of mutually exclusive rules of the form* $m(I, O, H_i, H_o, E, T) : -[\bar{G},]b_1, \ldots, b_n.$*, such that:*

(i) *I and O are the (possibly empty) lists of input and output arguments.*
(ii) *H_i and H_o are the input and output heaps.*
(iii) *E is an exception flag indicating whether the execution of m ends normally or with an uncaught exception.*
(iv) *If predicate m is defined by multiple rules, the guards in each one contain mutually exclusive conditions. We denote by m^k the k−th rule defining m.*
(v) *$\bar{G}$ is a sequence of constraints that act as execution guards on the rule.*
(vi) *$b_1, \ldots, b_n$ is a sequence of instructions, including arithmetic operations, calls to other predicates, and built-ins operations to handle the heap.*
(vii) *T is the trace term for m of the form $m(k, \langle T_{c_i}, \ldots, T_{c_m} \rangle)$, where k is the index of the rule and $T_{c_i}, \ldots, T_{c_m}$ are free logic variables representing the trace terms associated to the subsequence $c_i, \ldots, c_m$ of calls to other predicates in $b_1, \ldots, b_n$.*

Notice that the trace term T is a not a cardinal element in the translated program, but rather a supplementary argument with a central role in this paper.

2.2 Symbolic Execution

CLP-translated programs are symbolically executed using the standard CLP execution mechanism with special support for the use of dynamic memory [8].

Definition 2 (Symbolic Execution). *Let M be a method, m be its corresponding predicate from its associated CLP-translated program P, and P' be the union of P and a set of built-in predicates to handle dynamic memory. The symbolic execution of m is the CLP derivation tree, denoted as $\mathcal{T}_m$, with root $m(I, O, H_i, H_o, E, T)$ and initial constraint store $\theta = \{\}$ obtained using P'.*

2.3 Test Case Generation

The symbolic execution tree of programs containing loops or recursion is in general infinite. To guarantee termination of TCG it is therefore essential to impose a *termination criterion* that makes the symbolic execution tree finite:

Definition 3 (Finite symbolic execution tree, test case, and TCG). *Let m be the corresponding predicate for a method M in a CLP-translated program P, and let $\mathcal{C}$ be a termination criterion.*

- *$\mathcal{T}_m^{\mathcal{C}}$ is the finite and possibly incomplete symbolic execution tree of m with root $m(I, O, H_i, H_o, E, T)$ w.r.t. $\mathcal{C}$. Let B be the set of the successful (terminating) branches of $\mathcal{T}_m^{\mathcal{C}}$.*
- *A test case for m w.r.t. $\mathcal{C}$ is a tuple $\langle \theta, T \rangle$, where θ and T are, resp., the constraint store and the trace term associated to one branch $b \in B$.*
- *TCG is the process of generating the set of test cases associated to all branches in B.*

Each *test case* produced by TCG represents a class of inputs that will follow the same execution path, and its trace is the sequence of rules applied along such path. In a subsequent step, it is possible to produce actual values from the obtained constraint stores (e.g., by using labeling mechanisms in standard *clpfd* domains) therefore obtaining concrete and executable test cases. However, this is not an issue of this paper and we will comply with the above abstract definition of *test case*.

Example 1. Fig. 1 shows a Java program consisting of three methods: `lcm` calculates the least common multiple of two integers, `gcd` calculates the greatest common divisor of two integers, and `abs` returns the absolute value of an integer. The right side of the figure shows the equivalent CLP-translated program. Observe that each Java method corresponds to a set of CLP rules, e.g., method `lcm` is translated into predicates `lcm`, `cont`, `check` and `div`. The translation preserves the control flow of the program and transforms iteration into recursion (e.g. method `gcd`). Note that the example has been chosen deliberately small and simple to ease comprehension. For readability, the actual CLP code has been simplified, e.g., input and output heap arguments are not shown, since they do not affect the computation. Our current implementation [2] supports full sequential Java.

Coverage Criteria. By Def. 3, so far we have been interested in covering *all* feasible paths of the program under test w.r.t. a termination criterion. Now, our goal is to improve on efficiency by taking into account a *selection criterion* as well. First, let us define a *coverage criterion* as a pair of two components $\langle TC, SC \rangle$. TC is a *termination criterion* that ensures finiteness of symbolic execution. This can be done either based on execution steps or on loop iterations. In this paper, we adhere to loop-k, which limits to a threshold k the number of allowed loop iterations and/or recursive calls (of each concrete loop or recursive method). SC is a *selection criterion* that steers TCG to determine which paths of the symbolic execution tree will be explored. In other words, SC decides which test cases the TCG must produce. In the rest of the paper we focus on the following two coverage criteria:

- all-local-paths: It requires that all *local* execution paths within the method under test are exercised up to a loop-k limit. This has a potential interest in the context of unit testing, where each method must be tested in isolation.

```
int lcm(int a,int b) {
    if (a < b) {
        int aux = a;
        a = b;
        b = aux;
    }
    int d = gcd(a,b);
    try {
        return abs(a*b)/d;
    } catch (Exception e) {
(μ)     return -1;
    }
}
int gcd(int a,int b) {
    int res;
    while (b != 0) {
        res = a%b;
        a = b;
        b = res;
    }
    return abs(a);
}
int abs(int a) {
    if (a >= 0)
(κ)      return a;
    else
        return -a;
}
```

```
lcm([A,B],[R],_,_,E,lcm(1,[T])) :-
        A #>= B, cont([A,B],[R],_,_,E,T).
lcm([A,B],[R],_,_,E,lcm(2,[T])) :-
        A #<= B, cont([B,A],[R],_,_,E,T).
cont([A,B],[R],_,_,E,cont(1,[T,V])) :-
        gcd([A,B],[G],_,_,E,T),
        check([A,B,G],[R],_,_,E,V).
check([A,B,G],[R],_,_,E,check(1,[T,V])) :-
        M #= A*B, abs([M],[S],_,_,E,T),
        div([S,G],[R],_,_,E,V).
check([A,B,G],[R],_,_,exc,check(2,[])).
div([A,B],[R],_,_,ok,div(1,[])) :-
        B #\= 0, R #= A/B.
div([A,0],[-1],_,_,exc_caught,div(2,[])).   (μ)

gcd([A,B],[D],_,_,E,gcd(1,[T])) :-
        loop([A,B],[D],_,_,E,T).
loop([A,0],[F],_,_,E,loop(1,[T])) :-
        abs([A],[F],_,_,E,T).
loop([A,B],[E],_,_,G,loop(2,[T])) :-
        B #\= 0, body([A,B],[E],_,_,G,T).
body([A,B],[R],_,_,E,body(1,[T])) :-
        B #\= 0, M #= A mod B,
        loop([B,M],[R],_,_,E,T).
body([A,0],[R],_,_,exc,body(2,[])).

abs([A],[A],_,_,ok,abs(1,[])) :- A #>= 0.   (κ)
abs([A],[-A],_,_,ok,abs(2,[])) :- A #< 0.
```

Fig. 1. Motivating Example: Java (left) and CLP-translated (right) programs

- program-points(P): Given a set of program points P, it requires that all of them are exercised by at least one test case up to a loop-k limit. Intuitively, this criterion is the most appropriate choice for bug-detection and reachability verification purposes. A particular case of it is *statement coverage* (up to a limit), where all statements in a program or method must be exercised.

3 A Generic Framework for Guided TCG

The TCG framework as defined so far has been used in the context of coverage criteria only consisting of a termination criterion. In order to incorporate a selection criterion, one can employ a post-processing phase where only the test cases that are sufficient to satisfy the selection criterion are selected by looking at their traces. This is however not an appropriate solution in general due to the exponential explosion of the paths that have to be explored in symbolic execution.

In what follows, we develop a methodology where the TCG process is driven towards satisfying the selection criterion, stressing to avoid as much as possible the generation of irrelevant and/or redundant paths. The key idea that allows us to guide the TCG process is to use the trace terms of our CLP-translated program as input arguments. Let us observe also that we could either supply fully or partially instantiated traces, the latter ones represented by including free logic variables within the trace terms. This allows guiding, completely or partially, the symbolic execution towards specific paths.

Definition 4 (trace-guided TCG). *Given a method M, a termination criterion TC, and a (possibly partial) trace π, trace-guided TCG generates the set of test cases with traces, denoted* $\mathsf{tgTCG}(M, TC, \pi)$, *obtained for all successful branches in $\mathcal{T}_m^{TC}$ with root $m(Args_{in}, Args_{out}, H_{in}, H_{out}, E, \pi)$. We also define the* $\mathsf{firstOf\text{-}tgTCG}(M, TC, \pi)$ *to be the set corresponding to the leftmost successful branch in $\mathcal{T}_m^{TC}$.*

Observe that the TCG guided by one trace either generates: (a) exactly one test case if the trace is complete and corresponds to a feasible path, (b) none if it is unfeasible, or, (c) possibly several test cases if it is partial. In this case the traces of all test cases are instantiations of the partial trace.

Now, relying on trace-guided TCG and on the existence of a *trace generator* we define a generic scheme of *guided TCG*.

Definition 5 (guided TCG). *Given a method M; a coverage criterion $CC = \langle TC, SC\rangle$; and a trace generator TraceGen, that generates, on demand and one by one, (possibly partial) traces according to CC. Guided TCG is defined as the following algorithm:*

```
Input: M, ⟨TC, SC⟩, TraceGen
TestCases = {}
while TraceGen has more traces and TestCases does not satisfy SC
   Invoke TraceGen to generate a new trace in Trace
   TestCases ← TestCases ∪ firstOf-tgTCG(M, TC, Trace)
Output: TestCases
```

The intuition is as follows: The trace generator generates a trace satisfying SC and TC. If the generated trace is feasible, then the first solution of its trace-guided TCG is added to the set of test cases. The process finishes either when SC is satisfied, or when the trace generator has already generated all possible traces allowed by TC. If the trace generator is complete (see below), this means that SC cannot be satisfied within the limit imposed by TC.

Example 2. Let us consider the TCG for method `lcm` with program-points for points Ⓛ and Ⓚ as selection criterion. Observe the correspondence of these program points in both the Java and CLP code of Fig. 1. Let us assume that the trace generator starts generating the following two traces:

```
t1 : lcm(1,[cont(1,[G,check(1,[A,div(2,[])])])])
t2 : lcm(2,[cont(1,[G,check(1,[A,div(2,[])])])])
```

The first iteration does not add any test case since trace t_1 is unfeasible. Trace t_2 is proved feasible and a test case is generated. The selection criterion is now satisfied and therefore the process finishes. The obtained test case is shown in Example 7.

On Soundness, Completeness and Effectiveness: Intuitively, a concrete instantiation of the guided TCG scheme is *sound* if all test cases it generates satisfy the coverage criterion, and *complete* if it never reports that the coverage criterion is not satisfied when it is indeed satisfiable. *Effectiveness* is related to the number of iterations the algorithm performs. Those three features depend solely on the trace generator. We will refer to trace generators as being sound, complete or effective. The intuition is that a trace generator is sound if every trace it generates satisfies the coverage criterion, and complete if it produces an over-approximation of the set of traces satisfying it. Effectiveness is related to the number of unfeasible traces it generates, the larger the number, the less effective the trace generator.

4 Trace Generators for Structural Coverage Criteria

In this section we present a general approach for building sound, complete and effective trace generators for structural coverage criteria by means of program transformations. We then instantiate the approach for the all-local-paths and program-points coverage criteria and proposes Prolog implementations of the guided TCG scheme for both of them. Let us first define the notion of *trace-abstraction* of a program which will be the basis for defining our trace generators.

Definition 6 (trace-abstraction of a program). *Given a CLP-translated program with traces P, its trace-abstraction is obtained as follows: for every rule of P, (1) remove all atoms in the body of the rule except those corresponding to rule calls, and (2) remove all arguments from the head and from the surviving atoms of (1) except the last one (i.e., the trace term).*

Example 3. Fig. 2 shows the trace-abstraction of our CLP-translated program of Fig. 1. Let us observe that it essentially corresponds to its control-flow graph.

The trace-abstraction can be directly used as a trace-generator as follows: (1) Apply the termination criterion in order to ensure finiteness of the process. (2) Select, in a post-processing, those traces that satisfy the selection criterion. Such a trace generator produces on backtracking a superset of the set of traces of the program satisfying the coverage criterion. Note that this can be done as long as the criteria are structural. The obtained trace generator is by definition sound and complete. However, it can be very ineffective and inefficient due to the large number of unfeasible and/or unnecessary traces that it can generate. In the following, we develop two concrete, and more effective, schemes for the all-local-paths and program-points coverage criteria. In both cases, this is done by taking advantage of the notion of partial traces and the implicit information on the concrete coverage criteria.

```
lcm(lcm(1,[T])) :- cont(T).
lcm(lcm(2,[T])) :- cont(T).
cont(cont(1,[T,V])) :- gcd(T), check(V).
check(check(1,[T,V])) :- abs(T), div(V).
check(check(2,[])).
div(div(1,[])).
div(div(2,[])).
gcd(gcd(1,[T])) :- loop(T).
loop(loop(1,[T])) :- abs(T).
loop(loop(2,[T])) :- body(T).
body(body(1,[T])) :- loop(T).
body(body(2,[])).
abs(abs(1,[])).
abs(abs(2,[])).
```

Fig. 2. Trace-abstraction

```
lcm(lcm(1,[T])) :- cont(T).
lcm(lcm(2,[T])) :- cont(T).
cont(cont(1,[G,T])) :- check(T).
check(check(1,[A,T])) :- div(T).
check(check(2,[])).
div(div(1,[])).
div(div(2,[])).
```

```
lcm(1,[cont(1,[G,check(1,[A,div(1,[])])])])
lcm(1,[cont(1,[G,check(1,[A,div(2,[])])])])
lcm(1,[cont(1,[G,check(2,[])])])
lcm(2,[cont(1,[G,check(1,[A,div(1,[])])])])
lcm(2,[cont(1,[G,check(1,[A,div(2,[])])])])
lcm(2,[cont(1,[G,check(2,[])])])
```

Fig. 3. Slicing of method `lcm` for all-local-paths criterion

4.1 An Instantiation for the all-local-paths Coverage Criterion

Let us start from the trace-abstraction program and apply a syntactic program slicing which removes from it the rules that do not belong to the considered method.

Definition 7 (slicing for all-local-paths coverage criterion). *Given a trace-abstraction program P and an entry method M:*

1. *Remove from P all the rules that do not belong to method M.*
2. *For all remaining rules in P, remove from their bodies all the calls to rules which are not in P.*

The obtained sliced trace-abstraction, together with the termination criterion, can be used as a trace generator for the all-local-paths criterion for a method. The generated traces will have free variables in those trace arguments that correspond to the execution of other methods, if any.

Example 4. Fig. 3 shows on the left the sliced trace-abstraction for method `lcm`. On the right is the finite set of traces that is obtained from such trace-abstraction for any loop-K termination criterion. Observe that the free variables `G`, resp. `A`, correspond to the sliced away calls to methods `gcd` and `abs`.

Let us define the predicates: `computeSlicedProgram(M)`, that computes the sliced trace-abstraction for method M as in Def. 7; `generateTrace(M,TC,Trace)`, that returns in its third argument, on backtracking, all partial traces computed using such sliced trace-abstraction, limited by the termination criterion TC; and `traceGuidedTCG(M,TC,Trace,TestCase)`, which computes on backtracking the set tgTCG(M,Trace,TC) in Def. 4, failing if the set is empty, and instantiating on success `TestCase` and `Trace` (in case it was partial). The guided TCG scheme in Def. 5, instantiated for the all-local-paths criterion, can be implemented in Prolog as follows:

```
(1) guidedTCG(M,TC) :-
(2)     computeSlicedProgram(M),
(3)     generateTrace(M,TC,Trace),
(4)     once(traceGuidedTCG(M,Trace,TC,TestCase)),
(5)     assert(testCase(M,TestCase,Trace)),
(6)     fail.
(7) guidedTCG(_,_).
```

Intuitively, given a (possibly partial) trace generated in line (3), if the call in line (4) fails, then the next trace is tried. Otherwise, the generated test case is asserted with its corresponding trace which is now fully instantiated (in case it was partial). The process finishes when `generateTrace/3` has computed all traces, in which case it fails, making the program exit through line (7).

Example 5. The following test cases are obtained for the all-local-paths criterion for method `lcm`:

Constraint store	*Trace*
{A>=B}	lcm(1,[cont(1,[gcd(1,[loop(1,[abs(1,[])])]), check(1,[abs(1,[]),div(1,[])])])])
{A=B=0,Out=-1}	lcm(1,[cont(1,[gcd(1,[loop(1,[abs(1,[])])]), check(1,[abs(1,[]),div(2,[])])])])
{B>A}	lcm(2,[cont(1,[gcd(1,[loop(1,[abs(1,[])])]), check(1,[abs(1,[]),div(1,[])])])])

This set of three test cases achieves full code and path coverage on method `lcm` and is thus a perfect choice in the context of unit-testing. In contrast, the original, non-guided, TCG scheme with loop-2 as termination criterion produces nine test cases.

4.2 An Instantiation for the program-points Coverage Criterion

Let us first consider a simplified version of the program-points criterion so that only one program point is allowed, denoted as program-point. Starting again from the trace-abstraction program, we apply a syntactic bottom-up program slicing algorithm to filter away all the paths in the program that do not visit the program point of interest.

Definition 8 (slicing for program-point coverage criterion). *Given a trace-abstraction program P, a program point of interest pp, and an entry method M, the sliced program P' is computed as follows:*

1. *Initialize P' to be the empty program, and a set of clauses L with the clause corresponding to pp.*
2. *For each c in L which is not the clause for M, add to L all clauses in P whose body has a call to the predicate of clause c, and iterate until the set L stabilizes.*
3. *Add to P' all clauses in L.*
4. *Remove all calls to rules which are not in P' from the bodies of the rules in P'.*

The obtained sliced program, together with the termination criterion, can be used as a trace generator for the program-point criterion. The generated traces can have free variables representing parts of the execution which are not related (syntactically) to the paths visiting the program point of interest.

```
lcm(lcm(1,[T])) :- cont(T).
lcm(lcm(2,[T])) :- cont(T).
cont(cont(1,[G,T])) :- check(T).
check(check(1,[A,T])) :- div(T).
div(div(2,[])). ⓐ
```

```
lcm(1,[cont(1,[G,check(1,[A,div(2,[])])])])
lcm(2,[cont(1,[G,check(1,[A,div(2,[])])])])
```

Fig. 4. Slicing for program-point coverage criterion with *pp*=ⓐ from Fig. 1

Example 6. Fig. 4 shows on the left the sliced trace-abstraction program (using Def. 8) for method `lcm` and program point ⓐ from Fig. 1, i.e. the `return` statement within the `catch` block. On the right of the same figure, the traces obtained from such slicing using loop-2 as termination criterion.

Consider again predicates `computeSlicedProgram/2`, `generateTrace/4` and `traceGuidedTCG/4` with the same meaning as in Section 4.1, but being the first two now based on Def. 8 and extended with the program-point argument `PP`. The guided TCG scheme in Def. 5, instantiated for the program-points criterion, can be implemented in Prolog as follows:

```
guidedTCG(M,[],TC) :- !.
guidedTCG(M,[PP|PPs],TC) :-
    computeSlicedProgram(M,PP),
    generateTrace(M,PP,TC,Trace),
    once(traceGuidedTCG(M,Trace,TC,TestCase)), !,
    assert(testCase(M,TestCase,Trace)),
    removeCoveredPoints(PPs,Trace,PPs'),
    guidedTCG(M,PPs',TC).
guidedTCG(M,[PP|_],TC) :-
```

Intuitively, given the first remaining program point of interest `PP` (line `2`), a trace generator is computed and used to obtain a (possibly partial) trace that exercises `PP` (lines `3`–`4`). Then, if the call in line `5` fails, another trace for `PP` is requested on backtracking. When there are not more traces (i.e., line `4` fails) the process finishes through line `9` reporting that `PP` is not reachable within the imposed `TC`. If the call in line `5` succeeds, the generated test case is asserted with its corresponding trace (now fully instantiated in case it was partial), the remaining program points which are covered by `Trace` are removed obtaining `PPs'` (line `7`), and the process continues with `PPs'`. Note that a new sliced program is computed for each program point in `PPs'`. The process finishes through line `1` when all program points have been covered.

The above implementation is valid for the general case of program-points criteria with any finite set size. The trace generator, instead, has been deliberately defined for just one program point since this way the program slicing can be more aggressive, hence eluding the generation of unfeasible traces.

Example 7. The following test case is obtained for the program-points criterion for method `lcm` and program points Ⓜ and Ⓚ:

Constraint store	*Trace*
`{A=B=0,Out=-1}`	`lcm(1,[cont(1,[gcd(1,[loop(1,[abs(1,[])])]), check(1,[abs(1,[]),div(2,[])])])])`

This particular case exemplifies specially well how guided TCG can reduce the number of produced test cases through adequate control of the selection criterion.

5 Experimental Evaluation

We have implemented the guided TCG schemes for both all-local-paths and program-points coverage criteria as proposed in Section 4, and integrated them within PET [2,8], an automatic TCG tool for Java bytecode, which is available at `http://costa.ls.fi.upm.es/pet`. In this section we report on some experimental results which aim at demonstrating the applicability and effectiveness of guided TCG. The experiments have been performed using as benchmarks a selection of classes from the net.datastructures library [9], a well-known library of algorithms and data-structures for Java. In particular, we have used as "methods-under-test" the most relevant public methods of the classes *NodeSequence*, *SortedListPriorityQueue*, *BinarySearchTree* and *HeapPriorityQueue*, abbreviated respectively as *Seq*, *PQ*, *BST* and *HPQ*.

Table 1 aims at demonstrating the effectiveness of the guided TCG scheme for the all-local-paths coverage criterion. This is done by comparing it to standard way of implementing the all-local-paths coverage criterion, i.e., first generating all paths up to the termination criterion using standard TCG by symbolic execution, and then applying a filtering so that only the test cases that are necessary to meet the all-local-paths selection criterion are kept. Each row in the table corresponds to the TCG of one method using standard TCG vs. using guided TCG.

Table 1. Experimental results for the all-local-paths criterion

Method Info			Standard TCG			Guided TCG			
Class.Name	**BCs**	**Tt**	**T**	**N**	**CC**	**Tg**	**Ng**	**CCg**	**GT/UT**
Seq.elemAt	98	45	18	24	100%	9	5	100%	6/1
Seq.insertAt	220	85	41	39	100%	14	6	100%	8/2
Seq.removeAt	187	76	35	36	100%	10	4	100%	5/1
Seq.replaceAt	163	66	35	36	100%	9	4	100%	5/1
PQ.insert	357	144	148	109	100%	10	3	100%	4/1
PQ.remove	158	69	8	12	100%	20	7	100%	15/8
BST.addAll	260	125	1491	379	100%	22765	18	100%	151/133
BST.find	228	113	76	62	100%	82	5	100%	7/2
BST.findAll	381	178	1639	330	100%	1266	4	100%	6/2
BST.insert	398	184	2050	970	100%	1979	9	100%	18/9
BST.remove	435	237	741	365	98%	3443	26	98%	204/178
HPQ.insert	322	132	215	43	100%	26	5	100%	6/1
HPQ.remove	394	174	1450	40	100%	100	8	100%	19/11

For each method we provide: The number of reachable bytecode instructions (**BCs**) and the time of the translation of Java bytecode to CLP (**Tt**), including parsing and loading all reachable classes ; the time of the TCG process (**T**), the number of generated test cases before the filtering (**N**), and the code coverage achieved using standard TCG (**CC**); and the time of the TCG process (**Tg**), the number of generated test cases (**Ng**), the code coverage achieved (**CCg**), and the number of generated/unfeasible traces using guided TCG (**GT/UT**). All times are in milliseconds and are obtained as the arithmetic mean of five runs on an Intel(R) Core(TM) i5-2300 CPU at 2.8GHz with 8GB of RAM, running Linux Kernel 2.6.38. The code coverage measures, given a method, the percentage of its bytecode instructions which are exercised by the obtained test cases. This is a common measure in order to reason about the quality of the obtained test cases. As expected, the code coverage is the same in both approaches, and so is the number of obtained test cases. Otherwise, this would indicate a bug in the implementation.

Let us observe that the gains in time are significant for most benchmarks (column **T** vs. column **Tg**). There are however three notable exceptions for methods PQ.remove, BST.addAll and BST.remove, for which the guided TCG scheme behaves worse than the standard one, especially for BST.addAll. This happens in general when the control-flow of the method is complex, hence causing the trace generator to produce an important number of unfeasible traces (see last column). Interestingly, these cases could be statically detected using a simple syntactic analysis which looks at the control flow of the method. Therefore the system could automatically decide which methodology to apply. Moreover, Section 6 presents a trace-abstraction refinement that will help in improving guided TCG for programs whose control-flow is determine mainly by integer

Table 2. Experimental results for the program-points criterion

Method Info	Standard TCG			Guided TCG			
Class.Name	**T**	**N**	**CC**	**Tg**	**Ng**	**CCg**	**GT/UT**
Seq.elemAt	9	10	100%	6	3	100%	3/0
Seq.insertAt	39	36	100%	8	3	100%	3/0
Seq.removeAt	19	16	100%	8	3	100%	3/0
Seq.replaceAt	19	16	100%	8	3	100%	3/0
PQ.insert	149	109	100%	9	3	100%	3/0
PQ.remove	9	12	100%	5	3	100%	3/0
BST.addAll	1501	379	100%	284	2	100%	4/2
BST.find	77	62	100%	10	3	100%	3/0
BST.findAll	1634	330	100%	8	3	100%	3/0
BST.insert	2197	969	100%	35	3	100%	3/0
BST.remove	238	104	98%	61	3	98%	28/25
HPQ.insert	209	43	100%	24	3	100%	3/0
HPQ.remove	1385	38	100%	15	3	100%	3/0

linear constraints. Other classes of programs, e.g. BST.addAll, require a more sophisticated analysis, since their control-flow are strongly determined by object types and dynamic dispatch information. This discussion and further refinement is left out of the scope of this paper.

Table 2 aims at demonstrating the effectiveness of the guided TCG scheme for the program-points coverage criterion. For this aim, we have implemented the support in the standard TCG scheme to check the program-points selection criterion dynamically while the test cases are generated, in such a way that the process terminates when all program points are covered. Note that, in the worst case this will require generating the whole symbolic execution tree, as the standard TCG does. Table 2 compares the effectiveness of this methodology against that of the guided TCG scheme. Again, each row in the table corresponds to the TCG of one method using standard TCG vs. using guided TCG, providing for both schemes the time of the TCG process (**T** vs **Tg**), the number of generated test cases (**N** vs **Ng**), the code coverage achieved (**CC** vs **CCg**), and the number of generated/unfeasible traces using guided TCG (**GT/UT**). We have selected three program points for each method with the aim of covering as much code as possible. In all cases, such selection of program points allows obtaining the same code coverage as with the standard TCG even without the selection criterion (i.e. 100% coverage for all methods except 98% for BST.remove because of dead code). Let us observe that the gains in time are huge (column **T** vs. column **Tg**), ranging from one to two orders of magnitude, except for the simplest methods, for which the gain, still being significant, is not so notable. These results are witnessed by the low number of unfeasible traces that are obtained (column **GT/UT**), hence demonstrating the effectiveness of the trace-generator defined in Section 4.2.

Overall, we believe our experimental results support our initial claims about the potential interest of guiding symbolic execution and TCG by means of trace-abstractions. With the exception of some particular cases that deserve further study, our results demonstrate that we can achieve high code coverage without having to explore many unfeasible paths, with the additional advantage of discovering high quality (less in number and better selected) test cases.

6 Trace-Abstraction Refinement

As the above experimental results suggest, there are still cases where the trace-abstraction as defined in Def. 6 may still compromise the effectiveness of the guided TCG, because of the generation of too many unfeasible paths. This section discusses a complementary strategy to further optimize the framework. In particular, we propose a heuristics that aims to refine the trace-abstraction with information taken from the original program that will help reduce the number of unfeasible paths at symbolic execution. The goal is to reach a balanced level of refinement in between the original program (full refinement) and the trace-abstraction (empty refinement). Intuitively, the more information we include, the less unfeasible paths symbolic execution will explore, but the more costly it becomes.

The refinement algorithm consists of two steps: First, in a fixpoint analysis we approximate the instantiation mode of the variables in each predicate of the CLP-translated program. In other words, we infer which variables will be constrained or assigned a concrete value at symbolic execution time. In a second step, by program transformation, the trace-abstraction is enriched with clause arguments corresponding to the inferred variables, and with those goals in which they are involved.

6.1 Approximating Instantiation Modes

We develop a static analysis, similar to [5, 6], to soundly approximate the instantiation mode of the input argument variables in the program at symbolic execution time. The analysis is implemented as a fixpoint computation over the simple abstract domain $\{static, dynamic\}$. Namely, *dynamic* means that nothing was inferred about a variable and it will therefore remain a free unconstrained variable during symbolic execution; and *static* means that the variable will unify with a concrete value or will be constrained during symbolic execution. The analysis's result is a set of assertions in the form $\langle P, \mathcal{V}\rangle$ where P is a predicate name and $\mathcal{V}$ is the set of variables in P, each associated with an abstract value from the domain.

This analysis receives as input a CLP-translated program and a set of initial entries (predicate names). An event queue $\mathcal{Q}$ is initialized with this set of initial entries. The algorithm starts to process the events of $\mathcal{Q}$ until no more events are scheduled. In each iteration, an event p is removed from $\mathcal{Q}$ and processed as follows: Retrieve previously stored information $\psi \equiv \langle p, \mathcal{V}\rangle$ if any exists; else set $\psi \equiv \langle p, \emptyset\rangle$. For each rule r defining p, a new $\mathcal{V}_r$ is obtained by evaluating the body of r. The joint operation on the underlying abstract domain is performed to obtain $\mathcal{V}' \Leftarrow joint(\mathcal{V}, \mathcal{V}_r)$. If $\mathcal{V} \not\equiv \mathcal{V}'$ then set $\mathcal{V} \Leftarrow \mathcal{V}'$ and reschedule every predicate that calls p; else, if $\psi' \equiv \psi$ there is no need to recompute the calling predicates and the algorithm continues. That will ensure backward propagation of approximated instantiation modes. To propagate forward, the evaluation of r will schedule one event per call within its body. The process continues until a fixpoint is reached.

6.2 Constructing the Trace-Abstraction Refinement

This is a syntactic program transformation step of the refinement. It takes as input the original CLP-program and the instantiation information inferred in the first step and outputs a trace-abstraction refinement program. For each rule r of a predicate p in the program, the algorithm retrieves $\langle p, \mathcal{V}\rangle$. We denote $\mathcal{V}_s$ the projection of all variables in $\mathcal{V}$ whose inferred abstract value is *static*. The algorithm adds to the trace-abstraction refinement a new rule r' whose list of arguments is $\mathcal{V}_s$. The body of r' is constructed by traversing the body $b_1, \ldots, b_n$ of r and including 1) all guards and arithmetic operations b_i involving $\mathcal{V}_s$, and 2) all calls to other predicates, with the corresponding projection of $\mathcal{V}_s$ over the arguments of the calls.

Example 8. Consider the Java example of Fig. 5 (left side). Function `power` implements a exponentiation algorithm for positive integer exponents. Its CLP counterpart is shown at the right of the figure. The instantiation modes inferred by the first stage of our algorithm is presented at the right-bottom part of the figure. One can observe that variable `B` (the base of the exponentiation) remains *dynamic* all along the program, because it is never assigned any concrete value nor constrained by any guard. On the other hand, variable `E`'s final abstract value is *static*, since it is constrained by `0` and the also *static* variable `I` in rules `if` and `loop`. The following is the refined trace-abstraction that our algorithm constructs:

```
power([E],power(1,[T])) :- if([E],T).
if([E],if(1,[])) :- E #< 0.
if([E],if(2,[T])) :- E #>= 0, loop([E,1],T).
loop([E,I],loop(1,[])) :- I #> E.
loop([E,I],loop(2,[T])) :- I #=< E, Ip #= I+1, loop([E,Ip],T).
```

To illustrate how the trace-abstraction refinement can improve on effectiveness of the guided TCG, let us observe method `arraypower`. It iterates over the elements of an input array `a` and calls function `power` to update all even positions of the array by raising their values to the power of the integer input argument `e`. We report on the following performance results for this example and a coverage criterion $\langle$loop-2, $\{\}\rangle$:

- Standard non-guided TCG (i.e., full refinement) generates 11 test cases.
- Trace-abstraction guided TCG with the empty refinement generates 497 possibly (un)feasible traces.

```
void arraypower(int a[],int e) {
    int i=0;
    int n=a.length;
    for (i=0; i<n; i++)
        if (i%2==0)
            a[i]=power(a[i],e);
}
int power(int b, int e) {
    if (e >= 0) {
        int pow = 1;
        while (i <= e) {
            pow *= b;
            i++;
        }
        return pow;
    } else return -1;
}
```

```
power([B,E],[R],_,_,F,power(1,[T])) :-
    if([B,E],[R],_,_,F,T).
if([B,E],[-1],_,_,F,if(1,[])) :-
    E #< 0.
if([B,E],[R],_,_,F,if(2,[T])) :-
    E #>= 0), loop([B,E,1,1],[R],_,_,F,T).
loop([B,E,I,P],[P],_,_,ok,loop(1,[])) :-
    I #> E.
loop([B,E,I,P],[R],_,_,F,loop(2,[T])) :-
    I #=< E, Pp #= P*B, Ip #= I+1,
    loop([B,E,Ip,Pp],[R],_,_,F,T).
```

Inferred instantiation modes:
⟨power, {B= *dynamic*,E= *static*}⟩
⟨if,{B= *dynamic*,E= *static*}⟩
⟨loop, {B= *dynamic*,E= *static*,
I= *static*,P= *dynamic*}⟩

Fig. 5. Trace-abstraction refinement

- Trace-abstraction guided TCG with our trace-abstraction refinement reduces the number of possibly (un)feasible traces to 161.

These preliminary yet promising results, suggest the potential integrability of the trace-abstraction refinement algorithm presented in this section within the general guided TCG framework developed in this paper. The refinement is complementary to the slicings schemes presented in Section 4 without any modification. Unfortunately, the slicings could produce a loss of important information added by the refinement. This could be however improved by means of simple syntactic analyses on the sliced parts of the program. A deeper study of these issues remains as future work.

7 Related Work and Conclusions

Previous work also uses abstractions to guide symbolic execution and TCG by several means and for different purposes. Fundamentally, abstraction aims to reduce large data domains of a program to smaller domains [11]. One of the most relevant to ours is [3], where predicate abstraction, model checking and SAT-solving are combined to produce abstractions and generate test cases for C programs, with good code coverage, but depending highly on an initial set of predicates to avoid infeasible program paths. Rugta *et al.* [13] also proposes to use an abstraction of the program in order to guide symbolic execution and prune the execution tree as a way to scale up. Their abstraction is an under-approximation which tries to reduce the number of test cases that are generated in the context of concurrent programming, where the state explosion is in general problematic.

The main contribution of this paper is the development of a methodology for Guided TCG that allows to guide the process of test generation towards achieving more selective and interesting structural coverage. Implicit is the improvement in the scalability of TCG by guiding symbolic execution by means of trace-abstractions, since we gain more control over the symbolic execution state space to be explored. Moreover, whereas the main goal of our CLP-based TCG framework has been the exhaustive testing of programs, our new Guided TCG framework unveil new potential applications areas. Namely, the all-local-paths and program-points Guided TCG schemes we have presented in this paper, enable us to explore on the automation of other interesting software testing practices, such as selective and unit testing, goal-oriented testing and bug detection.

The effectiveness and applicability of Guided TCG is substantiated by an implementation within the PET system (`http://costa.ls.fi.upm.es/pet`), and encouraging experimental results. Nevertheless, our current and future work involves a more thorough experimental evaluation of the framework and the exploration of the new application areas in software testing. In a different line, a particularly challenging goal has been triggered which consists in developing static analysis techniques to achieve optimal refinement levels of the trace-abstraction programs. Last but not least, we plan to further study the generalization and integration of other interesting coverage criteria to our Guided TCG framework.

Acknowledgments. This work was funded in part by the Information & Communication Technologies program of the European Commission, Future and Emerging Technologies (FET), under the ICT-231620 *HATS* project, by the Spanish Ministry of Science and Innovation (MICINN) under the TIN2008-05624, TIN2012-38137 and PRI-AIBDE-2011-0900 projects, by UCM-BSCH-GR35/10-A-910502 grant and by the Madrid Regional Government under the S2009TIC-1465 *PROMETIDOS-CM* project.

References

1. Albert, E., Gómez-Zamalloa, M., Puebla, G.: Test Data Generation of Bytecode by CLP Partial Evaluation. In: Hanus, M. (ed.) LOPSTR 2008. LNCS, vol. 5438, pp. 4–23. Springer, Heidelberg (2009)
2. Albert, E., Cabañas, I., Flores-Montoya, A., Gómez-Zamalloa, M., Gutiérrez, S.: jPET: an Automatic Test-Case Generator for Java. In: Proc. of WCRE 2011. IEEE Computer Society (2011)
3. Ball, T.: Abstraction-guided test generation: A case study. Technical Report MSR-TR-2003-86, Microsoft Research (2003)
4. Cadar, C., Godefroid, P., Khurshid, S., Păsăreanu, C., Sen, K., Tillmann, N., Visser, W.: Symbolic execution for software testing in practice: preliminary assessment. In: Proc. of ICSE 2011. ACM (2011)
5. Craig, S.-J., Gallagher, J.P., Leuschel, M., Henriksen, K.S.: Fully Automatic Binding-Time Analysis for Prolog. In: Etalle, S. (ed.) LOPSTR 2004. LNCS, vol. 3573, pp. 53–68. Springer, Heidelberg (2005)
6. Debray, S.K.: Static inference of modes and data dependencies in logic programs. ACM Trans. Program. Lang. Syst. 11(3), 418–450 (1989)

7. Gómez-Zamalloa, M., Albert, E., Puebla, G.: Decompilation of Java Bytecode to Prolog by Partial Evaluation. Information and Software Technology 51(10), 1409–1427 (2009)
8. Gómez-Zamalloa, M., Albert, E., Puebla, G.: Test Case Generation for Object-Oriented Imperative Languages in CLP. Theory and Practice of Logic Programming, ICLP 2010 Special Issue 10(4-6) (2010)
9. Goodrich, M., Tamassia, R., Zamore, E.: The net.datastructures package, `http://net3.datastructures.net`
10. King, J.C.: Symbolic Execution and Program Testing. Communications of the ACM 19(7), 385–394 (1976)
11. Lakhnech, Y., Bensalem, S., Berezin, S., Owre, S.: Incremental verification by abstraction. In: Margaria, T., Yi, W. (eds.) TACAS 2001. LNCS, vol. 2031, pp. 98–112. Springer, Heidelberg (2001)
12. Păsăreanu, C.S., Visser, W.: A survey of new trends in symbolic execution for software testing and analysis. Int. J. Softw. Tools Technol. Transf. 11(4), 339–353 (2009)
13. Rungta, N., Mercer, E.G., Visser, W.: Efficient testing of concurrent programs with abstraction-guided symbolic execution. In: Păsăreanu, C.S. (ed.) SPIN 2009. LNCS, vol. 5578, pp. 174–191. Springer, Heidelberg (2009)

Simplifying the Verification of Quantified Array Assertions via Code Transformation*

Mohamed Nassim Seghir and Martin Brain

University of Oxford

Abstract. Quantified assertions pose a particular challenge for automated software verification tools. They are required when proving even the most basic properties of programs that manipulate arrays and so are a major limit for the applicability of fully automatic analysis. This paper presents a simple program transformation approach based on induction to simplify the verification task. The techniques simplifies both the program and the assertion to be verified. Experiments using an implementation of this technique show a significant improvement in performance as well as an increase in the range of programs that can be checked fully automatically.

1 Introduction

Arrays are a primitive data-structure in most procedural languages, so it is vital for verification tools to be able to handle them. Most of the properties of interest for array manipulating programs are stated in terms of *all* elements of the array. For example, all elements are within a certain range, all elements are not equal to a target value or all elements are in ascending order. Handling quantified assertions is thus a prerequisite for supporting reasoning about arrays. Proving them requires reasoning combining quantified assertions and loops. Unfortunately, handling the combination of loops and quantified assertions is difficult and poses considerable practical challenges to most of the existing automatic tools such as SLAM [1], BLAST [11], MAGIC [5], ARMC [18] and SATABS [6]. One approach requires the user provides loop invariants [2,7]. This is undesirable as writing loop invariants is a skilled, time intensive and error prone task. This paper aims to give a simple program transformation approach for handling some combinations of loops and quantified assertions, which include text book examples as well as real world examples taken from the Linux operating system kernel and the Xen hypervisor.

Transforming the program and assertions before applying existing verification techniques is a common and effective technique. It has a number of pragmatic advantages; it is often simple to implement, fast to use, re-uses existing technology and can be adapted to a wide range of tools. We present an approach that either removes loops or quantifiers and thus reduces the problem to a form solvable by existing tools. For example, if loops can be removed then decision procedures for

* Supported by the EU FP7 STREP PINCETTE (projectID ICT-257647) project.

E. Albert (Ed.): LOPSTR 2012, LNCS 7844, pp. 194–212, 2013.
© Springer-Verlag Berlin Heidelberg 2013

quantified formulas [4] can be applied to the loop-free code. If quantifiers can be removed then the previously mentioned automatic tools can be used as they are quite efficient in handling quantifier-free assertions. This work builds on the results presented in [19]; giving a formal basis for the technique, in particular:

1. Defining *recurrent fragments* and shows how they can be used to build inductive proofs of quantified assertions.
2. Giving an algorithm for locating recurrent fragments and transforming the program so that the resulting code is simpler than the original one.
3. Proving the soundness of the program transformation algorithm.
4. Illustrating how our approach extends naturally to multidimensional arrays and describing a scheme for handling loop sequences.

2 Overview

We illustrate our method by considering two sorting algorithms (insertion_sort and selection_sort) which represent challenging examples for automatic verification tools.

2.1 Pre-recursive Case (insertion_sort)

Let us consider the insertion_sort example illustrated in Figure 1(a). This program sorts elements of array a in the range $[0..n[$. We want to verify the assertion specified at line 18, which states that elements of array a are sorted in ascending order. Let us call $\mathcal{L}$ the outer `while` loop of example (a) in Figure 1 together with the assignment at line 4 which just precedes the loop. We also call φ the assertion at line 18. In the notation $\mathcal{L}(1,n)$ the first parameter represents the initial value of the loop iterator i and the second parameter is its upper bound (here upper bound means strictly greater). Also $\varphi(n)$ denotes the parameterized form of φ through n. In a Floyd-Hoare style [12], the verification task is expressed as a Hoare triple

$$\{\}\ \mathcal{L}(1,n)\ \{\varphi(n)\} \tag{1}$$

In (1) the initial value 1 for the loop iterator is fixed but its upper bound n can take any value. To prove (1) via structural induction on the iterator range, where the *well-founded order* is range inclusion, it suffices to use induction on n. We proceed as follows

- prove that $\{\}\ \mathcal{L}(1,1)\ \{\varphi(1)\}$ holds
- assume $\{\}\ \mathcal{L}(1,n-1)\ \{\varphi(n-1)\}$ and prove $\{\}\ \mathcal{L}(1,n)\ \{\varphi(n)\}$

By unrolling the outer `while` loop in Figure 1(a) backward, we obtain the code in Figure 1(b). One can observe that the code fragment from line 4 to 16 in Figure 1(b) represents $\mathcal{L}(1,n-1)$, we call it the *recurrent fragment*. If we denote by $\mathcal{C}$ the remaining code, i.e., from line 18 to 26 in Figure 1(b), then the Hoare triple (1) is rewritten to

$$\{\}\ \mathcal{L}(1,n-1);\mathcal{C}\ \{\varphi(n)\}.$$

```
void insertion_sort (int a [], int n)
{
  int i, j, index;
  i = 1;
  while(i < n)
  {
    index = a[i];
    j = i;
    while ((j > 0) && (a[j-1] > index))
    {
      a[j] = a[j-1];
      j = j - 1;
    }
    a[j] = index;
    i++;
  }

  assert(∀ x y. (0 ≤ x < n
         ∧ 0 ≤ y < n
         ∧ x < y)
         ⇒ a[x] ≤ a[y]);
}
```

(a)

```
void insertion_sort (int a [], int n)
{
  int i, j, index;
  i = 1;
  while(i < n-1)
  {
    index = a[i];
    j = i;
    while ((j > 0) && (a[j-1] > index))
    {
      a[j] = a[j-1];
      j = j - 1;
    }
    a[j] = index;
    i++;
  }

  index = a[n-1];
  j = n-1;
  while ((j > 0) && (a[j-1] > index))
  {
    a[j] = a[j-1];
    j = j - 1;
  }
  a[j] = index;
  i = n;

  assert(∀ x y. (0 ≤ x < n
         ∧ 0 ≤ y < n
         ∧ x < y)
         ⇒ a[x] ≤ a[y]);
}
```

(b)

Fig. 1. Insertion sort program (a) and the corresponding pre-recursive form (b)

$$\frac{\{\}\ \mathcal{L}(1,n-1)\ \{\varphi(n-1)\} \qquad \{\varphi(n-1)\}\ \mathcal{C}\ \{\varphi(n)\}}{\{\}\ \mathcal{L}(1,n-1);\mathcal{C}\ \{\varphi(n)\}}$$

Fig. 2. Hoare's rule of composition

Hoare's rule of composition can be applied in this context following the scheme illustrated in Figure 2. The first (left) premise of the rule represents the induction hypothesis, thus, assumed to be valid. It suffices then to prove the second premise to conclude the validity of (1). Hence, the verification problem in Figure 1(a) is reduced to the one illustrated in Figure 3, in addition to the verification of the basic case $\{\}\ \mathcal{L}(1,1)\ \{\varphi(1)\}$ which is trivial. The assumption at line 1 in Figure 3 represents the postcondition of the induction hypothesis. One can clearly see that the final code is much simpler than the original one. We start with a loop having two levels of nesting and we end up with a code containing a loop that has just one level of nesting, which means less loop invariants (fixed points) to compute.

```
assume(∀ x y. (0 ≤ x,y < n−1 ∧ x < y) ⇒ a[x] ≤ a[y]);
index = a[n−1];
j = n−1;
while ((j > 0) && (a[j−1] > index))
{
  a[j] = a[j−1];
  j = j – 1;
}
a[j] = index;
i = n;
assert(∀ x y. (0 ≤ x,y < n ∧ x < y) ⇒ a[x] ≤ a[y]);
```

Fig. 3. Result obtained by replacing the code fragment of the induction hypothesis with the corresponding post condition (which is here used as assumption)

2.2 Post-recursive Case (selection_sort)

Now, we consider the example selection_sort which is illustrated in Figure 4(a). Initially, instead of the two assignments at lines 5 and 6, we had just a single assignment $i := 0$. Also in the assertion at line 23 we had 0 instead of k. We introduced the fresh variable k to allow the application of induction on the iterator initial value as will be illustrated.

As previously, we write $\mathcal{L}(k,n)$ to denote the outer loop together with the assignment at line 6 in Figure 4(a). Unlike the previous example, here by unrolling $\mathcal{L}(k,n)$ backward, the remaining (first) $n-1$ iterations do not represent $\mathcal{L}(k,n-1)$. In fact, the iterator j of the inner loop in $\mathcal{L}(k,n)$ has n as upper bound in the first $n-1$ iterations of $\mathcal{L}(k,n)$, whereas j has $n-1$ as upper bound in $\mathcal{L}(k,n-1)$. However, by unrolling $\mathcal{L}(k,n)$ forward, we obtain the code in Figure 4(b). The code portion from line 20 to 35 represents $\mathcal{L}(k+1,n)$. In this case, the recursion occurs at the end, i.e., $\mathcal{L}(k,n) = \mathcal{C};\mathcal{L}(k+1,n)$, where $\mathcal{C}$ is the code fragment from line 5 to 18 in Figure 4(b).

For both $\mathcal{L}(k,n)$ and $\mathcal{L}(k+1,n)$ the iterator upper bound n is fixed, but the iterator initial value varies (k and $k+1$). Thus, this time we apply induction on the iterator initial value. Hence, we prove the basic case $\{\}\ \mathcal{L}(n,n)\ \{\varphi(n)\}$, then we assume that $\{\}\ \mathcal{L}(k+1,n)\ \{\varphi(k+1)\}$ holds and prove $\{\}\ \mathcal{L}(k,n)\ \{\varphi(k)\}$. Replacing $\mathcal{L}(k,n)$ with $\mathcal{C};\mathcal{L}(k+1,n)$ in the last formula, we obtain

$$\{\}\ \mathcal{C};\mathcal{L}(k+1,n)\ \{\varphi(k)\} \tag{2}$$

By introducing the assumption $\varphi(k+1)$ resulting from the induction hypothesis into (2), we get

$$\{\}\ \mathcal{C};\mathcal{L}(k+1,n)\ ;\mathsf{assume}(\varphi(k+1))\ \{\varphi(k)\} \tag{3}$$

By moving the assumption into the postcondition in (3), we obtain

$$\{\}\ \mathcal{C};\mathcal{L}(k+1,n)\ \{\varphi(k+1) \Rightarrow \varphi(k)\}.$$

```
void selection_sort (int a[], int n)
{
  int i, j, k, s;

  k = 0;
  i = k;
  while(i < n)
  {
    s = i;
    for(j = i+1; j < n; ++j)
    {
      if(a[j] < a[s])
      {
        s = j;
      }
    }
    t = a[i];
    a[i] = a[s];
    a[s] = t;
    i++;
  }

  assert(∀ x y. (k ≤ x < n
  ∧ k ≤ y < n
  ∧ x < y)
  ⇒ a[x] ≤ a[y]);
}
```

(a)

```
void selection_sort (int a[], int n)
{
  int i, j, k, s;

  k = 0;
  i = k;
  s = k;

  for(j = i+1; j < n; ++j)
  {
    if(a[j] < a[s])
    {
      s = j;
    }
  }
  t = a[k];
  a[k] = a[s];
  a[s] = t;

  i = k + 1;
  while(i < n)
  {
    s = i;
    for(j = i+1; j < n; ++j)
    {
      if(a[j] < a[s])
      {
        s = j;
      }
    }
    t = a[i];
    a[i] = a[s];
    a[s] = t;
    i++;
  }

  assert(∀ x y. (k ≤ x < n
  ∧ k ≤ y < n
  ∧ x < y)
  ⇒ a[x] ≤ a[y]);
}
```

(b)

Fig. 4. Selection sort (a) and the corresponding post-recursive form (b)

The last formula is simply written

$$\{\}\ \mathcal{L}(k,n)\ \{\varphi(k+1) \Rightarrow \varphi(k)\}.$$

We have

$$\varphi(k) \equiv \forall x\ y.(k \leq x < n \wedge k \leq y < n \wedge x < y) \Rightarrow a[x] \leq a[y]$$

and

$$\varphi(k+1) \equiv \forall x\, y.(k+1 \leq x < n \wedge k+1 \leq y < n \wedge x < y) \Rightarrow a[x] \leq a[y].$$

The formula $\varphi(k)$ can be split into two conjuncts $\varphi(k) \equiv \varphi(k+1) \wedge \varphi'(k)$ such that

$$\varphi'(k) \equiv \forall y.\ k+1 \leq y < n \Rightarrow a[k] \leq a[y]$$

thus

$$\varphi'(k) \Rightarrow (\varphi(k+1) \Rightarrow \varphi(k)).$$

Meaning that it suffices to prove

$$\{\}\ \mathcal{C}; \mathcal{L}(k+1, n)\ \{\varphi'(k)\}.$$

By repeating the steps applied to (2) we get

$$\{\}\ \mathcal{C}; \mathcal{L}(k+1, n)\ \{\varphi'(k+1) \Rightarrow \varphi'(k)\}.$$

One can check that the final postcondition $\varphi'(k+1) \Rightarrow \varphi'(k)$ is simply equivalent to $a[k] \leq a[k+1]$. We start with an assertion having two universally quantified variables and end up with an assertion which is quantifier-free. Many existing automatic tools can handle quantifier-free assertions. This time the simplification does not affect the code but weakens the target assertion. Here the application of induction is slightly different from the previous (pre-recursive) case. We assume that the target property $\varphi(k)$ holds for $k+1$ ($k+1 \leq n-1$) and prove it for k. We can always apply the classical reasoning scheme by rewriting $k+1$ to $n-p$ (p is fresh) as n is fixed. We then apply induction on p: we assume that $\varphi(p)$ holds for p ($p \geq 1$) and prove it for $p+1$.

Question: based on which criterion, we transform loops to pre- or post-recursive form? An answer to this question is given in section 4.

3 Preliminaries

3.1 Loops in Canonical Form

In our study, we consider loops $\mathcal{L}$ having one of the forms illustrated in Figure 5(a) and Figure 5(b). We say that they are in *canonical* form. The variable i is an *iterator*, it is incremented (decremented) by one at each loop iteration. In Figure 5(a), variable l represents the iterator initial value and variable u represents its upper bound (strictly greater). In Figure 5(b), variable u represents the iterator initial value and l its lower bound (strictly less). The iterator i is not modified within the loop body $\mathcal{B}$. If l and u are not numerical constants, they are also not modified within $\mathcal{B}$.

```
i := l;                 i := u;
while(i < u)            while(i > l)
{                       {
    B;                      B;
    i := i + 1;             i := i - 1;
}                       }

      (a)                     (b)
```

Fig. 5. Canonical form for loops

3.2 Notations

Given the loop $\mathcal{L}$ in canonical form, $\mathcal{L}.i$ refers to the loop iterator and $\mathcal{L}.s$ is the iterator sign, it can be "+" (increase) or "−" (decrease). The notation $\mathcal{L}(t,b)$ represents a parameterization of $\mathcal{L}$ with the iterator initial value t and the iterator bound b (upper or lower). The iterator range is abstractly given by $[t..b[$. This range is not oriented, which means that t is not necessarily less than b. We just know that the left element of the range is the initial value and the right one is the bound. E.g., $[5..0[$ models the range which is concretely defined by $]0..5]$. The body $\mathcal{B}$ of the loop $\mathcal{L}$ is denoted by $\mathcal{L}.B$. We have the generic operator next which takes the loop $\mathcal{L}$ and a value n for the iterator $\mathcal{L}.i$ as parameters and returns the next value of the iterator. I.e., $\mathsf{next}(\mathcal{L},n) = n+1$ if $\mathcal{L}.i$ increases and $\mathsf{next}(\mathcal{L},n) = n-1$ if $\mathcal{L}.i$ decreases. We also have the generic operator prev which computes the previous value of the loop iterator. Note that $\mathsf{next}(\mathcal{L},x)$ for $x \notin [t..b[$ is not defined as well as $\mathsf{prev}(\mathcal{L},x)$ for $x \notin]t..b]$. Finally, the projection of loop $\mathcal{L}$ on the range $[i'..b'[$, such that $[i'..b'[\subseteq [i..b[$, is denoted by $\mathcal{L}([i'..b'[)$. It represents the execution of $\mathcal{L}$ when the loop iterator takes its values in the range $[i'..b'[$. Note that for a given loop $\mathcal{L}(i,b)$, $\mathcal{L}(i',b')$ is different from $\mathcal{L}([i'..b'[)$. The first notation represents the loop obtained by replacing each occurrence of the symbols i and b with i' and b' respectively, which means that changes induced by the substitution can also affect the loop body $\mathcal{L}.B$. However, the loop projection is simply the iterative execution of the loop body $\mathcal{L}.B$ such that $\mathcal{L}.i$ ranges in $[i'..b'[$.

3.3 Array Quantified Assertions

Let us have the following grammar for predicates

$$\begin{aligned} \mathsf{pred} &::= \mathsf{pred} \wedge \mathsf{pred} \mid \mathsf{pred} \vee \mathsf{pred} \mid \neg\mathsf{pred} \mid \mathsf{e} > \mathsf{e} \mid \mathsf{e} = \mathsf{e} \\ \mathsf{e} &::= \mathsf{int} \mid \mathsf{id} \mid \mathsf{int} * \mathsf{e} \mid \mathsf{id}[\mathsf{i}] \mid \mathsf{e} + \mathsf{e} \\ \mathsf{i} &::= \mathsf{int} \mid \mathsf{id} \mid \mathsf{int} * \mathsf{i} \mid \mathsf{i} + \mathsf{i} \mid \mathsf{i}, \mathsf{i} \end{aligned}$$

In the grammar above, id stands for an identifier and int represents an integer constant.

We consider universally quantified assertions φ of the form

$$\varphi \equiv \forall x_1, \ldots, x_k \in [l..u[.\ \psi(x_1, \ldots, x_k) \tag{4}$$

such that ψ is generated via the above grammar under the condition that each of the variables $x_1, \ldots, x_k$ must occur in an expression generated by the last rule (head i) of the grammar. In other words, there must be at least one occurrence of each of the variables $x_1, \ldots, x_k$ in the index expression of an array that appears in ψ.

4 Source-to-Source Transformation

As seen in Section 2, the base of our program transformation is finding the recurrent fragment. In what follows, we formalize a criterion for identifying the recurrent fragment and thus transforming loops to pre- or post-recursive forms. We also present an algorithm which is based on the proposed criterion to soundly transform verification tasks.

4.1 Recurrent Fragments

Given a loop $\mathcal{L}(t, b)$ whose iterator has the range $[t..b[$ of length $|b - t|$, our goal is to identify the code fragment X in $\mathcal{L}(t, b)$ such that $\exists\ t', b' \in [t..b[$, $X = \mathcal{L}(t', b')$ and $|b' - t'| = |b - t| - 1$. It means that (t', b') is either $(\mathsf{next}(\mathcal{L}, t), b)$ or $(t, \mathsf{prev}(\mathcal{L}, b))$, which implies that X is either $\mathcal{L}[\mathsf{next}(\mathcal{L}, t)..b[$ or $\mathcal{L}[t..\mathsf{prev}(\mathcal{L}, b)[$. Thus, the relation between $\mathcal{L}(t, b)$ and X follows one of the forms in (5)

$$\mathcal{L}(t, b) = \begin{cases} \mathcal{C};\ X \\ \quad \text{or} \\ X;\ \mathcal{C} \end{cases} \tag{5}$$

where $\mathcal{C}$ is a code fragment and ";" is the sequencing operator. We call X the *recurrent* fragment.

Let us consider the case where $X = \mathcal{L}(\mathsf{next}(\mathcal{L}, t), b)$, it means that the iterator $\mathcal{L}.i$ in X ranges over $[\mathsf{next}(\mathcal{L}, t)..b[$. The code fragment in $\mathcal{L}(t, b)$ that potentially matches with X can only be the last $|b - t| - 1$ iterations, which corresponds to $\mathcal{L}([\mathsf{next}(\mathcal{L}, t)..b[)$ the projection of $\mathcal{L}(t, b)$ on $[\mathsf{next}(\mathcal{L}, t)..b[$. In this case, we have $\mathcal{L}(t, b) = \mathcal{C}; \mathcal{L}(\mathsf{next}(\mathcal{L}, t), b)$ such that $\mathcal{C}$ corresponds to the initial iteration where the loop iterator $\mathcal{L}.i$ is equal to t. It remains now to check whether the equation below holds

$$\mathcal{L}(\mathsf{next}(\mathcal{L}, t), b) = \mathcal{L}([\mathsf{next}(\mathcal{L}, t)..b[) \tag{6}$$

The code portions corresponding to the left and right side of the above equation are respectively represented in Figure 6(a) and Figure 6(b).

The symbol $\bowtie$ in Figure 6(a) and Figure 6(b) represents a relational operator which can be either " $<$ " or " $>$ ", depending on whether the loop iterator is increasing or decreasing. We want to find the condition under which the piece of code in Figure 6(a) and the one in Figure 6(b) are equivalent. A possible

(a)	(b)
$\mathcal{L}.i := \mathsf{next}(\mathcal{L}, t);$ $\mathsf{while}(\mathcal{L}.i \bowtie b)$ $\{$ $\quad \mathcal{B}[\mathsf{next}(\mathcal{L}, t)/t];$ $\quad \mathcal{L}.i := \mathsf{next}(\mathcal{L}, \mathcal{L}.i);$ $\}$	$\mathcal{L}.i := \mathsf{next}(\mathcal{L}, t);$ $\mathsf{while}(\mathcal{L}.i \bowtie b)$ $\{$ $\quad \mathcal{B};$ $\quad \mathcal{L}.i := \mathsf{next}(\mathcal{L}, \mathcal{L}.i);$ $\}$

Fig. 6. Code fragments corresponding to $\mathcal{L}(\mathsf{next}(\mathcal{L}, t), b)$ (left) and $\mathcal{L}([\mathsf{next}(\mathcal{L}, t)..b[)$ (right)

condition for equivalence is that t is equal to $\mathsf{next}(\mathcal{L}, t)$ the value with which it is replaced (Figure 6(a)). However, we know that the equality $t = \mathsf{next}(\mathcal{L}, t)$ is not possible, thus, we go for a stronger condition and simply choose identity as equivalence relation between programs. It means that the code fragments in Figure 6(a) and Figure 6(b) must be syntactically identical. This requires that changes induced by the substitution in Figure 6(a) must not affect the loop body, i.e., $\mathcal{B}[\mathsf{next}(\mathcal{L}, t)/t] = \mathcal{B}$. This is only true if $\mathcal{B}$ does not contain t. Hence a sufficient condition for the transformation to post-recursive form is given by the following proposition which is valid by construction.

Proposition 1. *(Transformation condition) $\mathcal{L}(t, b)$ is transformable to post-recursive form if the variable t representing the symbolic initial value of the loop iterator does not appear in $\mathcal{L}.B$ the body of $\mathcal{L}(t, b)$.*

A similar reasoning is applied to show that the loop $\mathcal{L}(t, b)$ is transformable to pre-recursive form under the condition that the *variable b representing the symbolic bound of the loop iterator must not appear in the body $\mathcal{L}.B$ of the loop* $\mathcal{L}(t, b)$.

4.2 Transformation Algorithm

Based on the previous result (proposition 1) concerning the criterion for loop transformation, we present algorithm Transform (Algorithm 1) which takes a Hoare triple (verification task) H as argument and returns another Hoare triple H', such that, if H' is proven to be valid then H is valid. The algorithm may also return H unchanged if neither the pre- nor the post-recursive transformation criterion is fulfilled. The induction is applied either on the iterator initial value t or its bound b. Therefore we first test, at line 1 of the algorithm, whether φ contains at least one of the symbols t or b, if not then the Hoare triple is not transformed. Function Ids takes an expression as parameter and returns the set of symbols in that expression. If the test at line 4 of the algorithm is true, the loop $\mathcal{L}$ is transformed to the pre-recursive form. In this case the reasoning scheme presented in section 2.1 is applied. The induction is then performed on the iterator bound b, that is why we have the additional condition (at line 4) that b must not be a numerical constant. Following the reasoning scheme of section 2.1, the code fragment in the returned Hoare triple represents the last iteration of the loop, i.e., the loop iterator is assigned $\mathsf{prev}(\mathcal{L}, b)$ (line 5) which is concretely $b-1$ if the iterator is increasing or $b+1$ if the iterator is decreasing. The precondition of the Hoare triple in both cases is simply the post condition of the induction hypothesis, i.e., b in φ is replaced with $\mathsf{prev}(\mathcal{L}, b)$. If the criterion for the transformation to post-recursive form is true (line 6), the reasoning scheme illustrated in section 2.2 is applied. In this case, the code in the Hoare triple remains unchanged but the postcondition is weakened using the postcondition of the induction hypothesis. This is illustrated in the return statements at line 7. Here the induction is applied on the initial value t of the iterator which must not be a numerical constant (line 6).

Back to the example of Figure 1(a), i represents an increasing iterator for the outer loop, its upper bound n does not appear in the loop body, thus, the obtained result (Figure 1(b)) corresponds to the return statement at line 5 of algorithm Transform. Concerning the example of Figure 4(a), the upper bound n of iterator i appears in the loop body, thus, the loop cannot be transformed to pre-recursive form. However the initial value k for i does not appear in the loop body, thus, the loop is transformed to post-recursive form (line 7 of algorithm Transform).

Algorithm 1. Transform

Input: $\{\ \}\ \mathcal{L}(t, b)\ \{\varphi\}$ Hoare triple
Output: Hoare triple

1 **if** $b \notin \mathsf{Ids}(\varphi) \land t \notin \mathsf{Ids}(\varphi)$ **then**
2   **return** $(\{\ \}\ \mathcal{L}(t, b)\ \{\varphi\})$;
3 **end**
4 **if** $((b \notin \mathsf{Ids}(\mathcal{L}.\mathsf{B})) \land (b\ \textit{is not constant}) \land (b \in \varphi))$ **then**
5   **return** $(\{\varphi[\mathsf{prev}(\mathcal{L}, b)/b]\}\ \mathcal{L}.i := \mathsf{prev}(\mathcal{L}, b);\ \mathcal{L}.B\ \{\varphi\})$;
6 **else if** $((t \notin \mathsf{Ids}(\mathcal{L}.\mathsf{B})) \land (t\ \textit{is not constant}) \land (t \in \varphi))$ **then**
7   **return** $(\{\ \}\ \mathcal{L}(t, b)\ \{\varphi[\mathsf{next}(\mathcal{L}, t)/t] \Rightarrow \varphi\})$;
8 **else**
9   **return** $(\{\ \}\ \mathcal{L}(t, b)\ \{\varphi\})$;
10 **end**

Proposition 2. *(Soundness) Let us have the Hoare triple $H = \{\ \}\ \mathcal{L}(t, b)\ \{\varphi\}$ and $H' = \mathsf{Transform}(H)$. If H' is true for $|b-t| \geq 1$ then H is true for $|b-t| \geq 1$, i.e., whenever H' is proven to be valid then H is also valid.*

Proof. Let us first consider the case where the loop iterator is increasing and the loop is transformable to pre-recursive form. According to the assumption, H' is true for $|b-t| = 1$, thus H' holds for $b = t+1$, i.e., $\{\}\ \mathcal{L}.i := t;\ \mathcal{L}.B\ \{\varphi[t+1/b]\}$. We also know that H' is valid for an arbitrary b ($|b-t| \geq 1$) such that

$$H' = \{\varphi[b-1/b]\}\ \mathcal{L}.i := b-1;\ \mathcal{L}.B\ \{\varphi\}$$

thus, all the following Hoare triples are valid

$$H'_1 = \{\varphi[t+1/b]\}\ \mathcal{L}.i := t+1;\ \mathcal{L}.B\ \{\varphi[t+2/b]\}$$
$$\ldots \quad \ldots$$
$$H'_{b-2-t} = \{\varphi[b-2/b]\}\ \mathcal{L}.i := b-2;\ \mathcal{L}.B\ \{\varphi[b-1/b]\}$$
$$H'_{b-1-t} = \{\varphi[b-1/b]\}\ \mathcal{L}.i := b-1;\ \mathcal{L}.B\ \{\varphi\}$$

in addition to

$$H'_0 = \{\}\ \mathcal{L}.i := t;\ \mathcal{L}.B\ \{\varphi[t+1/b]\}.$$

According to Hoare's rule of composition, we have

$$\frac{H'_0 \qquad H'_1 \qquad \ldots \qquad H'_{b-2-t} \qquad H'_{b-1-t}}{\{\}\ \mathcal{L}.i := t;\ \mathcal{L}.B\ ; \mathcal{L}.i := t+1;\ \mathcal{L}.B\ \ldots\ \mathcal{L}.i := b-1;\ \mathcal{L}.B\ \{\varphi\}}$$

The conclusion of the rule represents $\{\ \}\ \mathcal{L}(t,b)\ \{\varphi\}$, which is H. Hence, H is valid if H' is valid. In a similar way, we can prove the case where the iterator decreases.

Now, we assume that H is transformable to post-recursive form and the iterator is decreasing, i.e., $H' = \{\ \}\ \mathcal{L}(t,b)\ \{\varphi[t-1/t] \Rightarrow \varphi\}$.

The Hoare triple H' is assumed to be valid for an arbitrary value of t s.t., $|b-t| \geq 1$, thus we have $\{\ \}\ \mathcal{L}(t-1,b)\ \{\varphi[t-2/t] \Rightarrow \varphi[t-1/t]\}$ and we know that (according to post-recursive form) $\mathcal{L}(t,b) = \mathcal{L}.i := t;\ \mathcal{L}.B;\ \mathcal{L}(t-1,b)$, hence

$$\{\ \}\ \mathcal{L}(t,b)\ \{(\varphi[t-2/t] \Rightarrow \varphi[t-1/t]) \wedge (\varphi[t-1/t] \Rightarrow \varphi)\}$$

By reiterating the previous rewriting step (for post-recursive form)

$$\mathcal{L}(t,b) = \mathcal{L}.i := t;\ \mathcal{L}.B; \ldots;\ \mathcal{L}.i := b+1;\ \mathcal{L}.B$$

taking into account the assumption that H' holds for $|b-t| = 1$ i.e.,

$$\{\ \}\ \mathcal{L}.i := b+1;\ \mathcal{L}.B\ \{\varphi[b+1/t]\}$$

we obtain

$$\{\ \}\ \mathcal{L}(t,b)\ \{\varphi'\}$$

such that

$$\varphi' \equiv \varphi(b+1/t) \wedge (\varphi[b+1/t] \Rightarrow \varphi[b+2/t]) \wedge \ldots \wedge (\varphi[t-1/t] \Rightarrow \varphi).$$

Via modus ponens we obtain $\varphi' \Rightarrow \varphi$, thus $\{\ \}\ \mathcal{L}(t,b)\ \{\varphi\}$ which means that H is valid. Analogically, we can prove the case where the iterator increases. □

4.3 Multidimensional Case

Our approach extends naturally to multidimensional arrays. To illustrate this, we use the example matrix_init which is shown in Figure 7(a). The procedure matrix_init takes three arguments: the matrix (array) a, the number of matrix

```
1  void matrix_init(int a [][],  int n, int m)
2  {
3
4    int i,j;
5    for(i = 0; i < m; ++i)
6    {
7      for(j = 0; j < n; ++j)
8      {
9        a[i][j] = 0;
10     }
11   }
12
13   assert(∀ x y. (0 ≤ x < m
14   ∧ 0 ≤ y < n)
15   ⇒ a[x][y] = 0);
16 }
```

(a)

```
1  assume(∀ x y. (0 ≤ x < m−1
2  ∧ 0 ≤ y < n)
3  ⇒ a[x][y] = 0);
4
5  for(j = 0; j < n; ++j)
6  {
7    a[m−1][j] = 0;
8  }
9
10 assert(∀ x y. (0 ≤ x < m
11 ∧ 0 ≤ y < n)
12 ⇒ a[x][y] = 0);
```

(b)

```
1  assume(∀ x y. (0 ≤ x < m−1
2  ∧ 0 ≤ y < n)
3  ⇒ a[x][y] = 0);
4
5  assume(∀ x y. (0 ≤ x < m
6  ∧ 0 ≤ y < n−1)
7  ⇒ a[x][y]= 0);
8
9  a[m−1][n−1] = 0;
10
11 assert(∀ x y. (0 ≤ x < m
12 ∧ 0 ≤ y < n)
13 ⇒ a[x][y] = 0);
```

(c)

Fig. 7. Simple matrix initialization program (a) and the corresponding two successive pre-recursive form transformations (b) and (c)

rows `m` and the number of matrix columns `n`. The outer loop in the procedure ranges over matrix rows, while the inner one ranges over elements of each row and initializes every element to 0. The assertion at line 13 states that the elements $(0, 0)..(m-1, n-1)$ of the matrix are indeed initialized to 0. We can unroll the outer loop and apply the same reasoning used for the first example (Figure 1(a)). We choose the variable m to parametrize the code and the assertion, and transform the program to pre-recursive form. We obtain the result illustrated in Figure 7(b). This result can itself be transformed again by considering the code portion from line 5 to 12 (Figure 7(b)) and reiterating the process, however this time the program fragment as well as the assertion are parametrized by n. As result, we obtain the code fragment from line 5 to 13 in Figure 7(c). By adopting the reasoning scheme used in the first example, where the precondition of the Hoare triple is empty (true), the assumption (in Figure 7(b)) is ignored despite its relevance to the target assertion. Obviously, the assumption at line 5 in Figure 7(c) is too weak to prove the assertion. To avoid such a loss of information regarding the initial state, we propagate the precondition of the source Hoare triple to the resulting one by adopting the following reasoning scheme

$$\frac{\{\psi\}\ \mathcal{L}(t,n-1)\ \{\varphi(n-1)\wedge\psi\} \qquad \{\varphi(n-1)\wedge\psi\}\ \mathcal{C}\ \{\varphi(n)\}}{\{\psi\}\ \mathcal{L}(t,n-1);\mathcal{C}\ \{\varphi(n)\}}$$

Compared to the previous case (Figure 2), here we have the additional precondition ψ in the postcondition of the first premise of the rule. The induction hypothesis guarantees that $\varphi(n-1)$ is valid but does not say anything about the validity of ψ as postcondition. Thus, under the assumption that the validity of ψ is not affected by the execution of $\mathcal{L}(t,n-1)$, it suffices to prove that the second premise $\{\varphi(n-1)\wedge\psi\}\ \ \mathcal{C}\ \{\varphi(n)\}$ is valid to conclude the validity of the original verification task, namely $\{\psi\}\ \ \mathcal{L}(t,n)\ \{\varphi(n)\}$. Back to our example of Figure 7(b), we can see that the assumption is not affected by the code portion (line 5 to 8) as all array elements modified by the code have $m-1$ as first index while the assumption is about array elements whose first index is less than $m-1$. Hence, the assumption is preserved in the final result (at line 1) in Figure 7(c) and the second assumption (line 5) is the one coming from the induction hypothesis in the second transformation step. By combining the two assumptions, we can prove the target assertion. The question now is how to show that some quantified precondition is not affected by a given piece of code. For this we introduce the notion of *array access separation.*

Array Access Separation and Precondition Propagation. Given two array access (expressions) $a[i_1,\ldots,i_k]$ at program location ℓ_1 and $a[j_1,\ldots,j_k]$ at program location ℓ_2, we say that both access are separated if it exits a,b,c,d and x such that $1\leq x\leq k$, and the assertions $a\leq i_x\leq b$ and $c\leq j_x\leq d$ are valid at locations ℓ_1 and ℓ_2 respectively, and $[a..b]\cap[c..d]=\emptyset$. For each index expression we can compute the interval in which it ranges using a lightweight *interval analysis* [23]. However in practice the loop iterator often appears as array index, thus its interval is simply the iterator range.

Given a quantified array assertion φ of the form (4), we say that a piece of code $\mathcal{C}$ does not *affect* φ if

- Each expression $a[j_1,\ldots,j_k]$ appearing as left hand side of an assignment in $\mathcal{C}$ and each expression $a[i_1,\ldots,i_k]$ in φ are separated.
- Each simple variable x representing the left hand side of an assignment in $\mathcal{C}$ does not appear in φ.

Using the concept of array access separation, we propose a new version of the transformation algorithm (Algorithm 2). We first make call to the previous transformation algorithm (line 1). The precondition in the original Hoare triple is then propagated to the result if it is not affected by the code (line 2 and 3), otherwise we just get the Hoare triple which is obtained via the first transformation algorithm (Transform). The proof of the soundness of this algorithm, more precisely

Algorithm 2. TransformAndPropagate

Input: $\{\psi\}\ \mathcal{L}(t, b)\ \{\varphi\}$ Hoare triple
Output: Hoare triple
1 Let $\{\,\}\ \mathcal{C}\ \{\varphi'\}$ = Transform($\{\,\}\ \mathcal{L}(t, b)\ \{\varphi\}$);
2 **if** ($\mathcal{L}(t, b)$ *does not affect* ψ)) **then**
3   **return** ($\{\psi\}\ \mathcal{C}\ \{\varphi'\}$);
4 **else**
5   **return** ($\{\,\}\ \mathcal{C}\ \{\varphi'\}$);

the first case (line 3), is exactly similar to the one of Transform. It suffices just to include ψ in each pre- and postcondition of any Hoare triple used in the proof as ψ always holds (not affected).

Discussion. An assertion φ can be affected by an assignment of the form $*x = e$ even if $*x$ does not appear in φ. This is due to potential aliases that x may have. In this case, one can use *Morris' axiom of assignment* to make side effects explicit through a case split with respect to potential aliasing cases. E.g., if $*x$ and y are potential aliases, i.e, x may point to y, we explicitly add the statement $if(x == \&y)\ y = e$ after the assignment $*x = e$.

4.4 Loop Sequences

In some instances we have sequences of loops followed by assertions, i.e., $\mathcal{L}_1(t_1, b_1); \ldots; \mathcal{L}_k(t_k, b_k)\ \{\varphi\}$, and the assertion does not only depend on the final loop ($\mathcal{L}_k$) but also on other loops preceding it. Under some conditions we can still transform the original program to obtain a simpler one.

Pre-recursive Case. The original program can be transformed to pre-recursive form if the following conditions are fulfilled

1. The iterator interval is the same for all loops, thus we have $\mathcal{L}_1(t, b), \ldots, \mathcal{L}_k(t, b)$.
2. All loops are transformable to pre-recursive form.
3. For every m, s.t., $1 \le m < k$ and for every n, s.t., $m < n \le k$, the code fragments $\mathcal{L}_m.i :=$ prev$(\mathcal{L}_m, b); \mathcal{L}_m.B$ and $\mathcal{L}_n(t,$ prev$(\mathcal{L}_n, b))$ are *commutative.* Two code fragments C_1 and C_2 are commutative if for any state s the execution of either $C_1; C_2$ or $C_2; C_1$ starting at s gives the same state s'. This is the case if every (a) two array access a_1 and a_2 in C_1 and C_2 where one of them is a write are separated, and (b) if there is a write to a variable x in one fragment, then x must not appear in the second one.

To illustrate the transformation result, let us consider the following Hoare triple which fulfills the first condition

$$\{\}\ \mathcal{L}_1(t, b); \ldots; \mathcal{L}_k(t, b)\ \{\varphi\} \tag{7}$$

As all loops have the same parametrization, we can simply write

$$\{\}\ (\mathcal{L}_1; \ldots; \mathcal{L}_k)(t, b)\ \{\varphi\} \tag{8}$$

Assuming that all loops are transformable to pre-recursive form and that the iterator is increasing, (8) is rewritten to

$$\{\}\ \mathcal{L}_1(t, b-1); \mathcal{L}_1.i := b-1; \mathcal{L}_1.B; \ldots; \mathcal{L}_k(t, b-1); \mathcal{L}_k.i := b-1; \mathcal{L}_k.B\ \{\varphi\}.$$

Under the commutativity assumption, we obtain

$$\{\}\ \underbrace{(\mathcal{L}_1; \ldots; \mathcal{L}_k)(t, b-1)}_{\text{recurrent fragment}}; \mathcal{L}_1.i := b-1; \mathcal{L}_1.B; \ldots; \mathcal{L}_k.i := b-1; \mathcal{L}_k.B\ \{\varphi\}.$$

Using a reasoning analogous to the one applied for the pre-recursive case (section 2.1), we obtain the final verification task where the code part is composed of the last iterations from each loop

$$\{\varphi[b-1/b]\}\ \mathcal{L}_1.i := b-1; \mathcal{L}_1.B; \ldots; \mathcal{L}_k.i := b-1; \mathcal{L}_k.B\ \{\varphi\}.$$

The transformation for the case where the iterator is decreasing is straightforward, it suffices just to replace $b-1$ with $b+1$ in the last result.

Post-recursive Case. For the post-recursive case, the conditions are slightly different:

1. The iterator interval is the same for all loops, i.e., $\mathcal{L}_1(t, b), \ldots, \mathcal{L}_k(t, b)$.
2. All loops are transformable to post-recursive form.
3. For every m, s.t., $1 \leq m < k$ and for every n, s.t., $m < n \leq k$, the code fragments $\mathcal{L}_m(\mathsf{next}(\mathcal{L}_m, t), b)$ and $\mathcal{L}_n.i := t; \mathcal{L}_n.B$ are commutative.

For illustration, let us now assume that all loops are transformable to post-recursive form and the iterator is increasing. Under conditions 1 and 3, formula (8) is transformed to

$$\{\}\ \mathcal{L}_1.i := t; \mathcal{L}_1.B; \ldots; \mathcal{L}_k.i := t; \mathcal{L}_k.B\ ; \underbrace{(\mathcal{L}_1; \ldots; \mathcal{L}_k)(t+1, b)}_{\text{recurrent fragment}}\{\varphi\}.$$

Following the reasoning scheme applied in section 2.2, we obtain

$$\{\}\ \mathcal{L}_1(t, b); \ldots; \mathcal{L}_k(t, b)\ \{\varphi[t+1/t] \Rightarrow \varphi\}.$$

For the case where the iterator decreases, we just replace $t+1$ with $t-1$ in the postcondition of the Hoare triple above.

5 Implementation and Experiments

We have implemented our transformation technique in the software model checker ACSAR [20]. We recall the previously obtained results [19] to show the practical performance enhancement that our approach can bring. Moreover, this transformation allows to verify challenging examples which are out of the scope of many automatic verification tools, in particular sorting algorithms.

Table 1. Experimental results for academic and industrial examples

Program	Transform	Time (s)	
		S	T
string_copy	PS/PR$^\bullet$	0.39	0.41
scan	PS/PR$^\bullet$	0.27	0.14
array_init	PS/PR$^\bullet$	0.50	0.13
loop1	PS/PR$^\bullet$	0.51	0.21
copy1	PS/PR$^\bullet$	0.70	0.23
num_index	PS/PR$^\bullet$	0.68	0.21
dvb_net_feed_stop*	PS/PR$^\bullet$	3.41	0.30
cyber_init*	PS$^\bullet$/PR	9.47	5.60
perfc_copy_info**	PS/PR$^\bullet$	10.57	1.50
do_enoprof_op**	PS/PR$^\bullet$	8.9	0.54
selection_sort	PS	409.87	173.50
insertion_sort	PR	-	145.60
bubble_sort	PS	-	188.90

The results of our experiments are illustrated in Table 1. The column "Transform" indicates the type of transformation which is applicable, "PR" stands for pre-recursive and "PS" stands for post-recursive. If both transformations are applicable, we choose the one that delivers the best results and mark it with the superscript $^\bullet$ as illustrated in the table. The column "Time" is divided into two columns "S" which stands for the simple (our previous [21]) approach and "T" which represents the modular approach based on code transformation. Our benchmarks are classified in three categories. The first category (upper part of the table) concerns academic examples taken from the literature. The second class of examples covers typical use of arrays in real world system code. The programs are code fragments taken from the Linux kernel (superscript $*$) and drivers code as well as the Xen hypervisor[1] code (superscript $**$). The last category of benchmarks (sorting algorithms), represents the most challenging examples in terms of complexity. The selection_sort example is handled by both approaches (simple and modular). However, the difference in terms of execution time is considerable. For bubble_sort and insertion_sort, the simple technique seems to diverge as it is unable to terminate within a fixed time bound. However, the modular approach is able to handle both examples in a fairly acceptable time regarding the complexity of the property. To the best of our knowledge, apart

[1] A hypervisor is a software that permits hardware virtualization. It allows multiple operating systems to run in parallel on a computer. The Xen hypervisor is available at `http://www.xen.org/`

from the method presented in [22], no other technique in the literature can handle all these three sorting examples automatically. Please refer to [19] for more details about our experimental study.

6 Related Work

The verification of array quantified assertions received a lot of consideration in recent years. Various ideas and techniques have been developed to verify such properties. However, the idea of performing modular reasoning based on assertion or code decomposition has not yet been addressed. Moreover, the modularity aspect is orthogonal to issues investigated by methods discussed in this section, indeed our technique can also be combined with these methods. Lahiri and Bryant proposed an extension of predicate abstraction to compute universally quantified invariants [15]. Their technique is based on *index predicates* which are predicates that contain free index variables. These index variables are implicitly universally quantified at each program location. In a later work, they described heuristics for inferring index predicates using counterexamples [16]. This approach requires the implementation of adequate predicate transformers. Our method reuses the existing domain of quantifier-free predicates, therefore, it does not require the implementation of specialized domain transformers. In the same category, Jhala and McMillan proposed *range predicates* [13]: predicates that refer to an implicitly quantified variable that ranges over array intervals. An axiom-based algorithm is applied to infer new range predicates as Craig interpolants for the spurious counterexamples. This approach does not handle properties that require quantification over more than one variable. Our approach does not have this restriction. Template-based techniques [3,9,22] consist of providing templates that fix the form of potential invariants. The analysis then searches for an invariant that instantiates the template parameters. Srivastava and Gulwani combined this approach with predicate abstraction to verify properties of arrays with alternating quantifiers [22]. Such properties are out of the scope for our method. However, finding the appropriate template is itself a complicated task, therefore, the template is in general manually provided. Automatic methods to discover relevant templates have not yet been proposed. Recently, a new family of interesting methods based on first-order theorem provers has emerged. McMillan proposed an approach for the computation of universally quantified interpolants by adapting a saturation prover [17]. The technique presented by Kovacs and Voronkov allows the generation of first-order invariants over arrays that contain alternations of quantifiers [14]. Their analysis is based on extracting predicates that encode array updates within a loop. A saturation-based theorem prover is applied to these predicates to infer quantified invariants. The current state of these methods is still limited due to the lack of support for arithmetic theories in the underlying theorem provers. Abstract interpretation has also received its part of interest in the verification of quantified assertions. Gopan *et al.* [8] proposed an idea based on partitioning an array into several symbolic intervals and associating a symbolic variable to each interval. Halbwachs and Péron extended this technique by allowing relational properties between abstract variables which are associated to array intervals [10]. Despite restrictions, their technique seems to handle several interesting properties. However, as for many abstract interpretation based methods, their approach requires the implementation of the appropriate abstract domain as well as the corresponding abstract transformers. This makes their approach less flexible to integrate in state of the art software verification tools.

7 Conclusion

Improving the ability of tools to check quantified properties over arrays is a key issue in automated program verification. This paper defines the concept of *recurrent fragments* and shows, for the case of loops, how the recurrence scheme can be lifted to give an automatic inductive proof. Pattern matching can then be used to locate suitable loops and thus simplify the verification task. An algorithm based on this idea shows that it can be implemented as a source-to-source translation making it low cost (as no fixed points are computed) and allowing it to be used as a front-end for a variety of tools. Experimental results show that this reduces verification time and allows us to verify challenging programs which were out of scope for previous methods. Although this work focuses on simple loop iteration schemes and syntactic matching, the idea of recurrent fragments and linking the code structure to the induction scheme is much wider. Future work includes the implementation of the presented scheme for handling sequences of loops, having more flexible and robust matching, handling wider ranges of iteration patterns and generalizing recurrent fragments to handle recursive and parametric program fragments.

References

1. Ball, T., Rajamani, S.K.: The SLAM project: debugging system software via static analysis. In: POPL, pp. 1–3 (2002)
2. Barnett, M., Chang, B.-Y.E., DeLine, R., Jacobs, B., Leino, K.R.M.: Boogie: A modular reusable verifier for object-oriented programs. In: de Boer, F.S., Bonsangue, M.M., Graf, S., de Roever, W.-P. (eds.) FMCO 2005. LNCS, vol. 4111, pp. 364–387. Springer, Heidelberg (2006)
3. Beyer, D., Henzinger, T.A., Majumdar, R., Rybalchenko, A.: Invariant synthesis for combined theories. In: Cook, B., Podelski, A. (eds.) VMCAI 2007. LNCS, vol. 4349, pp. 378–394. Springer, Heidelberg (2007)
4. Bradley, A.R., Manna, Z., Sipma, H.B.: What's decidable about arrays? In: Emerson, E.A., Namjoshi, K.S. (eds.) VMCAI 2006. LNCS, vol. 3855, pp. 427–442. Springer, Heidelberg (2006)
5. Chaki, S., Clarke, E.M., Groce, A., Jha, S., Veith, H.: Modular verification of software components in C. In: ICSE, pp. 385–395 (2003)
6. Clarke, E., Kroning, D., Sharygina, N., Yorav, K.: SATABS: SAT-based predicate abstraction for ANSI-C. In: Halbwachs, N., Zuck, L.D. (eds.) TACAS 2005. LNCS, vol. 3440, pp. 570–574. Springer, Heidelberg (2005)

7. Dahlweid, M., Moskal, M., Santen, T., Tobies, S., Schulte, W.: Vcc: Contract-based modular verification of concurrent C. In: ICSE Companion, pp. 429–430 (2009)
8. Gopan, D., Reps, T.W., Sagiv, S.: A framework for numeric analysis of array operations. In: POPL, pp. 338–350 (2005)
9. Gulwani, S., McCloskey, B., Tiwari, A.: Lifting abstract interpreters to quantified logical domains. In: POPL, pp. 235–246 (2008)
10. Halbwachs, N., Péron, M.: Discovering properties about arrays in simple programs. In: PLDI, pp. 339–348 (2008)
11. Henzinger, T.A., Jhala, R., Majumdar, R., Sutre, G.: Lazy abstraction. In: POPL, pp. 58–70 (2002)
12. Hoare, C.A.R.: An axiomatic basis for computer programming. Commun. ACM 12(10), 576–580 (1969)
13. Jhala, R., McMillan, K.L.: Array abstractions from proofs. In: Damm, W., Hermanns, H. (eds.) CAV 2007. LNCS, vol. 4590, pp. 193–206. Springer, Heidelberg (2007)
14. Kovács, L., Voronkov, A.: Finding loop invariants for programs over arrays using a theorem prover. In: Chechik, M., Wirsing, M. (eds.) FASE 2009. LNCS, vol. 5503, pp. 470–485. Springer, Heidelberg (2009)
15. Lahiri, S.K., Bryant, R.E.: Constructing quantified invariants via predicate abstraction. In: Steffen, B., Levi, G. (eds.) VMCAI 2004. LNCS, vol. 2937, pp. 267–281. Springer, Heidelberg (2004)
16. Lahiri, S.K., Bryant, R.E.: Indexed predicate discovery for unbounded system verification. In: Alur, R., Peled, D.A. (eds.) CAV 2004. LNCS, vol. 3114, pp. 135–147. Springer, Heidelberg (2004)
17. McMillan, K.L.: Quantified invariant generation using an interpolating saturation prover. In: Ramakrishnan, C.R., Rehof, J. (eds.) TACAS 2008. LNCS, vol. 4963, pp. 413–427. Springer, Heidelberg (2008)
18. Podelski, A., Rybalchenko, A.: ARMC: The logical choice for software model checking with abstraction refinement. In: Hanus, M. (ed.) PADL 2007. LNCS, vol. 4354, pp. 245–259. Springer, Heidelberg (2007)
19. Seghir, M.N.: An assume guarantee approach for checking quantified array assertions. In: Johnson, M., Pavlovic, D. (eds.) AMAST 2010. LNCS, vol. 6486, pp. 226–235. Springer, Heidelberg (2011)
20. Seghir, M.N., Podelski, A.: ACSAR: Software model checking with transfinite refinement. In: Bošnački, D., Edelkamp, S. (eds.) SPIN 2007. LNCS, vol. 4595, pp. 274–278. Springer, Heidelberg (2007)
21. Seghir, M.N., Podelski, A., Wies, T.: Abstraction refinement for quantified array assertions. In: Palsberg, J., Su, Z. (eds.) SAS 2009. LNCS, vol. 5673, pp. 3–18. Springer, Heidelberg (2009)
22. Srivastava, S., Gulwani, S.: Program verification using templates over predicate abstraction. In: PLDI, pp. 223–234 (2009)
23. Zaks, A., Yang, Z., Shlyakhter, I., Ivancic, F., Cadambi, S., Ganai, M.K., Gupta, A., Ashar, P.: Bitwidth reduction via symbolic interval analysis for software model checking. IEEE Trans. on CAD of Integrated Circuits and Systems 27(8), 1513–1517 (2008)

Proving Properties of Co-logic Programs with Negation by Program Transformations

Hirohisa Seki*

Dept. of Computer Science, Nagoya Inst. of Technology,
Showa-ku, Nagoya, 466-8555 Japan
seki@nitech.ac.jp

Abstract. A framework for unfold/fold transformation of (constraint) co-logic programs has been proposed recently, which can be used to prove properties of co-logic programs, thereby allowing us to reason about infinite sequences of events such as behavior of reactive systems. The main problem with this approach is that only definite co-logic programs are considered, thus representing a rather narrow class of co-logic programs. In this paper we consider "negation elimination", a familiar program transformation method, tailored to co-logic programs; given a program for predicate $p(X)$, negation elimination derives a program which computes its negation $\neg p(X)$, when the program satisfies certain conditions. We show that negation elimination can be used for co-logic programs, and its application is correct under the alternating fixpoint semantics of co-logic programs. We show by examples how negation elimination, when incorporated into the previous framework for unfold/fold transformation, allows us to represent and reason about a wider class of co-logic programs. We also discuss the difference between negation elimination applied to co-logic programs and the conventional negative unfolding applied to stratified programs.

1 Introduction

Co-logic programming (co-LP) is an extension of logic programming recently proposed by Gupta et al. [5] and Simon et al. [22,23], where each predicate in definite programs is annotated as either *inductive* or *coinductive*, and the declarative semantics of co-logic programs is defined by an alternating fixpoint model: the least fixpoints for inductive predicates and the greatest fixpoints for coinductive predicates. Predicates in co-LP are defined over infinite structures such as infinite trees or infinite lists as well as finite ones, and co-logic programs allow us to represent and reason about properties of programs over such infinite structures. Co-LP therefore has interesting applications to reactive systems and verifying properties such as safety and liveness in model checking and so on.

A framework for unfold/fold transformation of (constraint) co-logic programs has been proposed recently [20]. The main problem with this approach is that

* This work was partially supported by JSPS Grant-in-Aid for Scientific Research (C) 24500171 and the Kayamori Foundation of Information Science Advancement.

E. Albert (Ed.): LOPSTR 2012, LNCS 7844, pp. 213–227, 2013.
© Springer-Verlag Berlin Heidelberg 2013

only definite co-logic programs are considered, thus representing a rather narrow class of co-logic programs.

In this paper we consider "negation elimination", a familiar program transformation method, tailored to co-logic programs. Given a program for predicate $p(X)$, negation elimination derives a program which computes its negation $\neg p(X)$, when the program satisfies certain conditions. Sato and Tamaki [18] first proposed a negation elimination method for definite programs, where they called it the "*negation technique*". We show that the negation technique can be used for co-logic programs, and its application is correct under the alternating fixpoint semantics of co-logic programs.

One of the motivations of this paper is to further study the applicability of techniques based on unfold/fold transformation not only to program development originally due to Burstall and Darlington [1], but also for proving properties of programs, which goes back to Kott [9] in functional programs. We show by examples how negation elimination, when incorporated into the previous framework for unfold/fold transformation, allows us to represent and reason about a wider class of co-logic programs. We also discuss the difference between negation elimination applied to co-logic programs and the conventional negative unfolding applied to stratified programs.

The organization of this paper is as follows. In Section 2, we summarise some preliminary definitions on co-logic programs and the previous framework for our unfold/fold transformation of co-logic programs. In Section 3, we present negation elimination for co-logic programs. In Section 4, we explain by examples how our transformation-based verification method proves properties of co-logic programs. Finally, we discuss about the related work and give a summary of this work in Section 5.[1]

Throughout this paper, we assume that the reader is familiar with the basic concepts of logic programming, which are found in [10].

2 A Framework for Transforming Co-logic Programs

In this section, we recall some basic definitions and notations concerning co-logic programs. The details and more examples are found in [5,22,23]. We also explain some preliminaries on constraint logic programming (CLP) (e.g., [8] for a survey), and our framework for transformation of co-LP.

Since co-logic programming can deal with infinite terms such as infinite lists or trees like $f(f(\dots))$ as well as finite ones, we consider the *complete* (or *infinitary*) Herbrand base [10,7], denoted by HB^*_P, where P is a program. Fig. 1 (left) shows an example of unfolding, one of basic transformations in our transformation system. The unfolding rule introduces a constraint consisting of equations $X = Y \wedge Y = [a|X]$, which shows the necessity of dealing with infinite terms. The theory of equations and inequations in this case is studied by Colmerauer [3].

[1] Due to space constraints, we omit most proofs and some details, which will appear in the full paper.

$$
\begin{array}{lll}
P_k : p \leftarrow q(X, [a|X]) & (1) \\
\quad q(Y, Y) \leftarrow r(Y) & (2) \\
\text{Unfold}: & \\
\quad P_{k+1} = (P_k \setminus \{(1)\}) \cup \{(3)\}, & \\
\quad p \leftarrow c \,[]\, r(Y) & (3) \\
\text{where} & \\
\quad c \equiv (X = Y \wedge Y = [a|X]) &
\end{array}
\qquad
\begin{array}{lll}
P'_k : p \leftarrow X_1 = X, X_2 = [a|X] \,[]\, q(X_1, X_2) & (1') \\
\quad q(Y_1, Y_2) \leftarrow Y_1 = Y_2, Y_1 = Y_3 \,[]\, r(Y_3) & (2') \\
\text{Unfold}: & \\
\quad P'_{k+1} = (P'_k \setminus \{(1')\}) \cup \{(3')\}, & \\
\quad p \leftarrow c' \,[]\, r(Y_3) & (3') \\
\text{where} & \\
\quad c' \equiv (X_1 = X \wedge X_2 = [a|X] \wedge X_1 = Y_1 \wedge & \\
\quad\quad X_2 = Y_2 \wedge Y_1 = Y_2 \wedge Y_1 = Y_3) &
\end{array}
$$

Fig. 1. Example of Unfolding (left) and its CLP form (right)

Throughout this paper, we assume that there exists at least one constant and one function symbol of arity ≥ 1, thus HB^*_P is non-empty.

Fig. 1 (right) shows the counterparts to the clauses in the figure (left) in standard CLP form. In general, let $\gamma : p(\tilde{t}_0) \leftarrow p_1(\tilde{t}_1), \ldots, p_n(\tilde{t}_n)$ be a (logic programming) clause, where $\tilde{t}_0, \ldots, \tilde{t}_n$ are tuples of terms. Then, γ is mapped into the following *pure* CLP clause:

$$p(\tilde{x}_0) \leftarrow \tilde{x}_0 = \tilde{t}_0 \wedge \tilde{x}_1 = \tilde{t}_1 \wedge \cdots \wedge \tilde{x}_n = \tilde{t}_n \,[]\, p_1(\tilde{x}_1), \ldots, p_n(\tilde{x}_n),$$

where $\tilde{x}_0, \ldots, \tilde{x}_n$ are tuples of new and distinct variables, and $[]$ means conjunction ("$\wedge$"). We will use the conventional representation of a (logic programming) clause as a shorthand for a pure CLP clause, for the sake of readability.

In the following, for a CLP clause γ of the form: $H \leftarrow c \,[]\, B_1, \ldots, B_n$, the head H and the body $B_1, \ldots, B_n$ are denoted by $hd(\gamma)$ and $bd(\gamma)$, respectively. We call c the *constraint* of γ. A conjunction $c \,[]\, B_1, \ldots, B_n$ is said to be a *goal* (or a *query*). The predicate symbol of the head of a clause is called the *head predicate* of the clause.

The set of all clauses in a program P with the same predicate symbol p in the head is called the definition of p and denoted by $Def(p, P)$. We say that a predicate p *depends on* a predicate q in P, iff either (i) $p = q$, (ii) there exists in P a clause of the form: $p(\ldots) \leftarrow c \,[]\, B$ such that predicate q occurs in B or (iii) there exists a predicate r such that p depends on r in P, and r depends on q in P. The *extended definition* [14] of p in P, denoted by $Def^*(p, P)$, is the conjunction of the definitions of all the predicates on which p depends in P.

2.1 Syntax and Semantics of Co-logic Programs

A *co-logic program* is a constraint definite program, where predicate symbols are annotated as either inductive or coinductive.[2] Let $\mathcal{P}$ be the set of all predicates in a co-logic program P, and we denote by $\mathcal{P}^{in}$ ($\mathcal{P}^{co}$) the set of inductive (coinductive) predicates in $\mathcal{P}$, respectively. There is one restriction on co-LP, referred

[2] We call an atom, A, an *inductive* (a *coinductive*) atom when the predicate of A is an inductive (a coinductive) predicate, respectively.

to as the *stratification restriction*: Inductive and coinductive predicates are not allowed to be mutually recursive. An example which violates the stratification restriction is $\{p \leftarrow q;\ q \leftarrow p\}$, where p is inductive, while q is coinductive. When P satisfies the stratification restriction, it is possible to decompose the set $\mathcal{P}$ of all predicates in P into a collection (called a *stratification*) of mutually disjoint sets $\mathcal{P}_0, \ldots, \mathcal{P}_r$ $(0 \leq r)$, called *strata*, so that, for every clause

$$p(\tilde{x}_0) \leftarrow c \,[\!]\, p_1(\tilde{x}_1), \ldots, p_n(\tilde{x}_n),$$

in P, we have that $\sigma(p) \geq \sigma(p_i)$ if p and p_i have the same inductive/coinductive annotations, and $\sigma(p) > \sigma(p_i)$ otherwise, where $\sigma(q) = i$, if the predicate symbol q belongs to $\mathcal{P}_i$. The following is an example of co-logic programs.

Example 1. [23]. Suppose that predicates *member* and *drop* are annotated as inductive, while predicate *comember* is annotated as coinductive.

$$member(H, [H|\_]) \leftarrow \qquad\qquad drop(H, [H|T], T) \leftarrow$$
$$member(H, [\_|T]) \leftarrow member(H, T) \qquad drop(H, [\_|T], T_1) \leftarrow drop(H, T, T_1)$$
$$comember(X, L) \leftarrow drop(X, L, L_1), comember(X, L_1)$$

The definition of *member* is a conventional one, and, since it is an inductive predicate, its meaning is defined in terms of the least fixpoint. Therefore, the prefix ending in the desired element H must be finite. The similar thing also holds for predicate *drop*.

On the other hand, predicate *comember* is coinductive, whose meaning is defined in terms of the greatest fixpoint (see the next section). Therefore, it is true if and only if the desired element X occurs an infinite number of times in the list L. Hence it is false when the element does not occur in the list or when the element only occurs a finite number of times in the list. □

Semantics of Co-logic Programs. The declarative semantics of a co-logic program is a stratified interleaving of the least fixpoint semantics and the greatest fixpoint semantics.

In this paper, we consider the complete Herbrand base HB_P^* as the set of elements in the domain of a *structure* $\mathcal{D}$ (i.e., a complete Herbrand interpretation [10]).

Given a structure $\mathcal{D}$ and a constraint c, $\mathcal{D} \models c$ denotes that c is true under the interpretation for constraints provided by $\mathcal{D}$. Moreover, if θ is a ground *substitution* (i.e., a mapping of variables on the domain $\mathcal{D}$, namely, HB_P^* in this case) and $\mathcal{D} \models c\theta$ holds, then we say that c is *satisfiable*, and θ is called a *solution* (or ground *satisfier*) of c, where $c\theta$ denotes the application of θ to the variables in c. We refer to [3] for an algorithm for checking constraint satisfiability.

Let P be a co-logic program with a stratification $\mathcal{P}_0, \ldots, \mathcal{P}_r$ $(0 \leq r)$. Let $\Pi_i\,(0 \leq i \leq r)$ be the set of clauses whose head predicates are in $\mathcal{P}_i$. Then, $P = \Pi_0 \uplus \ldots \uplus \Pi_r$. Similar to the "immediate consequence operator" T_P in the literature, our operator $T_{\Pi,S}$ assigns to every set I of ground atoms a new set $T_{\Pi,S}(I)$ of ground atoms as

$$T_{\Pi,S}(I) = \{A \in HB^*_\Pi \mid \text{there is a ground substitution } \theta \text{ and a clause in } \Pi \\ H \leftarrow c \,[\!]\, B_1, \cdots, B_n,\ n \geq 0, \text{ such that} \\ \text{(i) } A = H\theta, \text{ (ii) } \theta \text{ is a solution of } c, \text{ and} \\ \text{(iii) for every } 1 \leq i \leq n, \text{ either } B_i\theta \in I \text{ or } B_i\theta \in S\}.$$

In the above, the atoms in S are treated as facts. S is intended to be a set of atoms whose predicate symbols are in lower strata than those in the current stratum Π. We consider $T_{\Pi,S}$ to be the operator defined on the set of all subsets of HB^*_Π, ordered by standard inclusion. Next, two subsets $T^{\uparrow\alpha}_{\Pi,S}$ and $T^{\downarrow\alpha}_{\Pi,S}$ of the complete Herbrand base are defined as:

$$T^{\uparrow 0}_{\Pi,S} = \phi \text{ and } T^{\downarrow 0}_{\Pi,S} = HB^*_\Pi\,;$$

$$T^{\uparrow n+1}_{\Pi,S} = T_{\Pi,S}(T^{\uparrow n}_{\Pi}) \text{ and } T^{\downarrow n+1}_{\Pi,S} = T_{\Pi,S}(T^{\downarrow n}_{\Pi,S}) \text{ for a successor ordinal } n;$$

$$T^{\uparrow\alpha}_{\Pi,S} = \cup_{z<\alpha} T^{\uparrow z}_{\Pi,S} \text{ and } T^{\downarrow\alpha}_{\Pi,S} = \cap_{z<\alpha} T^{\downarrow z}_{\Pi,S}, \text{ for a limit ordinal } \alpha.$$

Finally, the model $M(P)$ of a co-logic program $P = \Pi_0 \uplus \ldots \uplus \Pi_r$ is defined inductively as follows: First, for the bottom stratum Π_0, let $M(\Pi_0) = T^{\uparrow\omega}_{\Pi_0,\emptyset}$, if $\mathcal{P}_0$ is inductive; $gfp(T_{\Pi_0,\emptyset})$ otherwise.

Next, for $k > 0$, let:

$$M(\Pi_k) = \begin{cases} T^{\uparrow\omega}_{\Pi_k,M_{k-1}}, & \text{if } \mathcal{P}_i \text{ is inductive,} \\ gfp(T_{\Pi_k,M_{k-1}}), & \text{if } \mathcal{P}_i \text{ is coinductive.} \end{cases}$$

where M_{k-1} is the model of lower strata than Π_k, i.e., $M_{k-1} = \cup_{i=0}^{k-1} M(\Pi_i)$. Then, the *model* of P is $M(P) = \cup_{i=0}^{r} M(\Pi_i)$, the union of all models $M(\Pi_i)$.

2.2 Transformation Rules for Co-logic Programs

We first explain our transformation rules, and then give some conditions imposed on the transformation rules which are necessary for correctness of transformation. Our transformation rules are formulated in the framework of CLP, following that by Etalle and Gabbrielli [4].

A sequence of programs $P_0, \ldots, P_n$ is said to be a *transformation sequence* with a given initial program P_0, if each P_{k+1} $(0 \leq k \leq n-1)$ is obtained from P_k by applying one of the following transformation rules R1-R4.

Our motivation of this paper is to use program transformation rules for proving properties of a given system represented in a co-logic program P_0. We thus assume that there exist two kinds of predicate symbols appearing in a transformation sequence: *base* predicates and *defined* predicates. A base predicate is defined in P_0, and it is intended to represent the structure and the behaviours of the given system. Its definition is therefore assumed to remain unchanged during program transformation. On the other hand, a defined predicate, which is introduced by the following *definition introduction* rule, is intended to represent a property of the given system such as safety and liveness properties. Its definition will be changed during program transformation so that its truth value in $M(P_n)$ will be easily known.

R1. Definition Introduction. Let δ be a clause of the form: $newp(\tilde{X}) \leftarrow c \,[]\, B$, where (i) *newp* is a *defined* predicate symbol not occurring in P_k, (ii) c is a constraint, and (iii) B is a conjunction of atoms whose predicate symbols are all base predicates appearing in P_0. Moreover, δ satisfies the following conditon called the *annotation rule*: *newp* is annotated as *inductive* iff there exists at least one inductive atom in B, while *newp* is annotated as *coinductive* iff every predicate symbol occurring in B is annotated as *coinductive*.

By *definition introduction*, we derive from P_k the new program $P_{k+1} = P_k \cup \{\delta\}$. For $n \geq 0$, we denote by $Defs_n$ the set of clauses introduced by the definition introduction rule during the transformation sequence $P_0, \ldots, P_n$. In particular, $Defs_0 = \emptyset$.

As mentioned above, the role of the predicate annotations such as "base" and "defined" is to specify the conditions for applying our transformation rules; they are orthogonal to the predicate annotations of "inductive" and "coinductive", which are used to represent the intended semantics of each predicate in co-logic programs. In particular, since the definition of $newp(\tilde{X})$ is non-recursive, its meaning is determined only by the base predicates B in P_k, irrelevant of whether its annotation is inductive or coinductive. However, some application of transformation rules to δ could possibly derive a clause in P_n $(n > k)$ of the form: $newp(\tilde{X}) \leftarrow newp(\tilde{X})$. Then, the annotation of predicate *newp* will matter, which is determined according to the above-mentioned annotation rule.

Our transformation rules include two basic rules: *unfolding* (R2) and *folding* (R3). They are the same as those in Etalle and Gabbrielli [4], which are CLP counterparts of those by Tamaki and Sato [24] for definite programs, so we omit these definitions here;

The following *replacement rule* allows us to make simple the definition of a defined predicate. The notion of *useless* predicates is originally due to Pettorossi and Proietti [14], where the rule is called *clause deletion rule*.

R4. Replacement Rule. We consider the following two rules depending on the annotation of a predicate.

- The set of the *useless* inductive predicates of a program P is the maximal set U of inductive predicates of P such that a predicate p is in U if, for the body of each clause of $Def(p, P)$, it has an inductive atom whose predicate is in U. By applying the *replacement rule* to P_k w.r.t. the useless inductive predicates in P_k, we derive the new program P_{k+1} from P_k by removing the definitions of the useless inductive predicates.
- Let $p(\tilde{t})$ be a coinductive atom, and γ be a clause (modulo variance $\simeq$) in P_k of the form: $p(\tilde{t}) \leftarrow p(\tilde{t})$. By applying the *replacement rule* to P_k w.r.t. $p(\tilde{t})$, we derive from P_k the new program $P_{k+1} = (P_k \setminus \{\gamma\}) \cup \{p(\tilde{t}) \leftarrow\}$.

2.3 Correctness of the Transformation Rules

In order to preserve the alternating fixpoint semantics of a co-logic program, we will impose some conditions on the application of folding rule. The conditions on folding are given depending on whether the annotation of the head predicate *newp* of a folded clause is inductive or not.

First, let *newp* be an *inductive* head predicate of a clause δ introduced by rule R1 (definition introduction). We call an inductive predicate p *primitive*, if, for some coinductive predicate q on which *newp* depends in $P_0 \cup \{\delta\}$, q depends on p in P_0; put simply, *newp* depends on an inductive predicate p through some coinductive predicate q. We denote by $\mathcal{P}_{pr}$ the set of all the primitive predicates, i.e., $\mathcal{P}_{pr} = \{p \in \mathcal{P}^{in} \mid \exists \delta \in Defs_n \exists q \in \mathcal{P}^{co}$, *newp* is the head predicate of δ s.t. *newp* depends on q in $P_0 \cup \{\delta\}$, and q depends on p in $P_0.\}$. We call an inductive predicate p *non-primitive*, if it is not primitive. We call an atom with non-primitive (primitive) predicate symbol a non-primitive (primitive) atom, respectively.

Let $P_0, \ldots, P_n$ be a transformation sequence, and δ be a clause first introduced by rule R1 in $Defs_i$ $(0 \leq i \leq n)$, Then, we mark it "*not TS-foldable*". Let γ be a clause in P_{k+1} $(0 \leq k < n)$. γ inherits its mark when γ is not involved in the derivation from P_k to P_{k+1}. γ is marked "*TS-foldable*", only if γ is derived from $\beta \in P_k$ by unfolding β of the form: $A \leftarrow c \,[]\, H, G$ w.r.t. H and H is a *non-primitive* inductive atom. Otherwise, γ inherits the mark of β. Similarly, γ inherits the mark of β, if γ is derived from $\beta \in P_k$ by folding.

When there are no coinductive predicates in $P_0 \cup Defs_n$, the above TS-foldable condition coincides with the one by Tamaki-Sato [24].

Next, let *newp* be an *coinductive* defined predicate of a clause δ introduced by rule R1. The next notion is due to [21], which is originally introduced to give a condition on folding to preserve the finite failure set.

Let $P_0, \ldots, P_n$ be a transformation sequence, and δ be a clause first introduced by rule R1 in $Defs_i$ $(0 \leq i \leq n)$. Then, each atom in the body $bd(\delta)$ is marked *inherited* in δ. Let γ be a clause in P_{k+1} $(0 \leq k < n)$. When γ is not involved in the derivation from P_k to P_{k+1}, each atom B in $bd(\gamma)$ is marked *inherited*, if so is it in γ in P_k. Otherwise, i.e., suppose that γ is derived by applying to some clause β in P_k either unfolding rule or folding rule. Then, each atom B in $bd(\gamma)$ is marked *inherited*, if it is not an atom introduced to $bd(\gamma)$ by that rule, and it is marked inherited in β in P_k. Moreover, the application of folding is said to be *fair*, if there is no folded atom in $bd(\beta)$ which is marked inherited. Intuitively, an atom marked inherited is an atom such that it was in the body of some clause in $Defs_i$ $(0 \leq i \leq n)$, and no unfolding has been applied to it.

We are now in a position to state the conditions imposed on folding and the correctness of our transformation rules.

Conditions on Folding. Let $P_0, \ldots, P_n$ be a transformation sequence. Suppose that P_k $(0 < k \leq n)$ is derived from P_{k-1} by folding $\gamma \in P_{k-1}$. The application of folding is said to be *admissible* if the following conditions are satisfied:
(1) P_k satisfies the stratification restriction,
(2) if $hd(\gamma)$ is an inductive atom, then γ is marked "TS-foldable" in P_{k-1}, and
(3) if $hd(\gamma)$ is a coinductive atom, then the application of folding is fair. □

Proposition 1. Correctness of Transformation [20]
Let P_0 be an initial co-logic program, and $P_0, \ldots, P_n$ $(0 \leq n)$ a transformation sequence, where every application of folding is admissible. Then, the transformation sequence is correct, i.e., $M(P_0 \cup Defs_n) = M(P_n)$. □

3 Negation Elimination in Co-logic Programs

To deal with a negative literal in a goal, we use the *negation technique* by Sato and Tamaki [18], a method for negation elimination for definite programs. The negation technique is a procedure to derive a set S' of definite clauses from a given definite program S such that (i) each predicate symbol p in S has one-to-one correspondence with a new predicate symbol, not_p, in S' with the same arity, and (ii) for any ground atom $p(\tilde{t})$ and $not\_p(\tilde{t})$,

$$M(S) \models \neg p(\tilde{t}) \text{ iff } M(S') \models not\_p(\tilde{t}) \qquad (*)$$

If S' satisfies the above conditions, it is called a *complementary program* of S. In [18], S is a definite program, and $M(S)$ is its least Herbrand model. In the following, we show the negation technique for co-logic programs under the alternating fixpoint semantics.

We explain the negation technique by using an example for saving space. Consider *drop* in Example 1, and let S be the definition of *drop*. We will derive the definition of *not_drop* (*not_d* for short) by applying the following steps:

(Step 1) Consider the completed definition of *drop*. In this case, we have:

$$drop(A,B,C) \leftrightarrow (\exists H,T)(\langle A,B,C\rangle = \langle H,[H|T],T\rangle) \vee (\exists H,X,T,T_1)(\langle A,B,C\rangle = \langle H,[X|T],T_1\rangle \wedge drop(H,T,T_1)) \qquad (1)$$

We denote by $\tilde{V} = \tilde{t}_1$ ($\tilde{V} = \tilde{t}_2$) the first (second) equation in (1), respectively.
(Step 2) Negate both sides of the completed definition, and every negative occurrence $\neg p(\tilde{t})$ is replaced by $not\_p(\tilde{t})$. In this case, we have:

$$not\_d(A,B,C) \leftrightarrow (\forall H,T)(\tilde{V} \neq \tilde{t}_1) \wedge (\forall H,X,T,T_1)(\tilde{V} \neq \tilde{t}_2 \vee not\_d(H,T,T_1)) \qquad (2)$$

(Step 3) Transform every conjunct on the right-hand side of the result of (Step 2) which is of the form: $(\forall \tilde{X})(\langle \tilde{A} \neq \tilde{t}\rangle \vee not\_p_1(\tilde{t}_1) \vee \cdots \vee not\_p_m(\tilde{t}_m))$ $(m \geq 1)$ to $(\forall \tilde{X})\langle \tilde{A} \neq \tilde{t}\rangle \vee (\exists \tilde{X})(\langle \tilde{A} = \tilde{t}\rangle \wedge not\_p_1(\tilde{t}_1)) \vee \cdots \vee (\exists \tilde{X})(\langle \tilde{A} = \tilde{t}\rangle \wedge not\_p_m(\tilde{t}_m))$. Note that, when some clause in S has an existential variable, this transformation is not valid. In this case we obtain from (2):

$$not\_d(A,B,C) \leftrightarrow (\forall H,T)(\tilde{V} \neq \tilde{t}_1) \wedge \{(\forall H,X,T,T_1)(\tilde{V} \neq \tilde{t}_2) \vee (\exists H,X,T,T_1)(\tilde{V} = \tilde{t}_2 \wedge not\_d(H,T,T_1))\} \qquad (3)$$

(Step 4) Transform the right-hand side to a disjunctive form. In this case, we obtain from (3):

$$not\_d(A,B,C) \leftrightarrow ((\forall H,T)(\tilde{V} \neq \tilde{t}_1) \wedge (\forall H,X,T,T_1)(\tilde{V} \neq \tilde{t}_2)) \vee \{(\forall H,T)(\tilde{V} \neq \tilde{t}_1) \wedge (\exists H,X,T,T_1)(\tilde{V} = \tilde{t}_2 \wedge not\_d(H,T,T_1))\} \quad (4)$$

(Step 5) Transform the completed definition given as the result of (Step 4) to a set of clauses, and then simplify constraints assuming that each predicate p is typed. Annotate the derived predicate as "coinductive" (resp. "inductive) if the annotation of the original predicate is inductive (resp. coinductive). In this case, we obtain from (4), assuming that B and C are lists:

$$not\_drop(A,[\,],C) \leftarrow$$
$$not\_drop(H,[X|T],T_1) \leftarrow X \neq H, not\_drop(H,T,T_1)$$
$$not\_drop(H,[H|T],T_1) \leftarrow T \neq T_1, not\_drop(H,T,T_1)$$

not_drop is annotated as "coinductive", since predicate *drop* is inductive.
(Step 6) Apply to all predicates in $Def^*(p,S)$ (Step 1) to (Step 5), and let S' is the set of the resulting clauses. We call S' the *result* of the negation technique applied to p in S.

The above transformations themselves are exactly the same as those in the original negation technique [18] for definite programs; the only difference is the necessity of predicate annotations in (Step 5). In the original negation technique, an extra condition such as the finiteness of SLD-tree for $p(\tilde{t})$ is necessary to obtain a complementary program, while such a condition is not needed here.

Proposition 2. Correctness of the Negation Technique
Let S be a co-logic program. If every clause in S has no existential variable, then the procedure of the negation technique (Step 1) to (Step 6) gives a complementary co-logic program S', i.e., for any ground term $\tilde{t}$,

$$M(S) \models \neg p(\tilde{t}) \text{ iff } M(S') \models not\_p(\tilde{t}).$$

Proof. (Sketch) Let S' be the definition of *not_drop* obtained in (Step 5) in the above. Since the general case will be shown similarly, we explain an outline of the proof that S' is a complementary program of S (i.e., the definition of *drop*).

We first note that, when S satisfies the stratification restriction, then so does S'. Let I' be the greatest fixpoint $gfp(S')$ of S', noting that *not_drop* is annotated as coinductive. Since I' is a fixpoint, it satisfies the IFF-definition of (4), and it also satisfies (3) and (2) in turn. We define:

$$I = \{drop(\tilde{t}) \mid not\_drop(\tilde{t}) \notin I'.\ \tilde{t} \text{ is a sequence of ground terms.}\}$$

Then, we can show that I satisfies (1) and is the least fixpoint model of S. □

Remark 1. In [18], Sato and Tamaki described an extension of the negation technique to the case where a clause in S has an existential variable. Namely, even when there are clauses with existential variables, (Step 3) in the negation technique is still valid, if the clauses have a *functional part* [18]: that is, let

γ be a clause in P of the form: $p_0(X) \leftarrow p_1(X,Y),\ldots,p_m(X,Y)$. Then, we say that γ with an existential variable Y has a *functional part* p_1 iff $p_1(X,Y)$ defines a partial function from X to Y (that is, whenever $M(P) \models p_1(a,b)$ and $M(P) \models p_1(a,b')$, then $b = b'$ holds for any ground terms a, b, b').

This is also true in co-logic programs, and we show it by the following example. The definition of predicate *comember* in Ex. 1 has an existential variable (i.e., L_1); we cannot apply the negation technique to $\neg comember(X,L)$. Moreover, it does not have a functional part, since $drop(X,L,L_1)$ does not define a partial function. Instead, we consider the following definition of *comember'*:

1. $comember'(X,L) \leftarrow drop'(X,L,L_1), comember'(X,L_1)$
2. $drop'(H,[H|T],T) \leftarrow$
3. $drop'(H,[H_1|T],T_1) \leftarrow H \neq H_1, drop'(H,T,T_1)$

where $drop'(X,L,L_1)$ now defines a partial function from (X,L) to L_1.

Applying the negation technique to $\neg comember'(X,L)$ and simple unfold/fold transformations, we obtain the following clauses:

4. $not\_comem(X,L) \leftarrow not\_d(X,L)$
5. $not\_comem(X,L) \leftarrow drop'(X,L,L_1), not\_comem(X,L_1)$
6. $not\_d(H,[]) \leftarrow$
7. $not\_d(H,[H_1|T]) \leftarrow H \neq H_1, not\_d(H,T)$

where we denote *not_comemember'* by *not_comem* for short, and *not_d* is a coinductive predicate. We will use this in Example 3. □

In conventional program transformation for logic programs, *negative unfolding* ([14,19] for example) is applied to a negative literal in the body of a clause. The following example will show the difference between the negation technique in co-LP and negative unfolding in (locally) stratified programs, which reflects the differences underlying the two different semantics.

Example 2. Consider the following stratified program $P_0 = \{1,2,3\}$:

1. $p \leftarrow \neg q(X)$	4. $p \leftarrow \neg q(X), \neg r$	(neg. unfold 1)
2. $q(X) \leftarrow q(X)$	5. $p \leftarrow p, \neg r$	(fold 4)
3. $q(X) \leftarrow r$		

We first consider a *wrong* transformation sequence [14]: we apply to clause 1 negative unfolding w.r.t. $\neg q(X)$, followed by folding the resulting clause. Let $P_2 = (P_0 \setminus \{1\}) \cup \{5\}$. Then, we have that $PERF(P_0) \models p$, while $PERF(P_2) \not\models p$. Thus, the above transformation does not preserve the perfect model semantics; put somewhat simply, folding immediately after negative unfolding is not allowed as noted in [14].

Next, we consider the use of the negation technique in unfold/fold transformation of co-LP. We apply the negation technique to $\neg q(X)$ in P_0, obtaining $P_0' = \{1',2',3'\}$:

1'. $p \leftarrow not\_q(X)$	4'. $p \leftarrow not\_q(X), not\_r$	(unfold 1')
2'. $not\_q(X) \leftarrow not\_q(X), not\_r$	5'. $p \leftarrow p, not\_r$	(fold 4')
3'. $not\_r \leftarrow$		

The annotations of $not\_q(X)$ and not_r are coinductive, since the semantics of $q(X)$ and r are given by the least fixpoints in the perfect model semantics.

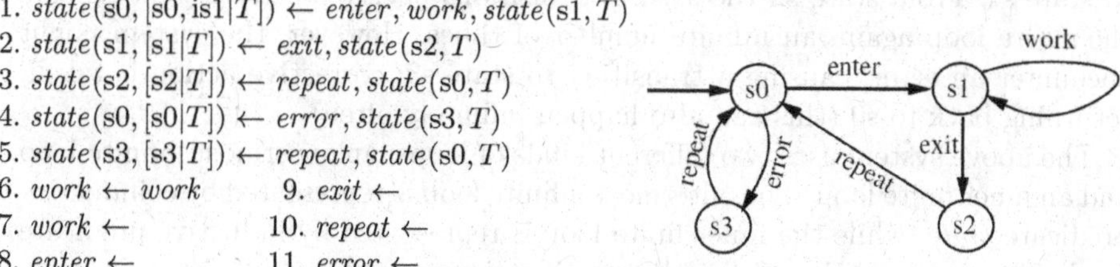

1. state(s0, [s0, is1|T]) ← enter, work, state(s1, T)
2. state(s1, [s1|T]) ← exit, state(s2, T)
3. state(s2, [s2|T]) ← repeat, state(s0, T)
4. state(s0, [s0|T]) ← error, state(s3, T)
5. state(s3, [s3|T]) ← repeat, state(s0, T)
6. work ← work
7. work ←
8. enter ←
9. exit ←
10. repeat ←
11. error ←
work
enter
s0
s1
exit
repeat
error
repeat
s3
s2

Fig. 2. Example: a self-correcting system [22]

On the other hand, we consider p a defined predicate, annotated "coinductive" according to the annotation rule (R1). Then, we consider the above transformation sequence: we apply to clause $1'$ unfolding w.r.t. $not\_q(X)$, followed by folding the resulting clause. Let $P_2' = (P_0' \setminus \{1'\}) \cup \{5'\}$. Then, we have that $M(P_2') \models p$. The above transformation thus preserves the alternating fixpoint semantics. □

The above example suggests that the negation technique in co-LP, when used with unfold/fold transformations, would circumvent a restriction imposed on the use of negative unfolding, thereby making amenable to subsequent transformations, which will hopefully lead to a successful proof.

4 Proving Properties of Co-logic Programs with Negation

In this section, we explain by examples how negation elimination in Sect. 3 will be utilized to prove properties of co-logic programs.

Let P be a co-logic program and *prop* be a predicate specifying a property of interest which is defined in terms of the predicates in P. Then, in order to check whether or not $M(P) \models \exists X\, prop(X)$, our transformation-based verification method is simple: we first introduce a defined predicate f defined by clause δ of the form: $f \leftarrow prop(X)$, where the annotation of predicate f is determined according to the annotation rule in R1. We then apply the transformation rules for co-logic programs given in Sect. 2.2 to $P_0 = P$ as an initial program, constructing a transformation sequence $P_0, P_1 = P_0 \cup \{\delta\}, \ldots, P_n$ so that the truth value of f in $M(P_n)$ will be easily known. In particular, if the definition of f in P_{n-1} consists of a single self-recursive clause $f \leftarrow f$, we will apply the replacement rule to it, obtaining P_n from P_{n-1}, where $Def(f, P_n) = \emptyset$ (i.e., $M(P_n) \models \neg f$) if f is inductive, $Def(f, P_n) = \{f \leftarrow\}$ (i.e., $M(P_n) \models f$) otherwise.

The first example due to [22] is on proving a liveness property of a self-correcting system in Fig. 2.

Example 3. Let P_s be the clauses $1 - 11$ in Fig. 2, which encodes the self correcting system in the figure. The system consists of four states s0, s1, s2 and s3. It starts in state s0, enters state s1, performs a finite amount of work in state s1. This inner loop state in s1 is denoted by is1 (clause 1). The system then exits to state s2. From state s2 the system transitions back to state s0, and repeats the entire loop again, an infinite number of times. However, the system might encounter an error, causing a transition to state s3; corrective action is taken, returning back to s0 (this can also happen infinitely often).

The above system uses two different kinds of loops: an outermost infinite loop and an inner finite loop. The outermost infinite loop is represented by coinductive predicate *state*, while the inner finite loop is represented by inductive predicate *work*. The program P_s satisfies the stratification restriction.

Suppose that we want to prove a property of the system, ensuring that the system must traverse through the work state s2 infinitely often. The counterexamples to the property can be specified as: $\exists T\ state(\mathrm{s0}, T), \neg comember(\mathrm{s2}, T)$, meaning that the state s2 is not present infinitely often in the infinite list T.

To express the counterexample, we first introduce the following clause:
12. $f \leftarrow state(\mathrm{s0}, T), not\_comem(\mathrm{s2}, T)$
where f is an inductive defined predicate, and *not_comem* is an inductive predicate defined in Remark 1. Let P_0 be P_s together with the extended definition of *not_comem*. Then, we can consider the following transformation sequence:

13. $f \leftarrow state(\mathrm{s0}, T'), not\_d(\mathrm{s2}, T')$	(unfold$^+$ 12)
14. $g \leftarrow state(\mathrm{s0}, T), not\_d(\mathrm{s2}, T)$	(def. intro.), $g \in \mathcal{P}^{co}$
15. $f \leftarrow g$	(fold 13)
16. $g \leftarrow state(\mathrm{s0}, T'), not\_d(\mathrm{s2}, T')$	(unfold$^+$ 14)
17. $g \leftarrow g$	(fold 14)
18. $g \leftarrow$	(replacement 17)
19. $f \leftarrow$	(unfold 15)

This means that $M(P_0 \cup \{12, 14\}) \models f$, which implies that there exists a counterexample to the original property; in fact, $T = [\mathrm{s0}, \mathrm{s3}|T]$ satisfies the body of clause 12.

In the previous approach [20], we introduced a predicate $absent(X, T)$ in advance, which corresponds to $\neg comember(X, T)$. Then, we perform a sequence of transformations similar to the above. By contrast, we derive predicate $not\_comem(X, T)$ here by using the negation technique. The current approach is therefore more amenable to automatic proof of properties of co-LP. □

Example 4. Adapted from [16]. We consider regular sets of infinite words over a finite alphabet. These sets are denoted by ω-regular expressions whose syntax is defined as follows:

$e ::= a \mid e_1e_2 \mid e_1 + e_2 \mid e^\omega$ with $a \in \Sigma$ (regular expressions)

$f ::= e^\omega \mid e_1e_2^\omega \mid f_1 + f_2$ (ω-regular expressions)

Given a regular (or an ω-regular) expression r, by $\mathcal{L}(r)$ we indicate the set of all words in Σ^* (or Σ^ω, respectively) denoted by r. In particular, given a regular expression e, we have that $\mathcal{L}(e^\omega) = \{w_0w_1 \cdots \in \Sigma^\omega \mid \text{for } i \geq 0, w_i \in \mathcal{L}(e) \subseteq \Sigma^*\}$.

Now we introduce a co-logic program, called P_f, which defines the predicate ω-*acc* such that for any ω-regular expression f, for any infinite word w, $\omega\text{-}acc(f, w)$ holds iff $w \in \mathcal{L}(f)$. The ω-program P_f consists of the clauses in Fig. 3, together with the clauses defining the predicate *symb*, where $symb(a)$ holds iff $a \in \Sigma$. Clauses 1-6 specify that, for any finite word w and regular expression e, $acc(e, w)$

1. $acc(E, [E]) \leftarrow symb(E)$
2. $acc(E_1E_2, X) \leftarrow$
   $app(X_1, X_2, X), acc(E_1, X_1), acc(E_2, X_2)$
3. $acc(E_1 + E_2, X) \leftarrow acc(E_1, X)$
4. $acc(E_1 + E_2, X) \leftarrow acc(E_2, X)$
5. $acc(E^*, []) \leftarrow$
6. $acc(E^*, X) \leftarrow$
   $app(X_1, X_2, X), acc(E, X_1), acc(E^*, X_2)$
7. $\omega\text{-}acc(F_1 + F_2, X) \leftarrow \omega\text{-}acc(F_1, X)$
8. $\omega\text{-}acc(F_1 + F_2, X) \leftarrow \omega\text{-}acc(F_2, X)$
9. $\omega\text{-}acc(E^\omega, X) \leftarrow$
   $app(X_1, X_2, X), acc(E, X_1), \omega\text{-}acc(E^\omega, X_2)$
10. $app([], Y, Y) \leftarrow$
11. $app([S|X], Y, [S|Z]) \leftarrow app(X, Y, Z)$

Fig. 3. Example: A Program P_f Which Accepts ω-languages

holds iff $w \in \mathcal{L}(e)$. Similarly, clauses 7-9 specify that, for any infinite word w and ω-regular expression f, $\omega\text{-}acc(f, w)$ holds iff $w \in \mathcal{L}(f)$. The annotation of ω-*acc* is coinductive, while the other predicates are inductive.

Now, let us consider the ω-regular expressions $f_1 \equiv_{def} a^\omega$ and $f_2 \equiv_{def} (b^*a)^\omega$. The following two clauses:

12. $expr_1(X) \leftarrow \omega\text{-}acc(a^\omega, X)$
13. $expr_2(X) \leftarrow \omega\text{-}acc((b^*a)^\omega, X)$

together with program P_f, define the predicates $expr_i$ $(i = 1, 2)$ such that, for every infinite word w, $expr_i(w)$ holds iff $w \in \mathcal{L}(f_i)$. Moreover, we introduce predicate $not\_expr_2(X)$ defined as: for any ground term t,

14. $M(P_f \cup \{13\}) \models \neg expr_2(t)$ iff $M(P_f \cup \{13\}) \models not\_expr_2(t)$.

If we introduce a clause:

15. $not\_contained(X) \leftarrow expr_1(X), not\_expr_2(X)$,

then we have that $\mathcal{L}(f_1) \subseteq \mathcal{L}(f_2)$ iff $M(P_f \cup \{12, 13, 14, 15\}) \not\models \exists X not\_contained$ (X). We note that the negation technique is not applicable to $\omega\text{-}acc(E^\omega, T)$; clause 9 has existential variables and, unlike the previous example, it does not have a functional part. The definition of $not\_\omega\text{-}acc$ is therefore not given in an explicit clausal form at this moment. Instead, we will apply the negation technique 'on the fly' in the following.

To check the above containment, we introduce the following clause:

16. $g \leftarrow not\_contained(X)$

where g is an inductive predicate. We can then consider the following transformation sequence:

17. $g \leftarrow \omega\text{-}acc(a^\omega, X), not\_\omega\text{-}acc((b^*a)^\omega, X)$ (unfold$^+$ 16)
18. $g_1 \leftarrow \omega\text{-}acc(a^\omega, X), not\_\omega\text{-}acc((b^*a)^\omega, X)$ (def. intro.), $g \in \mathcal{P}^{in}$
19. $g \leftarrow g_1$ (fold 17)
20. $g_1 \leftarrow \omega\text{-}acc(a^\omega, T), not\_\omega\text{-}acc((b^*a)^\omega, [a|T])$ (unfold$^+$ 18)

From P_f and a goal $\omega\text{-}acc((b^*a)^\omega, [a|T])$, we get the following resultant [11]: $\omega\text{-}acc((b^*a)^\omega, [a|T]) \leftarrow \omega\text{-}acc((b^*a)^\omega, T)$.

Since the above clause does not contain an existential variable, we have now from the negation technique:

21. $not\_\omega\text{-}acc((b^*a)^\omega, [a|T]) \leftarrow not\_\omega\text{-}acc((b^*a)^\omega, T)$

Using the above clause, we continue the above-mentioned transformation:

22. $g_1 \leftarrow \omega\text{-}acc(a^\omega, T), not\_\omega\text{-}acc((b^*a)^\omega, T)$ (unfold 20)
23. $g_1 \leftarrow g_1$ (fold 22)
24. def. of g_1 (clause 23) removed (replacement 23)
25. clause 19 removed (unfold 19)

This means that $M(P_f \cup \{12, 13, 14, 15\}) \not\models g$, namely $\mathcal{L}(f_1) \subseteq \mathcal{L}(f_2)$.

The above example is originally due to Pettorossi, Proietti and Senni [16], where (i) the given problem is encoded in an *ω-program* P, a locally stratified program on infinite lists, then (ii) their transformation rules in [16] are applied to P, deriving a *monadic* ω-program T, and finally (iii) the decision procedure in [15] is applied to T to check whether or not $PERF(T) \models \exists X\, prop(X)$.

On the other hand, our approach uses co-LP, which allows us to make the representation more succinct, about the half the lines of the ω-program in this particular example. □

5 Related Work and Concluding Remarks

We have shown that negation elimination based on the negation technique (NT) can be used to co-logic programs, and its application is correct under the alternating fixpoint semantics of co-logic programs. In co-LP, NT allows us to derive a complementary program under a weaker condition than in the original NT for definite programs. We have explained in Sect. 4 how NT, when incorporated into the previous framework for unfold/fold transformation, allows us to represent and reason about a wider class of co-logic programs which our previous framework [20] cannot deal with.

Simon et al. [22] have proposed an operational semantics, co-SLD resolution, for co-LP, which has been further extended by Min and Gupta [12] to co-SLDNF resolution. On the other hand, as explained in Example 3, NT derives a program which simulates failed computations of a given program. When NT is applicable, our approach is simpler than co-SLDNF resolution in that we can do without an extra controlling mechanism such as a positive/negative context in co-SLDNF resolution.

Pettorossi, Proietti and Senni [16] have proposed another framework for transformation-based verification based on ω-programs, which can also represent infinite computations. As explained in Example 4, our approach can prove the given property in a simpler and more succinct manner, as far as this particular example is concerned. However, the detailed comparison between our approach based on co-logic programs and their approach based on ω-programs is left for future work.

One direction for future work is to extend the current framework to allow a more general class of co-logic programs. Gupta et al. [6], for example, have discussed such an extension, where they consider co-LP without the stratification restriction.

Acknowledgement. The author would like to thank anonymous reviewers for their constructive and useful comments on the previous version of the paper.

References

1. Burstall, R.M., Darlington, J.: A Transformation System for Developing Recursive Programs. J. ACM 24(1), 144–167 (1977)
2. Clarke, E.M., Grumberg, O., Peled, D.A.: Model Checking. MIT Press (1999)

3. Colmerauer, A.: Prolog and Infinite Trees. In: Logic Programming, pp. 231–251. Academic Press (1982)
4. Etalle, S., Gabbrielli, M.: Transformations of CLP Modules. Theor. Comput. Sci., 101–146 (1996)
5. Gupta, G., Bansal, A., Min, R., Simon, L., Mallya, A.: Coinductive logic programming and its applications. In: Dahl, V., Niemelä, I. (eds.) ICLP 2007. LNCS, vol. 4670, pp. 27–44. Springer, Heidelberg (2007)
6. Gupta, G., Saeedloei, N., DeVries, B., Min, R., Marple, K., Kluźniak, F.: Infinite computation, co-induction and computational logic. In: Corradini, A., Klin, B., Cîrstea, C. (eds.) CALCO 2011. LNCS, vol. 6859, pp. 40–54. Springer, Heidelberg (2011)
7. Jaffar, J., Stuckey, P.: Semantics of infinite tree logic programming. Theoretical Computer Science 46, 141–158 (1986)
8. Jaffar, J., Maher, M.J.: Constraint Logic Programming: A Survey. J. Log. Program. 19/20, 503–581 (1994)
9. Kott, L.: Unfold/fold program transformations. In: Algebraic Methods in Semantics, pp. 411–434. Cambridge University Press (1985)
10. Lloyd, J.W.: Foundations of Logic Programming, 2nd edn. Springer (1987)
11. Lloyd, J.W., Shepherdson, J.C.: Partial Evaluation in Logic Programming. J. Logic Programming 11, 217–242 (1991)
12. Min, R., Gupta, G.: Coinductive Logic Programming with Negation. In: De Schreye, D. (ed.) LOPSTR 2009. LNCS, vol. 6037, pp. 97–112. Springer, Heidelberg (2010)
13. Pettorossi, A., Proietti, M.: Transformation of Logic Programs: Foundations and Techniques. J. Logic Programming 19/20, 261–320 (1994)
14. Pettorossi, A., Proietti, M.: Perfect Model Checking via Unfold/Fold Transformations. In: Palamidessi, C., Moniz Pereira, L., Lloyd, J.W., Dahl, V., Furbach, U., Kerber, M., Lau, K.-K., Sagiv, Y., Stuckey, P.J. (eds.) CL 2000. LNCS (LNAI), vol. 1861, pp. 613–628. Springer, Heidelberg (2000)
15. Pettorossi, A., Proietti, M., Senni, V.: Deciding Full Branching Time Logic by Program Transformation. In: De Schreye, D. (ed.) LOPSTR 2009. LNCS, vol. 6037, pp. 5–21. Springer, Heidelberg (2010)
16. Pettorossi, A., Proietti, M., Senni, V.: Transformations of logic programs on infinite lists. Theory and Practice of Logic Programming 10, 383–399 (2010)
17. Przymusinski, T.C.: On the Declarative and Procedural Semantics of Logic Programs. J. Automated Reasoning 5(2), 167–205 (1989)
18. Sato, T., Tamaki, H.: Transformational Logic Program Synthesis. In: Proc. FGCS 1984, Tokyo, pp. 195–201 (1984)
19. Seki, H.: On Inductive and Coinductive Proofs via Unfold/fold Transformations. In: De Schreye, D. (ed.) LOPSTR 2009. LNCS, vol. 6037, pp. 82–96. Springer, Heidelberg (2010)
20. Seki, H.: Proving Properties of Co-logic Programs by Unfold/Fold Transformations. In: Vidal, G. (ed.) LOPSTR 2011. LNCS, vol. 7225, pp. 205–220. Springer, Heidelberg (2012)
21. Seki, H.: Unfold/Fold Transformation of Stratified Programs. Theoretical Computer Science 86, 107–139 (1991)
22. Simon, L., Mallya, A., Bansal, A., Gupta, G.: Coinductive Logic Programming. In: Etalle, S., Truszczyński, M. (eds.) ICLP 2006. LNCS, vol. 4079, pp. 330–345. Springer, Heidelberg (2006)
23. Simon, L.E.: Extending Logic Programming with Coinduction, Ph.D. Dissertation, University of Texas at Dallas (2006)
24. Tamaki, H., Sato, T.: Unfold/Fold Transformation of Logic Programs. In: Proc. 2nd Int. Conf. on Logic Programming, pp. 127–138 (1984)

Program Analysis and Manipulation to Reproduce Learners' Erroneous Reasoning

Claus Zinn

Department of Computer Science
University of Konstanz
Box D188, 78457 Konstanz, Germany
claus.zinn@uni-konstanz.de

Abstract. Pedagogical research shows that learner errors are seldom random but systematic. Good teachers are capable of inferring from learners' input the erroneous procedure they are following, and use the result of such deep cognitive diagnoses to repair its incorrect parts. We report a method for the automatic reconstruction of such erroneous procedures based on learner input and the analysis and manipulation of logic programs. The method relies on an iterative application of two algorithms: an innovative use of algorithmic debugging to identify learner errors by the analysis of (initially) correct (*sic*) Prolog-based procedures, and a subsequent program manipulation phase where errors are introduced into (initially) correct procedures. The iteration terminates with the derivation of an erroneous procedure that was followed by the learner. The procedure, and its step-wise reconstruction, can then be used to inform remedial feedback.

1 Introduction

The diagnosis of learner input is central for the provision of effective scaffolding and remedial feedback in intelligent tutoring systems. A main insight is that errors are rarely random, but systematic. Either learners have acquired an erroneous procedure, which they execute in a correct manner, or learners attempt to execute a correct procedure but encounter an *impasse* when executing one of its steps. To address the impasse, learners are known to follow a small set of *repair* strategies such as skipping the step in question, or backing-up to a previous decision point where performing the step can be (erroneously) avoided.

Effective teaching depends on deep cognitive diagnosis of such learner behaviour to identify erroneous procedures or learners' difficulties with executing correct ones. State-of-the-art intelligent tutoring systems, however, fail to give a full account of learners' erroneous skills. In *model tracing* tutors (*e.g.*, the Lisp tutor [2]; the Algebra Tutor [4]), appropriately designed user interfaces and tutor questions invite learners to provide their answers in a piecemeal fashion. It is no longer necessary to reproduce a student's line of reasoning from question to (final) answer; only the student's next step towards a solution is analyzed,

E. Albert (Ed.): LOPSTR 2012, LNCS 7844, pp. 228–243, 2013.
© Springer-Verlag Berlin Heidelberg 2013

and immediate feedback is given. Model tracing tutors thus keep learners close to ideal problem solving paths, hence preventing learners to fully exhibit erroneous behaviour. *Constraint-based tutors* (*e.g.*, the SQL tutor [7]) perform student modelling based on constraints [9]. Here, diagnostic information is not derived from an analysis of learner actions but of problem states the student arrived at. With no representation of actions, the constraint-based approach makes it hard to identify and distinguish between the various (potentially erroneous) procedures learners follow to tackle a given problem.

None of the two approaches attempt to explain buggy knowledge or skills. There is no explicit and machine-readable representation to mark deviations of an expert rule from the buggy skill; and also, there is no mechanism for automatically deriving buggy skills from correct ones. In this paper, we report a method capable of reconstructing erroneous procedures from expert ones. The method is based on an iterative analysis and manipulation of logic programs. It relies on an innovative use of algorithmic debugging to identify learner error by the analysis of (initially) correct (*sic*) Prolog-based procedures (modelling expert skills), and a subsequent program manipulation to introduce errors into the correct procedure to finally produce the erroneous procedure followed by the learner. The method extends our previous work [14] by having algorithmic debugging now qualify the irreducible disagreement with its cause, *i.e.*, by specifying those elements in the learner's solution that are missing, incorrect, or superfluous. Moreover, we have defined and implemented a perturbation algorithm that can use the new information to transform Prolog programs into ones that can reproduce the observed error causes.

The remainder of the paper is structured as follows. Sect. 2 introduces the domain of instruction (multi-column subtraction), typical errors and how they manifest themselves in this domain, and our previous work on adapting Shapiro's algorithmic debugging to support diagnosis in intelligent tutoring systems. Sect. 3 first describes our new extension to Shapiro's technique; it then explains how we interleave the extended form of algorithmic debugging with automatic code perturbation, and how this method can be used iteratively to track complex erroneous behaviour. In Sect. 4, we give examples to illustrate the effectiveness of the method. In Sect. 5, we discuss related work, and Sect. 6 lists future work and concludes.[1]

2 Background

Our approach to cognitive diagnosis of learner input is applicable for any kind of domain that can be encoded as a logic program. For illustration purposes, we focus on the well-studied example domain of multi-column subtraction.

[1] For our research, we have chosen basic school arithmetics, a subject that most elementary schools teach in 3rd. and 4th. grade. In the paper, we use the terms "pupil", "learner" and "student" to refer to a school child.

```
subtract(PartialSum, Sum) :-
      length(PartialSum, LSum),
      mc_subtract(LSum, PartialSum, Sum).

mc_subtract(_, [], []).
mc_subtract(CurrentColumn, Sum, NewSum) :-
      process_column(CurrentColumn, Sum, Sum1),
      shift_left(CurrentColumn, Sum1, Sum2, ProcessedColumn),
      CurrentColumn1 is CurrentColumn - 1,
      mc_subtract(CurrentColumn1, Sum2, SumFinal),
      append(SumFinal, [ProcessedColumn], NewSum).

process_column(CurrentColumn, Sum, NewSum) :-
      last(Sum, LastColumn),        allbutlast(Sum,RestSum),
      subtrahend(LastColumn, S),    minuend(LastColumn, M),
      S > M, !,
      add_ten_to_minuend(CurrentColumn, M, M10),
      CurrentColumn1 is CurrentColumn - 1,
      decrement(CurrentColumn1, RestSum, NewRestSum),
      take_difference(CurrentColumn, M10, S, R),
      append(NewRestSum,[(M10, S, R)],NewSum).

process_column(CurrentColumn, Sum, NewSum) :-
      last(Sum, LastColumn),        allbutlast(Sum,RestSum),
      subtrahend(LastColumn, S),    minuend(LastColumn, M),
      % S =< M,
      take_difference(CurrentColumn, M, S, R),
      append(RestSum,[(M, S, R)], NewSum).

shift_left( _CurrentColumn, SumList, RestSumList, Item ) :-
      allbutlast(SumList, RestSumList),
      last(SumList, Item).

add_ten_to_minuend( _CC, M, M10)   :- irreducible, M10 is M + 10.
take_difference(    _CC, M, S, R) :- irreducible, R is M - S.

decrement(CurrentColumn, Sum, NewSum ) :- irreducible,
      last( Sum, (M, S, R) ), allbutlast( Sum, RestSum),
      M == 0, !,
      CurrentColumn1 is CurrentColumn - 1,
      decrement(CurrentColumn1, RestSum, NewRestSum),
      NM is M + 10, NM1 is NM - 1,
      append( NewRestSum,  [ (NM1, S, R)], NewSum).

decrement(CurrentColumn, Sum, NewSum ) :- irreducible,
      last( Sum, (M, S, R) ), allbutlast( Sum, RestSum),
      % \+ (M == 0),
      M1 is M - 1,
      append( RestSum,  [ ( M1, S, R) ], NewSum ).

minuend( (M, _S, _R), M).    subtrahend( (_M, S, _R), S).    irreducible.
```

Fig. 1. Multi-column subtraction in Prolog

2.1 Multi-column Subtraction

Fig. 1 gives an implementation of multi-column subtraction in Prolog. Sums are processed column by column, from right to left. The predicate **subtract/2** determines the sum's length and passes the arguments to **mc_subtract/3**, which implements the recursion.[2] The predicate **process_column/3** gets a partial sum, processes its right-most column and takes care of borrowing (**add_ten_to_minuend/3**) and payback (**decrement/3**) actions. Sums are encoded as Prolog lists of columns, where a column is represented as 3-element term `(M, S, R)` representing minuend, subtrahend, and result cell.[3] The program code implements the *decomposition method*. If the subtrahend `S` is larger than the minuend `M`, then `M` is increased by 10 (borrow) before the difference between `M` and `S` is taken. To compensate, the minuend in the column left to the current one is decreased by one (pay-back).

2.2 Error Analysis in Multi-column Subtraction

While some errors may be caused by a simple oversight (usually, pupils are able to correct such errors as soon as they see them), most errors are *systematic* errors (those keep re-occurring again and again). It is the systematic errors that we aim at diagnosing as they indicate a pupil's wrong understanding about some subject matter. Fig. 2 lists the top four errors of the Southbay study. [11, Chapter 10]; the use of squares is explained below.

```
                          3  □                7             □
    5  2  □4              4  ¹0  ¹4           8̶  □           3  ¹2
  - 2  9  8             - 1   8   7           9̶  ¹0  ¹4     - 1   7
  = 3  7  4             = 2   2   7         - 2   3   7     = 2   5
                                            = 5   7   7
```

(a) smaller-from-larger (b) stops-borrow-at-zero (c) borrow-across-zero (d) borrow-no-decrement

Fig. 2. The top four most frequent subtraction errors, see [11]

Errors can be classified according to the following scheme: *errors of omission* – forget to do something; *errors of commission* – doing the task incorrectly; and *sequence errors* – doing the task not in the right order. These errors occur because learners may have acquired an incorrect procedure, or they "know" the correct procedure but are unable to execute it. In [11], VanLehn explains the origin of errors in terms of *impasses* and *repairs*. According to his theory, an impasse occurs when an individual fails to execute a step in a procedure; to overcome the impasse, a meta-level activity – a repair action – is performed

[2] Note that the variable `CurrentColumn` is not required to specify our algorithm for multi-column subtraction; it is rather necessary for the mechanisation of the Oracle (see below), and thus passed on as argument to most of the predicates.

[3] The predicates **append/3** (append two lists), **last/2** (get last list element), and **allbutlast/2** (get all list elements but the last) are defined as usual.

4 6 3	8 3 5	7 0 2	8 0 7	6 0 4
- 2 5 1	- 2 1 8	- 3 4 7	- 2 0 8	- 8 7
=	=	=	=	=
(a)	(b)	(c)	(d)	(e)

Fig. 3. Exercises on multi-column subtraction [5, page 59f]

to cope with the impasse. While some repairs result in correct outcomes, others result in buggy procedures. Consider the error exhibited in Fig. 2(a). In VanLehn's theory, this can be explained as follows. The learner cannot perform a complete borrow action; to cope with this impasse, a repair is executed by "backing-up" to a situation where no borrowing is required: the smaller number is always subtracted from the larger number. In Fig. 2(d), the same impasse is encountered with a different repair. The borrow is executed, with correct differences computed for all columns, but the corresponding decrements (paybacks) are omitted. In Fig. 2(b,c), the learner does not know how to borrow from zero. In (b), this is repaired by a "no-op", that is, a *null operation* with no effect: for minuends equal to zero, the decrement operation is simply omitted (note that the learner performed a correct borrow/payback pair in the other columns). In (c), the learner repairs the situation by borrowing from the hundreds rather than the tens column; this can be modelled as "partial no-op".

An accurate diagnosis of a learner's competences can only be obtained when the learner's performance is observed and studied across a well-designed set of exercises demanding the full set of domain skills. Fig. 3 depicts exercises taken from a German textbook on third grade school mathematics [5]. While none of the errors given in Fig. 2 show-up in the exercise Fig. 3(a), the errors `borrow-no-decrement` and `smaller-from-larger` are observable in each of the tests Fig. 3(b)–3(e). The exercises Fig. 3(c)–3(e) can be used to check for the bugs `stops-borrow-at-zero` or `borrow-across-zero`, but also for others not listed here.

In the sequel, given a learner's performance across a set of exercises, we aim to diagnose his errors by reconstructing their underlying erroneous procedures *automatically* using algorithmic debugging and program transformation techniques. Our innovative method relies only on expert programs and learner answers.

2.3 Shapiro's Algorithmic Debugging

Shapiro's algorithmic debugging technique for logic programming prescribes a systematic manner to identify bugs in programs [10]. In the top-down variant, the program is traversed from the goal clause downwards. At each step during the traversal of the program's AND/OR tree, the programmer is taking the role of the *oracle*, and answers whether the currently processed goal holds or not. If the oracle and the buggy program agree on a goal, then algorithmic debugging passes to the next goal on the goal stack. If the oracle and the buggy program disagree on the result of a goal, then this goal is inspected further. Eventually an *irreducible disagreement* will be encountered, hence locating the program's clause where the buggy behavior is originating from. Shapiro's algorithmic debugging method extends, thus, a simple meta-interpreter for logic programs.

Shapiro devised algorithmic debugging to systematically identify bugs in incorrect programs. Our Prolog code for multi-column subtraction in Fig. 1, however, presents the expert model, that is, a presumably correct program. Given that cognitive modelling seeks to reconstruct students' erroneous procedure by an analysis of their problem-solving behavior, it is hard to see – at least at first sight – how algorithmic debugging might be applicable in this context. There is, however, a simple, almost magical trick, which we first reported in [14]. Shapiro's algorithm can be turned on its head: instead of having the Oracle specifying how the assumed incorrect program should behave, we take the expert program to take the role of the buggy program, and the role of the Oracle is filled by students' potentially erroneous answers. An irreducible disagreement between program behavior and student answer then pinpoints students' potential misconception(s).

An adapted version of algorithmic debugging for the tutoring context is given in [14]. Students can "debug" the expert program in a top-down fashion. Only clauses declared as being *on the discussion table* are subjected to Oracle questioning so that rather technical steps (such as `last/2` and `allbutlast/2`) do not need to be discussed with learners. Moreover, a Prolog clause whose body starts with the subgoal `irreducible/2` is subjected to Oracle questioning; but when program and Oracle disagree on such a clause, this disagreement is irreducible so that the clause's (remaining) body is not traversed. In [14], we have also shown how we can relieve learners from answering questions. The answers to all questions posed by algorithmic debugging are automatically reconstructed from pupils' exercise sheets, given their solution to a subtraction problem. With the mechanisation of the Oracle, diagnoses are obtained automatically.

3 Combining Algorithmic Debugging with Program Manipulation

Our goal is to reconstruct learners' erroneous procedures by only making use of an expert program (encoding the skills to be learned) and learners' performances when solving exercises. For this, the expert program is transformed in an iterative manner until a buggy program is obtained that reproduces all observed behaviour. The program transformation is informed by the results of algorithmic debugging as the source of the irreducible disagreement points to the part in the program that requires perturbation. To better inform program transformation, we augment algorithmic debugging to also return the nature of the disagreement.

3.1 Causes of Disagreements

Whenever expert and learner disagree on a goal, one of the following cases holds:

- the learner solution *misses* parts that are present in the expert solution, *e.g.*, a result cell in the multi-column subtraction table has not been filled out;

– the learner solution has *incorrect* parts with regard to the expert solution, *e.g.*, a result cell is given a value, albeit an incorrect one; and
– the learner solution has *superfluous* parts not present in the expert solution, *e.g.*, the learner performed a borrowing operation that was not necessary.

We have extended our variant of algorithmic debugging, and its cooperating mechanised Oracle, to enrich irreducible disagreements with their nature (*missing*, *incorrect*, or *superfluous*). Moreover, we have extended algorithmic debugging to record all agreements until irreducible disagreements are encountered. The program perturbation code has access to these past learner performances and uses this information to steer its perturbation strategy.

3.2 Main Algorithm

Fig. 4 gives a high-level view of the algorithm for the reconstruction of learners' erroneous procedure. The function `ReconstructErroneousProcedure/3` is recursively called until a program is obtained that reproduces learner behaviour, in which case there are no further disagreements. Note that multiple perturbations may be required to reproduce single bugs, and that multiple bugs are tackled by iterative applications of algorithmic debugging and code perturbation.

```
function ReconstructErroneousProcedure(Program, Problem, Solution)
    (Disagr, Cause) ← AlgorithmicDebugging(Program, Problem, Solution)
    if Disagr = nil then
        return Program
    else
        NewProgram ← Perturbation(Program, Disagr, Cause)
        ReconstructErroneousProcedure(NewProgram, Problem, Solution)
    end if
end function

function Perturbation(Program, Clause, Cause)
    return chooseOneOf(Cause)
        DeleteCallToClause(Program, Clause)
        DeleteSubgoalOfClause(Program, Clause)
        ShadowClause(Program, Clause)
        SwapClauseArguments(Program, Clause)
end function
```

Fig. 4. Pseudo-code: compute variant of *Program* to reproduce a learner's *Solution*

The irreducible disagreement resulting from the algorithmic debugging phase locates the code pieces where perturbations must take place; its cause determines the kind of perturbation. The function `Perturbation/3` can invoke various kinds of transformations: the deletion of a call to the clause in question, or the deletion of one of its subgoals, or the shadowing of the clause in question by a more specialized instance, or the swapping of the clause' arguments. These perturbations reproduce errors of omission and commission, and reflect the repair strategies learners use when encountering an impasse. Future transformations may involve the consistent change of recursion operators (reproducing sequence errors), and the insertion of newly created subgoals to extend the body of the clause in question (reproducing irreducible disagreements with cause *superfluous*).

The generic program transformations that we have implemented require an annotation of the expert program clauses with *mode* annotations, marking their arguments as input and output arguments. Our algorithm for clause call deletion, for instance, traverses a given program until it identifies a clause whose body contains the clause in question; once identified, it removes the clause in question from the body and replaces all occurrences of its output argument by its input argument in the adjacent subgoals as well as in the clause's head, if present. Then, `DeleteCallToClause/2` returns the modified program.

3.3 Problem Sets

A befitting diagnosis of a learner's knowledge and potential misconceptions requires an analysis of his actions across a set of exercises. Our code for algorithmic debugging and code perturbation is thus called for each exercise of a test set. We start the diagnosis process with the expert program; the program will get perturbated whenever there is a disagreement between program and learner behaviour. At the end of the test set, we obtain a program that mirrors learner performance across all exercises. As we will see, some perturbations might seem ad hoc; here, the program is perturbated with base clauses that only reproduce behavior specific to a given exercise. A post-processing step is thus invoked to generalize over exercise-specific perturbations.

Note that we expect learners to exhibit a problem-solving behaviour that is consistent across exercises. To simulate this idealized behaviour – and to test our approach – we have hand-coded buggy variations of the multi-column subtraction routine to reproduce the top-ten errors of the Southbay study [11, Chapter 10]. Consistent learner behaviour is thus simulated by executing these buggy routines. We use the following "driver" predicate to diagnose learners:

Program ← *ExpertProcedure*
for all Problem of the test set **do**
  Solution ← *BuggyProcedure*(*Problem*)
  Program ← *ReconstructErroneousProcedure*(*Program*, *Problem*, *Solution*)
  ▷ side effect: agreements and disagreements are recorded and used.
end for
GeneralisedProgram ← *Generalise*(*Program*)

4 Examples

We now explain our approach for the typical errors given in Fig. 2.

4.1 Simple Omission Error: Omit Call to Clause in Question

In Fig. 2(d), the learner forgets to honor the pay-back operation, following the borrowing that happened in the first (right-most) column. The execution of the adapted version of algorithmic debugging produces the following dialogue (with all questions automatically answered by the mechanised Oracle):

```
algorithmic_debugging(subtract([(3, 1, S1), ( 2, 7, S2)],
                               [(3, 1, 2), (12, 7, 5)],
                    IrreducibleDisagreement).

do you agree that the following goal holds:
   subtract( [(3, 1, R1), ( 2, 7, R2)],
             [(2, 1, 1), (12, 7, 5)])
|: no.

do you agree that the following goal holds:
   mc_subtract(2, [(3, 1, R1), ( 2, 7, R2)],
                  [(2, 1,  1), (12, 7,  5)])
|: no.

   process_column(2,[(3, 1, R1), ( 2, 7, R2)],[(2, 1, R1), (12, 7,  5)])
|: no.

   add_ten_to_minuend(2, 2, 12)
|: yes.

   decrement(1, [( 3, 1, R1)], [( 2, 1, R1)])
|: no.
=> IrreducibleDisagreement=( decrement(1,[ (3,1, R1), (2,1, R1) ]),
                             missing )
```

With the indication of error (the location is marked by □ in Fig. 2(d)), program transformation now attempts various manipulations to modify the expert program. Given the cause "missing", the perturbation heuristics chooses to delete the call to the indicated program clause, which succeeds: in the body of the first clause of `process_column/3`, we eliminate its subgoal `decrement(CurrentColumn1, RestSum, NewRestSum)` and subsequently replace its output variable `NewRestSum` with its input variable `RestSum`.[4] With this program manipulation, we achieve the intended effect; the resulting buggy program reproduces the learner's answer; both program and learner agree on the top `subtract/2` goal.

4.2 Complex Error of Omission and Commission

An iterative execution of algorithmic debugging and program manipulation to the problem in Fig. 2(a) shows how a complex error is attacked step by step.

First Run. Running algorithmic debugging on the expert program and the learner's solution concludes the dialogue (now omitted) with the disagreement

```
add_ten_to_minuend( 3, 4, 14)
```

[4] The subgoal introducing `CurrentColumn1` becomes obsolete, and is removed too.

and the cause "missing". The code perturbation algorithm makes use of `DeleteCallToClause/2` to delete the subgoal `add_ten_to_minuend/3` from the first program clause of `process_column/3`; the single occurrence of its output variable `M10` is replaced by its input variable `M` in the subsequent calls to `take_difference/4` and `append/3`.

Second Run. Algorithmically debugging the modified program yields

```
decrement(2, [ (5, 2, R3), (2, 9, R2)], [ (5, 2, R3), (1, 9, R2)])
```

a disagreement with cause "missing". Again, `DeleteCallToClause/2` is invoked, now to delete the subgoal `decrement/3` from the first clause of `process_column/3`. The occurrence of its output variable `NewRestSum` in the subsequent call to `append/3` is replaced by its `decrement/3` input variable `RestSum`; also the subgoal introducing `CurrentColumn1` is deleted. With these changes, we obtain a new program that is closer in modelling the learner's erroneous behaviour.

Third Run. Re-entering algorithmic debugging with the modified expert program now yields an irreducible agreement in the ones column with cause "incorrect":

```
take_difference(3, 4, 8, 4)
```

A mere deletion of a call to the clause in question is a bad heuristics as the result cell must obtain a value; moreover, past learner performances on the same problem test set indicate that the learner successfully executed instances of the skill `take_difference/4`. We must thus perturbate the clause's body, or shadow the clause with an additional, more specialized clause. A simple manipulation to any of the clause's subgoals fails to achieve the intended effect. To accommodate the result provided by the learner, we shadow the existing clause with:

```
take_difference( _CC, 4, 8, 4) :- irreducible.
```

Note that this new clause can be taken directly from the Oracle's analysis. While the new clause covers the learner's input, it is rather specific.

Fourth Run. We now obtain an irreducible disagreement in the tens column:

```
take_difference(2, 2, 9, 7)
```

with cause "incorrect". Similar to the previous run, we add another clause for `take_difference/4` to capture the learner's input:

```
take_difference( _CC, 2, 9, 7) :- irreducible.
```

With these changes to the expert program, we now obtain a buggy program that entirely reproduces the learner's solution in Fig. 2(a).

Generalization. When we run the driver predicate with the buggy learner simulation "smaller-from-larger" for the exercises given in Fig. 3, we obtain a buggy program which includes the following base clauses for `take_difference/4`:

```
take_difference(3, 5,8,3).      take_difference(3, 2,7,5).
take_difference(2, 0,4,4).      take_difference(2, 7,8,1).
take_difference(2, 4,7,3).      take_difference(2, 0,8,8).
```

complementing the existing general clause (see Fig. 1).

We use the inductive logic program Progol [8] to yield a more general definition for taking differences. For this, we construct an input file for Progol in which we re-use the mode declarations used to annotate the expert program. Progol's background knowledge is defined in terms of the subtraction code (see Fig. 1), and Progol's positive examples are the base cases given above. Moreover, the expert's view on the last two irreducible disagreements, namely, `take_difference(4,8,-4)` and `take_difference(2,9,-7)` are taken as negative examples to support the generalisation of the learner's `take_difference/3` clause.[5]

Running Progol (version 4.2) on the input file yields a new first clause of taking differences that generalizes its base cases. Taken together, we obtain:

```
take_difference(A,B,C) :- leq(A,B), sub(B,A,C).
take_difference(A,B,C) :- irreducible, C is A-B.
```

4.3 Error Analysis: Stops-Borrow-at-Zero and Borrow-across-Zero

The solutions given in Fig. 2(b) and Fig. 2(c) indicate two different repair strategies when encountering the impasse "unable to borrow from zero". We give a detailed discussion for the first, and an abridged one for the second error type.

Error: Stops-Borrow-at-Zero. A first run of algorithmic debugging returns the following irreducible disagreement in the tens column with the cause "incorrect":

```
decrement(2, [ (4, 1, R1), (0, 8, R2)], [ (3, 1, R1), (9, 8, R2)]).
```

Note that the learner is executing the `decrement/3` operation successfully in the hundreds (and in prior exercises with minuends different from zero). A program perturbation other than the deletion of the call to `decrement/3` – as in the previous two examples – is needed. The clause `decrement/3` must thus be called, but its corresponding clause body altered. For this, code perturbation calls `Delete-SubgoalOfClause/2`, which first locates the correct clause in question, given the irreducible disagreement. Given that the minuend of the last column is zero,

[5] At the time of writing, the input file for the ILP system is crafted manually. For technical reasons, we remove the argument `CurrentColumn` from all clauses; also we added to Progol's background knowledge the clauses: `sub( X, Y, Z) :- Z is X - Y.` and `leq(X,Y) :- X =< Y.` They also appear in Progol's `modeb` declarations.

the first clause of decrement/3 is selected. The perturbation DeleteSubgoalOf-Clause/2 now tries repairs to mirror those of the learner. Either the decrement/3 step is skipped entirely, or some of its subgoals are deleted.

The error stops-borrow-at-zero can be reproduced with an entire "no-op". The following steps are carried out to modify the body of the respective decrement/3 clause: (i) the input argument of the clause is mapped to its output argument; (ii) all clauses rendered superfluous by this perturbation are removed (see discussion of borrow-across-zero); and (iii) the existing first clause of decrement/3 is replaced by the new perturbated version, which is:

```
decrement(CurrentColumn, Sum, Sum ) :- irreducible,
      last( Sum, (M, S, R) ), allbutlast( Sum, RestSum),
      M == 0, !.
```

The modified program reproduces the observed learner behaviour.

Error: Borrow-across-Zero. The error borrow-across-zero follows the main line of perturbation than the previous error. But rather than repairing the impasse with a "no-op", an output argument different from the input argument is constructed. For this, one or more steps that are involved in its construction are deleted. Similar to the last error example, a dependency network of variables is created and exploited, as we now explain.

In the given clause, the output argument is NewSum; it has an occurrence in append/3, where it depends on the variables NewRestSum as well as NM1, S, and R, which are all input variables, given the mode declaration of append/3. To modify NewSum, we consider all subgoals that feature the aforementioned input variables as output variables. Consider NM1, which is constructed by the subgoal NM1 is NM - 1. When we delete this subgoal, and replace the occurrence of NM1 by NM in append/3, we obtain a program that reproduces the error don't-decrement-zero, see [11, page 226].[6] Another call to DeleteSubgoalOfClause/2 is required to obtain borrow-across-zero. Now, as the variable NM depends on the value of M, we can replace the occurrence of NM in append/3 by M to yield the intended effect.

4.4 Summary

In all examples, algorithmic debugging correctly indicated the program clause that required manipulation (this is not always the case, see future work on multiple models). Moreover, the cause of the disagreement as well as prior learner performances correctly informed the perturbation actions. Note that our perturbations DeleteCallToClause/2 and DeleteSubgoalOfClause/2 mirror the repair strategies that learners perform when encountering an impasse; they often omit or only partially execute steps relevant to the impasse. Our approach proved to be a general, effective, and also computationally inexpensive method. In this

[6] When we only delete the recursive call in the body of decrement/3, we obtain borrow-from-zero, the fifth most frequent learner error.

paper, we have illustrated the method being able to reproduce the top five subtraction errors of the Southbay study, which together account for 45% of all errors in this study.

5 Related Work

There is only little research in the intelligent tutoring systems community that builds upon logic programming and meta-level reasoning techniques. In [1], Beller & Hoppe use a fail-safe meta-interpreter to identify student error. A Prolog program, modelling the expert knowledge for doing subtraction, is executed by instantiating its output parameter with the student answer. While standard Prolog interpretation would fail, a fail-safe meta-interpreter can recover from execution failure, and can also return an execution trace. Beller & Hoppe then formulate *error patterns* which they then match against the execution trace; each successful match is indicating a plausible student bug.

In Looi's "Prolog Intelligent Tutoring System" [6], Prolog programs written by students are debugged with the help of the automatic derivation of mode specifications, dataflow and type analysis, and heuristic code matching between expert and student code. Looi also makes use of algorithmic debugging techniques borrowed from Shapiro [10] to test student code with regard to termination, correctness and completeness. The Oracle is mechanised by running expert code that most likely corresponds to given learner code, and simple code perturbations are carried out to correct erroneous program parts.

Most closely related to our research is the work of Kawai et al. [3]. Expert knowledge is represented as a set of Prolog clauses, and Shapiro's Model Inference System (MIS) [10], following an inductive logic programming (ILP) approach, is used to synthesize learners' (potentially erroneous) procedure from expert knowledge and student answers. Once the procedure to fully capture learner behaviour is constructed, Shapiro's Program Diagnosis System (PDS), based upon *standard* algorithmic debugging, is used to identify students' misconceptions, that is, the bugs in the MIS-constructed Prolog program.

While Kawai et al. use similar logic programming techniques, there are substantial differences to our approach. By turning Shapiro's algorithm on its head, we are able to identify the first learner error with no effort – for this, Kawai et al. require an erroneous procedure, which they need to construct first using ILP.[7] To analyse a learner's second error, we overcome the deviation between expert behaviour and learner behaviour by introducing the learner's error into the expert program. This "correction" is performed by small perturbations within existing code, or by the insertion of elements of little code size. We believe these changes to the expert program to be less complex than the synthesis of entire erroneous procedures by inductive methods alone.

[7] Moreover, in our approach, given the mechanization of the Oracle, no questions need to be answered by the learner.

6 Future Work and Conclusion

In our approach, an irreducible disagreement corresponds to an erroneous aspect of a learner answer not covered by the expert program (or a perturbated version thereof). We have seen that most program transformations are clause or goal deletions, mirroring the repair strategies of learners when encountering an impasse. We have also seen an example where program perturbation meant complementing existing program clauses with more specialized instances. We demonstrated the successful use of Progol to perform a post-processing generalization step to capture learner performance in a general manner.

Our perturbation algorithm could profit from program splicing [12]. Once algorithmic debugging has located an irreducible disagreement between expert performance and learner performance, the respective clause could serve as a *slicing criterion* for the computation of a subprogram (the *program slice*), whose execution may have an effect on the values of the clause's arguments. The program slice could support or replace the variable dependency network we use.

At the time of writing, the input files to Progol are manually written. In the future, we aim at obtaining a better understanding of Progol's inductive mechanism and its tweaking parameter to generate the input file automatically, and to better integrate Progol into the program perturbation process. In the light of [3], we are also considering a re-implementation of Shapiro's MIS and follow-up work to better integrate inductive reasoning into our approach. This includes investigating the generation of system-learner interactions to obtain additional positive or negative examples for the induction process. These interactions must not distract students from achieving their learning goals; ideally, they can be integrated into a tutorial dialogue that is beneficial to student learning.

In practise, the analysis of learner performances across a set of test exercises may sometimes yield multiple diagnoses. Also, some learners may show an inconsistent behaviour (which we have idealised in this paper). In the future, we would like to investigate the automatic generation of follow-up exercises that can be used to disambiguate between several diagnoses, and to address observed learner inconsistencies.

Expert models with representational requirements or algorithmic structure different to the subtraction algorithm given in Fig. 1 might prove beneficial for errors not discussed here. At the time of writing, we have implemented four different subtraction methods: the Austrian method, its trade-first variant, the decomposition method, and an algorithm that performs subtraction from left to right. We have also created variants to these four methods (with differences in sequencing subgoals, or the structure of clause arguments). When a learner follows a subtraction method other than the decomposition method, we are likely to give a wrong diagnosis of learner action whenever the decomposition method is the only point of reference to expert behaviour. To improve the quality of diagnosis, it is necessary to compare learner performance to *multiple* models. We have developed a method to identify the algorithm the learner is most likely following. The method extends algorithmic debugging by counting the number of *agreements* before and after the first irreducible disagreement; also, the "code size" that is being agreed upon is taken into account (for details, see [13]). In the future, we will combine the choice of expert model (selecting the model closest to observed behaviour) with the approach presented in this paper.

Once an erroneous procedure has been constructed by an iterative application of algorithmic debugging and automatic code perturbation, it will be necessary to use the procedure as well as its construction history to inform the generation of remedial feedback. Reconsider our example Fig. 2(a), where the learner always subtracted the smaller from the larger digit. Its construction history shows that the following transformations to the correct subtraction procedure were necessary to reproduce the learners' erroneous procedure: (i) deletion of the goal `add_ten_to_minuend/3`, (ii) deletion of the goal `decrement/3`, (iii) shadowing of the goal `take_difference/4` with more specialized instances, and (iv) a generalization step. Some reasoning about the four elements of the construction history is necessary to translate the perturbations into a compound diagnosis to generate effective remediation. For this, we would like to better link the types of disagreements obtained from algorithmic debugging and the class of our perturbations with the kinds of impasses and repair strategies of VanLehn's theory [11].

There is good evidence that a sound methodology of cognitive diagnosis in intelligent tutoring can be realized in the framework of declarative programming languages such as Prolog. In this paper, we reported on our work to combine an innovative variant of algorithmic debugging with program transformation to advance cognitive diagnosis in this direction.

Acknowledgments. Thanks to the reviewers whose comments helped improve the paper. Our work is funded by the German Research Foundation (ZI 1322/2/1).

References

1. Beller, S., Hoppe, U.: Deductive error reconstruction and classification in a logic programming framework. In: Brna, P., Ohlsson, S., Pain, H. (eds.) Proc. of the World Conference on Artificial Intelligence in Education, pp. 433–440 (1993)
2. Corbett, A.T., Anderson, J.R., Patterson, E.J.: Problem compilation and tutoring flexibility in the lisp tutor. In: Intell. Tutoring Systems, Montreal (1988)
3. Kawai, K., Mizoguchi, R., Kakusho, O., Toyoda, J.: A framework for ICAI systems based on inductive inference and logic programming. New Generation Computing 5, 115–129 (1987)
4. Koedinger, K.R., Anderson, J.R., Hadley, W.H., Mark, M.A.: Intelligent tutoring goes to school in the big city. Journal of Artificial Intelligence in Education 8(1), 30–43 (1997)
5. Maier, P.H.: Der Nussknacker. Schülerbuch 3. Schuljahr. Klett Verlag (2010)
6. Looi, C.-K.: Automatic debugging of Prolog programs in a Prolog Intelligent Tutoring System. Instructional Science 20, 215–263 (1991)
7. Mitrović, A.: Experiences in implementing constraint-based modeling in SQL-tutor. In: Goettl, B.P., Halff, H.M., Redfield, C.L., Shute, V.J. (eds.) ITS 1998. LNCS, vol. 1452, pp. 414–423. Springer, Heidelberg (1998)
8. Muggleton, S.: Inverse Entailment and Progol. New Generation Computing Journal (13), 245–286 (1995)

9. Ohlsson, S.: Constraint-based student modeling. Journal of Artificial Intelligence in Education 3(4), 429–447 (1992)
10. Shapiro, E.Y.: Algorithmic Program Debugging. ACM Distinguished Dissertations. MIT Press (1983); Thesis (Ph.D.) – Yale University (1982)
11. VanLehn, K.: Mind Bugs: the origin of procedural misconceptions. MIT Press (1990)
12. Weiser, M.: Program Slicing. IEEE Trans. Software Eng. 10(4), 352–357 (1984)
13. Zinn, C.: Identifying the closest match between program and user behaviour (unpublished manuscript), `http://www.inf.uni-konstanz.de/~zinn`
14. Zinn, C.: Algorithmic debugging to support cognitive diagnosis in tutoring systems. In: Bach, J., Edelkamp, S. (eds.) KI 2011. LNCS, vol. 7006, pp. 357–368. Springer, Heidelberg (2011)

Author Index